Polymeric Drug Delivery I

ACS SYMPOSIUM SERIES **923**

Polymeric Drug Delivery I

Particulate Drug Carriers

Sönke Svenson, Editor
Dendritic NanoTechnologies, Inc.

Sponsored by the
ACS Division of Polymeric Materials: Science and
Engineering, Inc.

American Chemical Society, Washington, DC

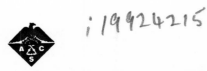

; 19924215

Library of Congress Cataloging-in-Publication Data

Polymeric drug delivery / Sönke Svenson, editor ; sponsored by the ACS Division of Polymeric Materials: Science and Engineering, Inc.

    p. cm.—(ACS symposium series ; v. 923-)

    "Developed from a symposium sponsored by the Division of Polymeric Materials: Science and Engineering, Inc., at the 226th National Meeting of the American Chemical Society, New York, September 7–11, 2003"—T.p. verso.

    Includes bibliographical references and indexes.

    Contents: v. 1. Particulate drug carriers

    ISBN 13: 978–0–8412–3918–0 (alk. paper)

    1. Polymeric drug delivery systems—Congresses. 2. Polymeric drugs—Congresses.

    I. Svenson, Sönke, 1956- II. American Chemical Society. Division of Polymeric Chemistry: Science and Engineering, Inc. III. American Chemical Society. Meeting (226th : 2003 : New York, N.Y.) IV. Series.

RS201 P65P6412   2005
615'.3—dc22   2005053686

The paper used in this publication meets the minimum requirements of American National Standard for Information Sciences—Permanence of Paper for Printed Library Materials, ANSI Z39.48–1984.

Distributed by Oxford University Press

Todd Emrick supplied the images for the front cover.

ISBN 10: 0-8412-3918-5

PRINTED IN THE UNITED STATES OF AMERICA

Foreword

The ACS Symposium Series was first published in 1974 to provide a mechanism for publishing symposia quickly in book form. The purpose of the series is to publish timely, comprehensive books developed from ACS sponsored symposia based on current scientific research. Occasionally, books are developed from symposia sponsored by other organizations when the topic is of keen interest to the chemistry audience.

Before agreeing to publish a book, the proposed table of contents is reviewed for appropriate and comprehensive coverage and for interest to the audience. Some papers may be excluded to better focus the book; others may be added to provide comprehensiveness. When appropriate, overview or introductory chapters are added. Drafts of chapters are peer-reviewed prior to final acceptance or rejection, and manuscripts are prepared in camera-ready format.

As a rule, only original research papers and original review papers are included in the volumes. Verbatim reproductions of previously published papers are not accepted.

ACS Books Department

Contents

ix

Preface

Achieving the therapeutic drug concentration at the desired location within a body during the required length of time without causing local over-concentrations and their adverse side effects is the main challenge in the development of a successful drug delivery system. Key factors to be considered in this effort are drug bioavailability, biocompatibility (or toxicity for that matter), targeting, and drug-release profile. The bioavailability of class II and III drugs (by the Biopharmaceutical Classification System of Drugs) is challenged by their poor water solubility and poor membrane permeability, respectively. Poor water solubility results in the rejection of about 40% of newly developed drugs by the Pharmaceutical Industry, while about 20% of marketed drugs exhibit suboptimal performance due to this challenge. Strategies to overcome this problem include the use of particular drug carriers such as liposomes, (polymeric) micelles, dendrimers, emulsion droplets, and engineered micro- and nanoparticles. These carriers encapsulate a drug, this way enhancing their water solubility (bioavailability) and reducing their toxicity (enhanced biocompatibility). In addition, carriers can transport drugs to the desired location by either passive targeting (enhanced permeability and retention, EPR, effect) or by active targeting through ligands that interact with receptors that are overexpressed at the surface of tumor cells. Alternatively, problem drugs can be encapsulated into depot matrices, whose release profile, defined by diffusion and matrix degradation, determines the concentration level of the free drug. The third strategy does not employ auxiliaries such as carriers or matrices but focuses on particle engineering. For example, reducing the size of a drug particle will increase its surface area and enhance its solubilization rate. All three strategies are explored in university laboratories as well as by the Pharmaceutical Industry.

Thus the main motivation for organizing the *Polymeric Drug*

Delivery: Science and Application symposium during the 226th National Meeting of the American Chemical Society (ACS) in New York, New York, in September 2003, was to provide a forum for both communities, academia and industry, to discuss progress in these three strategic approaches. Seventy-five well-recognized international experts, equally representing both communities, had been invited to present their research being conducted in the United States (57), the United Kingdom (4), Canada (3), Germany (3), France (3), the Netherlands (3), Italy (1) and Spain (1). Highlights from this symposium are presented in two volumes within the ACS Symposium Series: *Polymeric Drug Delivery I:Particulate Drug Carriers* (ACS Symposium Series 923) and *Polymeric Drug Delivery II:Polymeric Matrices and Drug Particle Engineering* (ACS Symposium Series 924). To provide an even broader overview, two contributions from Japanese researchers have been included in the books.

The forty-four chapters selected for publication within the Symposium Series are divided into three main sections. Following an overview, the first section describes in twenty chapters the use of carriers such as liposomes, micelles, dendrimers, emulsion droplets, nanoparticles, and yeast cells in the (targeted) delivery of poorly water-soluble drugs, small organic molecules, macromolecules such as proteins and nucleic acids, and metal ions for molecular imaging purposes. Different routes of application are described, including oral and transepithelial delivery (Volume I).

The second section, with thirteen chapters, is devoted to the use of polymeric matrices. The application of polymer solutions, nanogels, hydrogels, and millirods in gene and drug delivery, as well the prediction of drug solubility in polymeric matrices, the *in situ* study of drug release from t ablets b y F ourier-Transform i nfrared i maging, a nd a ntimicrobial release coatings are presented.

The third section details in nine chapters the use of supercritical fluids, controlled precipitation processes, and application of excipients in particle engineering and size control. The final chapter describes the preparation of fast dissolving tablets, another approach to increase the availability of drugs. These two sections, supplemented by an overview of the topics, are presented in Volume II.

These books are intended for readers in the chemical and pharmaceutical industry and academia who are interested or involved in drug delivery research as well as for advanced students who are interested in this active and rapidly developing research area.

I am using this opportunity to congratulate three presenters at t he symposium and coauthors of this book, whose contributions received special recognition during the ACS National Meeting. Dr. Theresa Reineke and her co-worker Yemin Liu (University of Cincinnati), received the Arthur K. Doolittle Award for Fall 2003 for the best paper presented at the ACS Division of Polymeric Materials: Science and Engineering, Inc., (PMSE) meeting for their presentation on "Synthesis and Characterization of Polyhydroxyamides for DNA Delivery". Dr. Brian Johnson (Princeton University), received the 2003 ICI Student Award in Applied Polymer Science for his presentation on "Nanoprecipitation of Organic Actives Using Mixing and Block Copolymer Stabilization".

I deeply appreciate the willingness of the authors to contribute to this important overview of drug delivery technologies. I also greatly appreciate the help of many researchers, who have devoted their time to this project by reviewing these contributions, ensuring clarity and technical accuracy of the manuscripts. I thank the PMSE for the opportunity to hold the symposium as a part of their program and especially Ms. Eileen Ernst, the PMSE Administrative Assistant and "Florence Nightingale" to me, for her invaluable help with placing all abstracts and preprints into the system. I appreciate the patience and support of the ACS Symposium Series acquisitions and production team during the production of these books.

Last but not least I thank the ACS Corporation Associates (CA) as the "gold sponsor"; The Dow Chemical Company as the "silver sponsor"; and Elan NanoSystems; Epic Therapeutics, Inc.; Guilford Pharmaceuticals, Inc.; Johnson and Johnson, Center for Biomaterials and Advanced Technologies; and Thar Technologies, Inc. as "bronze sponsors" for their financial support of the symposium. It would have been impossible to assemble this outstanding group of researchers without these contributions.

Sönke Svenson

Dendritic NanoTechnologies, Inc.
2625 Denison Drive
Mount Pleasant, MI 48858
(989) 774–1179 (telephone)
(989) 774–1194 (fax)
Svenson@dnanotech.com (email)

Overview

Chapter 1

Advances in Particulate Polymeric Drug Delivery

Sönke Svenson

Dendritic NanoTechnologies Inc., 2625 Denison Drive, Mount Pleasant, MI 48858 (email: svenson@dnanotech.com)

An o verview is given presenting various strategies to deliver small molecule and macromolecule drugs. Micelles, liposomes, emulsions, dendrimers and micro and nanocontainers are the carrier types employed in these strategies. Targeting of specific organs and triggered release of the drugs from their carriers are being highlighted.

Introduction

The development of molecular nanostructures with well-defined particle size and shape is of eminent interest in biomedical applications such as the delivery of active pharmaceuticals, imaging agents, or gene transfection. For example, structures utilized as carriers in drug delivery generally should be in the nanometer range and uniform in size to enhance their ability to cross cell membranes and reduce the risk of undesired clearance from the body through the liver or spleen. Two traditional routes to produce particles that will meet these requirements to some extent have been widely investigated. The first route takes advantage of the ability of amphiphilic molecules (i.e., molecules consisting of a hydrophilic and hydrophobic moiety) to self-assemble in water above a system-specific critical micelle concentration (CMC) to form micelles. Size and shape of these micelles depend on the geometry of the constituent monomers, intermolecular interactions, and conditions of the bulk solution (i.e., concentration, ionic strength, pH, and temperature). Spherical micelles are monodisperse in size; however, they are highly dynamic in nature with monomer exchange rates in millisecond to microsecond time ranges. Micelles have the ability to encapsulate and carry lipophilic actives within their hydrocarbon cores. Depending on the specific system, some micelles either spontaneously rearrange to form liposomes after a change of solution conditions, or when exposed to external energy input such as agitation, sonication, or extrusion through a filter membrane. Liposomes consist of bilayer lipid membranes (BLM) enclosing an aqueous core, which can be utilized to carry hydrophilic actives. Furthermore, liposomes with multilamellar membranes provide cargo space for lipophilic actives as well. However, most liposomes are considered energetically metastable and eventually will rearrange to form planar bilayers. (1,2)

The second route relies on engineering the well-defined particles through processing protocols. Examples for this approach include (i) shearing or homogenization of oil-in-water (o/w) emulsions or w/o/w double emulsions to produce stable and monodisperse droplets, (ii) extrusion of polymer strands or viscous gels through nozzles of defined size to manufacture stable and monodisperse micro and nanospheres, and (iii) layer-by-layer (LbL) deposition of polyelectrolytes and other polymeric molecules around colloidal cores, resulting in the formation of monodisperse nanocapsules after removal of the templating core. Size, degree of monodispersity, and stability of these structures depend on the systems that are being used in these applications. (3)

Currently, a new third route to create very well-defined, monodisperse, stable molecular level nanostructures is being studied based on the "dendritic state" architecture. (4,5) The challenge is to develop critical structure-controlled methodologies to produce appropriate nanoscale modules that will allow cost-effective synthesis and controlled assembly of more complex nanostructures in a very routine manner. Dendritic architecture is undoubtedly one of the most pervasive topologies observed throughout biological systems at virtually all dimensional length scales. This architecture offers unique interfacial and functional performance advantages because of the high level of control over its size, shape, branching density and composition, utilizing well established organic synthesis protocols.

The fourth route utilizes natural carriers, for example viruses and microorganisms such as yeast cells. Adenovirus is a natural carrier studied for gene transfection applications because of its high efficiency in crossing cell membranes. However, the risk of potential adverse health effects limits the regulatory and public acceptance of viral carriers. As a viable alternative, in recent years microorganisms such as live *lactobacillus* species and common baker's yeast have been considered as novel vectors for the delivery of bioactive proteins and peptides via the gastrointestinal tract. *(6,7)* The potential risk of using live organisms for drug delivery have resulted in a shift to using non-viable bacteria, yeast and yeast cell walls, where much of the site-specific targeting remains possible without problematic issues in controlling the activity of the microorganism. *(8)*

Several of these systems and their utilization in drug delivery and gene transfection are being highlighted in this overview and some will be discussed in detail in the following chapters of this book.

Results and Discussion

Micelles and Liposomes

Regardless of the specific type, carriers utilized in drug delivery and gene transfection must fulfill two requirements to avoid non-specific capture at non-tumor sites. First, drug carriers must be smaller than approximately 200 nm to evade uptake by the reticuloendothelial system, and their molecular weights should be larger than the critical value of approximately 40,000 Da to prevent renal filtration. Second, drug carriers should not strongly interact or randomly being taken up by organs, especially the reticuloendothelial system. Therefore, hydrophobic and cationic carrier surfaces should be avoided, while hydrophilic surfaces with neutral or weak negative overall charge will reduce or even prevent random uptake.

The average size (i.e., 10 to 100 nm in diameter) of micelles makes them suitable carriers for these medical applications. The use of polymeric constituents, formed from block or graft copolymers, will reduce the aforementioned high exchange rate between micelles and monomers and, thus, increase micelle stability. Most studies of polymeric micelles have employed AB or ABA-type block copolymers because the close relationship between micelle-forming behavior and structure of the polymers can be evaluated more easily with these copolymers than with graft or multi-segmented block copolymers. A more detailed discussion of these considerations is presented by *Yokoyama* in chapter 3 of this volume.

The hydrophobic core – hydrophilic shell structure of micelles can be achieved through various routes, generally utilizing the amphiphilic character of the constituent polymeric monomers. Pluronics® (ABA block copolymers), poly(ethylene glycol)-*b*-poly(ester)s, poly(ethylene glycol)-*b*-poly(L-amino acid) and poly(ethylene glycol)-phospholipid micelles have been studied extensively for drug delivery. *(9,10)* An interesting approach is the employment of poly(ethylene glycol)-phospholipid micelles for the delivery of the antifungal, amphotericin B (AmB), described by *Kwon et al.* in chapter 2. Advantages of this system are the presumed compatibility between drug and phospholipid, the proven safety profile of poly(ethylene glycol)-phospholipid, which are generally regarded as safe materials (GRAS), and perhaps the ability to disaggregate AmB to reduce its toxicity.

Another approach to combine safe materials with hydrophobic character relies on polysaccharide-based micelles, where hydrophobicity is induced through alkyl chains that connect to the polysaccharides via poly(oxyethylene) linkers (POE_y-C_n). Size, stability, and colloidal properties of these hydrophobically-modified (HM) polysaccharide micelles depend on their chemical composition, the number of saccharide units, and the architecture of the polymer. (11) While a number of fundamental studies of HM-polysaccharides have been reported, their use as carriers of poorly water-soluble drugs in oral delivery has only been studied very recently. (12) The stability of micellar carriers can be enhanced beyond hydrophobic interactions through chemical cross-linking between monomers, either within the core or the shell domain. (13,14)

The other major group of drug carriers based on self-assembly of amphiphilic monomers is comprised of liposomes. Liposomes can be con-structed from biodegradable and nontoxic constituents and are able to non-covalently encapsulate molecules (i.e., chemotherapeutic agents, hemoglobin, imaging agents, drugs, and genetic material) within their 100-200 nm diameter interior. Liposomes are typically categorized in one of four categories: (i) conventional liposomes, which are composed of neutral and/or negatively-charged lipids and often cholesterol; (ii) long-circulating liposomes, which incorporate PEG covalently bound to a lipid; (iii) immunoliposomes, used for targeting where antibodies or antibody fragments are bound to the surface, and (iv) cationic liposomes, that are used to condense and deliver DNA. Liposomes are characterized by their surface charge, size, composition, and number and fluidity of their lamellae. (15,16) Immune recognition results in liposome clearance from the body and accumulation in liver, kidney, and spleen. For delivery to tumors, increased circulation time and evasion of the immune system is vital. The attachment of biocompatible groups onto the surface of liposome such as PEG chains ("Stealth liposomes") is a common approach to increase the circulation time. These PEG chains produce a steric barrier to protein binding. However, detachment of the PEG chains with time results in deprotection of the liposomes and constitutes a serious limitation to this approach. A very elegant solution to this dilemma is presented by *Auguste et al.* in chapter 8, using multiply attached polymers as a means for constructing polymer-protected liposomes. The concept is established by a series of PEG-based comb copolymers with concatenated PEG chains having hydrophobic anchoring groups between the linked PEG chains.

In a related approach, PEG-lipid conjugates, connected via acid-labile linkers, have been inserted into the surface of planar lamellae and used to force curvature into these lamellae, resulting in the formation of pH-sensitive liposomes. The intrinsic low pH within the endosomal compartment results in hydrolysis of the PEG chains and re-transformation of the liposome into its parent lamellae, releasing the drug content within the cell. Eight acid-labile poly(ethylene glycol) conjugated vinyl ether lipids have been synthesized to test the efficiency of this approach. *(17,18)*

Another dilemma in particulate drug delivery is manifested in the fact that cationic carriers are very efficient in DNA compaction and transfection but, at the same time, form aggregates upon contact with serum that in its worst block capillaries *in vivo* or, in milder forms, restrict transfection to the immediate vicinity of the injection site. Anionic or neutral liposomes, on the other hand, are unable to incorporate large pieces of DNA due to electrostatic repulsion. One proposed solution to this dilemma is the use of liposomes composed of lipids that have an anionic charge under physiological conditions but become cationic at a pH<6. Charge reversible liposomes containing lipids with carboxylate and imidazole head groups are under investigation as carriers for nucleic acids. *(19)*

Sustained delivery of certain drugs will avoid the need for repeated administration of these actives. One approach of sustained delivery has been realized utilizing multivesicular liposomes (DepoFoam®). The multivesicular particles containing the drugs were prepared by a two-step double emulsification process. The first step is the emulsification of an aqueous solution containing the drug with a lipid solution in chloroform to produce a "water-in-oil" emulsion. A subsequent emulsification with a second aqueous solution results in a "water-in-oil-in-water" double emulsion. Chloroform is removed and the resulting multivesicular liposomes are washed and resuspended in appropriate buffer. All DepoFoam formulations tested had characteristic multivesicular morphology and narrow, monomodal particle size distribution in the range 12-20 μm. *(20,21)*

Emulsions

Many macromolecular and biomolecular drugs, for example most proteins, polypeptides, oligonucleotides and plasmid DNA, degrade under extreme pH or in the presence of digestive enzymes. In order to successfully deliver these drugs, formulations that are able to protect and deliver them to the targeted sites are mandated. A novel method for the encapsulation of plasmid DNA into a compatible polymer has been developed based on a double emulsion formulation process, that (i) produces particles smaller than those made by conventional methods, (ii) encapsulates plasmid DNA with greater than 90% efficiency, (iii) protects plasmid DNA from degradation, and (iv) has release profiles similar to conventional nonviral delivery methods. *(22)* Details of its recent progress are presented by *Niedzinski et al.* in chapter 16. To even further reduce the stress for proteins during the encapsulation process, recently a microencapsulation method has been proposed that applies a solvent exchange method to generate reservoir-type microcapsules with reduced contact between encapsulated drugs and the potentially damaging environments (i.e., extensive exposure of proteins to a large water/organic solvent (w/o) interfacial area during microencapsulation; mechanical stresses such as emulsification or homogenization; and extended contact with hydrophobic polymers and their degradation products). The details of this solvent exchange process are being discussed by *Yeo et al.* in chapter 17.

Oil-water interfacial assembly of amphiphilic graft copolymers has been employed in another approach to form hollow microcapsules that can be stabilized by cross-linking. Poly(cyclooctene)-g-poly(ethylene glycol) (PEG) copolymers, synthesized by ring-opening metathesis copolymerization of PEG-functionalized cyclooctene macromonomers with other cyclooctene derivatives, are observed to segregate to the toluene-water interface. Covalent cross-linking by ring-opening cross-metathesis with a *bis*-cyclooctene PEG derivative imparts mechanical integrity to these hollow capsules. This strategy can be utilized for the preparation of encapsulants and carriers for hydrophobic molecules, such as the anti-cancer drug doxorubicin (DOX). These novel capsules may be well suited for controlled release therapies, where the transport of drugs can be regulated by factors such as crosslink density, hydrolytic stability, and enzymatic stability of the polymer. *(23,24)*

Emulsions or, in the following example, microemulsions can not only be employed to deliver drugs but also to remove drugs from systemic circulation in case of an overdose. One microemulsion that proved effective in lowering the concentration of model toxins in saline or blood was composed of ethyl butyrate oil with Pluronic and caprylate cosurfactants. This system exhibited *in vitro* lowering of toxin concentrations and the ability to suppress the *in vivo* elongation of the QRS heart beat time span when injected into a rat exposed to a lethal dose of amitriptyline. In preliminary experiments the mortality rate was 25% with microemulsion treatment compared to 75% with saline treatment. *(25)*

Micro and Nanoparticles Based on Small Molecules, Polymers and Peptides

Polymeric molecules have been tailored for specific delivery applications. For example, poly(ethylene imine) (PEI) is the leading nonviral gene carrier described in the literature today. Moreover, it has been modified with a number of groups and modalities, including chemical groups for shielding of the cationic charge, targeting groups for specific cells, and biodegradable linkers for increased biocompatibility. Although the number of clinical trials with viral carriers overshadows that of nonviral carriers due to their inherently high degree of transfection and their increased persistence of gene expression, these carriers are prone to host immunogenic responses and are limited in the transgene size. The non-viral gene carrier PEI consists of a wide-range of varying molecular weights, branched and linear architectures, chemical modifications, and targeting modalities. T he i mpetus t owards m odifying P EI is to primarily increase gene transfer efficiency without accepting a loss in biocompatibility. An overview over this important topic is being presented by *Furgeson et al.* in chapter 13.

Another strategy to synthesize a polymeric carrier tailored to the need of its guest drug employs cyclodextrins. These polymers are capable of forming inclusion complexes with small molecules and the side-chains of larger compounds. Numerous types of cyclodextrin-polymer conjugates have been prepared to include species that are neutral, positive or negatively charged and deliver therapeutics ranging from small molecules (>1 nm in size) to oligonucleotides (1-10 nm) and plasmids (30-200 nm when condensed). *(26)*

In an effort to create nontoxic and highly effective synthetic transfection reagents, a tartarate comonomer has been polymerized with a series of amine co-monomers to yield a new family of copolymers. Four new poly(L-tartar-amidoamine)s have been designed and studied. Results of gel shift assays indicated that the polymers can bind plasmid DNA (pDNA) at polymer nitrogen to pDNA phosphate (N/P) ratios higher than one. Dynamic light scattering experiments revealed that each polymer compacted pDNA into nanoparticles (polyplexes) in the approximate size range to be endocytosed by cultured cells. These polyplexes exhibited high delivery efficiency without cytotoxic effects, indicating that these polymers have great promise as new gene delivery vehicles. The details of this study by *Liu et al.* are being presented in chapter 15.

Crossing biological barriers, a main obstacle in drug delivery and gene transfection, has triggered a mechanistic hypothesis for how water-soluble guani-dinium-rich transporters, attached to small cargos (MW <3000 Da), can migrate across the non-polar lipid membrane of a cell and enter the cytosol. Positively-charged and water-soluble arginine oligomers can associate with negatively-charged, bidentate hydrogen bond acceptor groups of endogenous membrane constituents, leading to the formation of membrane-soluble ion pair complexes. The resultant, less polar, ion pair complexes partition into the lipid bilayer and migrate in a direction and with a rate influenced by the membrane potential. The complex dissociates on the inner leaf of the membrane, and the transporter conjugate enters the cytosol. Several molecules have been studied to evaluate the structural requirements for these peptides to penetrate cell membranes. Even highly branched guanidinium-rich oligosaccharides and dendrimers are efficient transporters. *(27,28)* Similarly, a new peptide-based strategy (Pep-1) for protein transduction into cells has been developed, utilizing an amphipathic peptide consisting of a hydrophobic domain and a hydrophilic lysine-rich domain. Pep-1 efficiently delivers a variety of fully biologically active peptides and proteins into cells, without the need for prior chemical cross-linking or denaturation steps. *(29)* Both approaches are being presented in chapters 11 and 12.

Besides polymers, certain small molecules such as surfactants and other amphiphilic compounds have been utilized in the delivery of plasmid DNA and macromolecular drugs. Examples of these molecules as well as the advantages and disadvantages of this approach are being discussed in chapters 6 and 14.

Targeting and Release

A widely used targeting ligand for tumor-specific delivery is folic acid or vitamin B_{12}. The high affinity folate receptor (hFR) is overexpressed on many solid tumor surfaces. Other ligands for active targeting include whole monoclonal antibodies (mAb) or their Fab' fragments against the internalizing epitope, CD19 or the non-internalizing epitope, CD20, as well as the tumour-specific antibody, CC52, which is directed against a surface antigen on CC531 colon adenocarcinoma cells. Some aspects of active targeting are being discussed by *Kamps et al.* and *Thurmond et al.* in chapters 7 and 10. *(30,31)*

Drug release can be triggered by several means, including changes in pH as discussed earlier or temperature. Alternatively, the conversion of polymers from a hydrophobic to a more hydrophilic state by a chemical or enzymatic process is a way to destabilize micelles or vesicles. For example, micelles containing amphiphilic block copolymers of PEG and the thermosensitive block 2-hydroxy-propylmethacrylamide lactate are destabilized upon hydrolysis of the lactate esters in the side chains. This process causes swelling and ultimately disso-lution of the micelles due to increasing hydrophilicity of the core (see *van Nostrum et al.* in chapter 4).

References

1. Svenson, S. Controlling surfactant self-assembly. *Curr. Opin. Colloid Interface Sci.* **2004**, *9*, 201-212.
2. Svenson, S. Self-assembly and self-organization: Important processes – but can we predict them?. *J. Dispersion Sci. Technol.* **2004**, *25*, 101-118.
3. Svenson, S. (ed.), *Carrier-based Drug Delivery*, ACS Symposium Series, Vol. 879, American Chemical Society, Washington, DC, **2004**.
4. Tomalia, D.A. The dendritic state. *Materials Today* **2005**, *8(3)*, 34-46.
5. Tomalia, D.A. Birth of a new macromolecular architecture: Dendrimers as quantized building blocks for nanoscale synthetic organic chemistry. *Aldrichimica Acta* **2004**, *37*, 39-57.
6. Blanquet, S.; Antonelli, R.; Laforet, L.; Denis, S.; Marol-Bonnin, S.; Alric, M. Living recombinant Saccharomyces cerevisiae secreting proteins or peptides as a new drug delivery system in the gut. *J. Bio-technol.* **2004**, *110*, 37-49.
7. Stubbs, A.C.; Martin, K.S.; Coeshott, C.; Skaates, S.V.; Kuritzkes, D.R.; Bellgrau, D.; Franzusoff, A.; Duke, R.C.; Wilson, C.C. Whole

recombinant yeast vaccine activates dendritic cells and elicits protective cell-mediated immunity. *Nature Medicine*, **2001**, *7*, 625-629.

8. Paukner, S.; Kohl, G.; Jlava; K.; Lubitz, W. Sealed bacterial ghosts - Novel targeting vehicles for advanced drug delivery of water-soluble substances. *J. Drug Targeting*, **2003**, *111*, 151-161.

9. Torchilin, V.P. Structure and design of polymeric surfactant-based drug delivery systems. *J. Controlled Release* **2001**, *73*, 137-172.

10. Kwon, G.S. Polymeric micelles for delivery of poorly water-soluble compounds. *Crit. Rev. Ther. Drug Carrier Syst.* **2003**, *20*, 357-404.

11. Akiyoshi, K.; Kang, E.C.; Kurumada, S.; Sunamoto, J. Controlled association of amphiphilic polymers in water: Thermosensitive nanoparticles formed by self-assembly of hydrophobically modified pullulans and poly(N-isopropylacrylamides). *Macromolecules* **2000**, *33*, 3244-3249.

12. Francis, M.F.; Piredda, M.; Winnik, F.M. Solubilization of poorly water soluble drugs in micelles of hydrophobically modified hydroxypropyl-cellulose copolymers. *J. Controlled Release* **2003**, *93*, 59-68.

13. Tian, L.; Zhou, N.; Yam, L.; Wang, J.; Uhrich, K.E. The core cross-linkable micelles from PEG-lipid type amphiphiles. *Polym Mater. Sci. Eng. Preprints*, 226th National Meeting, Am. Chem. Soc., New York, NY, **2003**.

14. Thurmond, K.B.; Kowalewski, T.; Wooley, K.L. Shell cross-linked knedels: A synthetic study of the factors affecting the dimensions and properties of amphiphilic core-shell nanospheres. *J. Am. Chem. Soc.* **1997**, *119*, 6656-6665.

15. Jang, S.H.; Wientjes, M.G.; Lu, D.; Au, J.L.S. Drug delivery and transport to solid tumors. *Pharm. Res.* **2003**, *20*, 1337-1350.

16. Storm, G.; Crommelin, D.J.A. Liposomes: Quo vadis? *Pharm. Sci. Technol. Today* **1998**, *1*, 19-31.

17. Shin, J.; Shum, P.; Patri, V.S.; Kim, J.-M.; Thompson, D.H. Stereo-electronic and surface charge effects on the performance of DOPE liposomes containing acid-labile vinyl ether PEG-lipids. *Polym Mater. Sci. Eng. Preprints*, 226th National Meeting, Am. Chem. Soc., New York, NY, **2003**.

18. Boomer, J.A.; Inerowicz, H.D.; Zhang, Z.-Y.; Bergstrand, N.; Edwards, K.; Kim, J.-M.; Thompson, D.H. Acid-triggered release from sterically stabilized fusogenic liposomes via a hydrolytic dePEGylation strategy. *Langmuir* **2003**, *19*, 6408-6415.

19. Endert, G.; Fankhaenel, S.; Panzner, S. A charge reversible lipid for vector construction. *Polym Mater. Sci. Eng. Preprints*, 226th National Meeting, Am. Chem. Soc., New York, NY, **2003**.

20. Katre, N.V. Multivesicular liposomes (DepoFoam) for sustained delivery of therapeutics. *Polym Mater. Sci. Eng. Preprints*, 226th National Meeting, Am. Chem. Soc., New York, NY, **2003**.

21. Langston M.V.; Ramprasad, M.P.; Kararli, T.T.; Galluppi, G.R.; Katre, N.V. Modulation of the sustained delivery of myelopoietin (Leridistim) encapsulated in multivesicular liposomes (DepoFoam). *J. Controlled Release* **2003**, *89*, 87-99.

22. Fattal, E.; Roques, B.; Puisieux, F.; Blanco-Prieto, M.J.; Couvreur, P. Multiple emulsion technology for the design of microspheres containing peptides and oligopeptides. *Adv Drug Del Rev* **1997**, *28*, 85-96.

23. Breitenkamp, K.; Simeone, J.; Jin, E.; Emrick, T. Novel amphiphilic graft copolymers prepared by ring-opening metathesis polymerization of poly(ethylene glycol)-substituted cyclooctene macromonomers. *Macromolecules* **2002**, *35*, 9249-9252.

24. Breitenkamp, K.; Emrick, T. Novel polymer capsules from amphiphilic graft copolymers and cross-metathesis. *J. Am. Chem. Soc.* **2003**, *125*, 12070-12071.

25. Partch, R.; Powell, E.; Shah, D.; Varshney, M.; Dennis, D.; Morey, T.; Lee, Y.-H. Chemical selectivity of dispersed phases for removal of overdosed toxins from blood. *Polym Mater. Sci. Eng. Preprints*, 226th National Meeting, Am. Chem. Soc., New York, NY, **2003**.

26. Cheng, J.; Bellocq, N.; Pun, S.H.; Khin, K.T.; Liu, A.; Jensen, G.S.; Dartt, C.B.; Davis, M.E. Linear, cyclodextrin-based polymers for the delivery of broad ranging therapeutics. *Polym Mater. Sci. Eng. Preprints*, 226th National Meeting, Am. Chem. Soc., New York, NY, **2003**.

27. Luedtke, N.W.; Carmichael, P.; Tor, Y. Cellular Uptake of Amino-glycosides, Guanidinoglycosides, and Poly-Arginine. *J. Am. Chem. Soc.* **2003**, *125*, 12374-12375.

28. Chung, H.H.; Harms, G.; Seong, C.M.; Choi, B.H.; Min, C.; Taulane, J.P.; Goodman, M. Dendritic Oligoguanidines as Intracellular Translocators. *Biopolymers* **2004**, *76*, 83-96.

29. Morris, M.C.; Depollier, J.; Mery, J.; Heitz, F.; Divita, G. A peptide carrier for the delivery of biologically active proteins into mammalian cells. *Nat Biotechnol.*, **2001**, *19*, 1173-1176.

30. Allen, T.M.; Charrois, G.J.R.; Sapra, P. Recent advances in passively and actively targeted liposomal drug delivery systems for the treatment of cancer. *Polym Mater. Sci. Eng. Preprints*, 226th National Meeting, Am. Chem. Soc., New York, NY, **2003**.

31. Pan, D.; Turner, J.L.; Wooley, K.L. Folic acid-conjugated nanostructured materials designed for cancer cell targeting. *Chem. Commun.* **2003**, 2400-2401.

Particulate Drug Carriers

Chapter 2

Polymeric Micelles for the Delivery of Polyene Antibiotics

Ronak Vakil[1], Anuj Kuldipkumar[1], David Andes[2], Yvonne Tan[3], and Glen S. Kwon[1,*]

[1]School of Pharmacy, University of Wisconsin, Madison, WI 53705
[2]School of Medicine, University of Wisconsin, Madison, WI 53705
[3]School of Pharmaceutical Sciences, University of Sains Malaysia, Penang, Malaysia
*Corresponding author: gskwon@pharmacy.wisc.edu

Poly(ethylene glycol)-phospholipids assemble into nanoscopic supramolecular core-shell structures (micelles) that are being studied for drug solubilization. In this study, we demonstrate that poly(ethylene glycol)-phospholipid micelles effectively solubilize and deaggregate a membrane-acting antifungal drug, amphotericin B, based on size-exclusion chromatography and absorption spectroscopy. Deaggregated amphotericin B is encapsulated in 1,2-distearoyl-*SN*-glycero-3-phosphoethanol-amine-N-[methoxy (polyethylene glycol)] (M_n = 5,800 g/mol) micelles at a lipid to drug molar ratio of 4. Thus, amphotericin B is expected to be selective for fungal cells at a membrane level and less toxic than its standard formulation, where it is highly aggregated.

Introduction

Amphiphilic block copolymers (ABC) assemble into nanoscopic supramolecular core-shell structures, micelles, which are under extensive study for drug delivery (*1-3*). The shell of ABC micelles consists of poly(ethylene glycol), a non-toxic, water-soluble polymer, which prevents protein adsorption and cellular adhesion, and the core of ABC micelles consists of a smaller hydrophobic block or moiety, varying in drug delivery due to requirements of biocompatibility, compatibility with the companion drug, usually a poorly water-soluble compound, and stability of the ABC micelle in blood. In drug delivery, a goal is to study the potential of ABC micelles as 'nanocarriers' for potent, toxic and poorly water-soluble drugs.

Many existing drugs used in the treatment of life-threatening diseases and many compounds coming out of massive drug discovery efforts have poor water solubility due to hydrophobicity, low hydrogen bond capacity and large size. Adjustments in pH, cosolvents, cyclodextrins and surfactants are the primary methods used for drug solubilization (*4*). However, drug solubilization by these methods is often unable to realize adequate levels of drug in solution for parenteral administration at target doses. In addition, parenteral excipients for drug solubilization often exert toxicity, particularly in combination with anti-tumor drugs. Cremophor® EL, a nonionic surfactant commonly used for drug solubilization, causes hemolysis, hypersensitivity reactions, neurotoxicity and hyperlipidemia after intravenous administration (*5*).

Pluronics® (ABA block copolymers), poly(ethylene glycol)-*b*-poly(ester)s, poly(ethylene glycol)-*b*-poly(L-amino acid) (p(L-AA)) and poly(ethylene glycol)-phospholipid micelles have been studied extensively for drug delivery (*1-3*). Though poly(ethylene glycol)-phospholipid micelles are not true ABC micelles, they have been shown to possess similar properties for drug delivery, such as structural stability in blood (*6*). For drug solubilization, a primary consideration is the compatibility between the drug and the core-forming block (*3,7*). Considerations of structural similarity or polarity, e.g., solubility parameters, gives some insight into compatibility. A goal is to better understand the structural factors that dictate drug loading and drug release of ABC micelles, and this includes consideration of ABC assembly.

Poorly water-soluble drugs are often released rapidly from ABC micelles due to the break-up of ABC micelles in blood or rapid diffusion of drug from their cores, which are small and easily traversed. Increasing the core dimensions of ABC micelles by increasing the molecular weight of the hydrophobic block can increase drug loading and slow drug release. However, varying molecular weight of ABCs is synthetically challenging; water solubility of an ABC puts an upper limit on the molecular weight of the hydrophobic block; and ABCs may assemble into supramolecular structures besides micelles (such as vesicles) as the molecular weight of the hydrophobic block approaches the molecular weight of poly(ethylene glycol) (8,9). The diffusivity of drugs in cores of ABC micelles depends on the size of the drug, core viscosity and compatibility between drug and core of the ABC micelles. For drug targeting, ABC micelles that have covalently bound or complexed drug in their cores have been considered (10,11).

Paclitaxel incorporated in poly(ethylene glycol)-b-poly(lactic acid) micelles is less toxic in rodents than its standard formulation, Taxol®, which contains Cremophor® EL, without major changes in its pharmacokinetics (3). Apparently, paclitaxel is released fairly rapidly from poly(ethylene glycol)-b-poly(lactic acid) micelles. Poly(ethylene glycol) and poly(lactic acid) are biocompatible and approved for use in humans. Thus, PEG-b-PLA micelles may permit higher doses of paclitaxel for the treatment of ovarian and breast cancer and superior anti-tumor activity. Phase II clinical trials are underway to test this hypothesis.

Poly(ethylene glycol)-b-p(L-AA) micelles increase the circulation time of doxorubicin, another anti-tumor drug, reduce its accumulation in the heart (site of dose-limiting cardiotoxicity) and increase its accumulation at tumors due to preferential extravasation at leaky vasculature (1-3). Tumors that recruit their own blood supply by angiogenesis possess leaky blood vessels, which serve as a conduit for drug delivery (12). Drug delivery systems that are small (< 100 nm), long-circulating and capable of sustained or controlled drug release target solid tumors through leaky vasculature. Poly(ethylene glycol)-b-poly(aspartic acid)-doxorubicin conjugate micelles that carry physically-bound doxorubicin in their cores are being evaluated in Phase I clinical trials in Japan.

In efforts on poly(ethylene glycol)-b-p(L-AA) micelles, we have shown that the side chains of a p(L-AA) block can be easily adjusted in order to vary the properties of ABC micelles, enhance drug incorporation, and gain some measure of controlled drug release (3). In these studies, the length of an acyl side chain attached on poly(ethylene glycol)-b-p(L-AA) was found to govern the extent of interaction, incorporation, and release of an antifungal, amphotericin B (AmB) (Figure 1). As a result, the toxicity of AmB (hemolysis) was reduced without a loss of antifungal effects.

We hypothesize that poly(ethylene glycol)-b-p(L-AA) micelles increase the therapeutic index of AmB by sustained release or changes in the self-aggregation state of this drug. The latter is because of the membrane activity of AmB, and its preferential interaction with ergosterol in fungal membranes versus cholesterol in mammalian cell membranes when it exists in a monomeric or deaggregated state (Figure 2) (13). AmB is highly aggregated in Fungizone®, its standard formulation, which contains sodium deoxycholate. Thus, AmB is toxic, causing acute reactions, such as fever, chills and rigor, and its renal toxicity is dose-limiting. AmB is the primary drug used to treat life-threatening systemic fungal diseases, which are common among immuno-compromised patients, e.g., AIDS patients, organ transplant patients. Fungal pathogens are now the fourth most common cause of bloodstream infection in hospitals with nearly a 40% mortality rate (14).

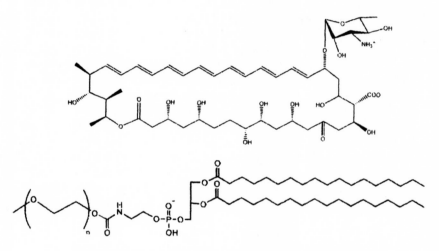

Figure 1. Chemical structures of AmB and PEG-DSPE.

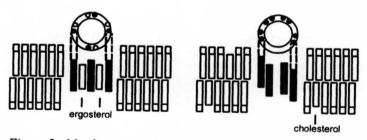

Figure 2. Membrane activity of monomeric and aggregated AmB.

18

More recently, we have studied poly(ethylene glycol)-phospholipid micelles for the delivery of AmB, owing to the presumed compatibility between the drug and phospholipid, proven safety profile of poly(ethylene glycol)-phospholipid, which are approved for use in humans and perhaps the ability to deaggregate AmB to reduce its toxicity. Thus, we hypothesize that poly(ethylene glycol)-phospholipid micelles deaggregate AmB and lower its toxicity without a loss of antifungal activity. Because poly(ethylene glycol)-phospholipids are generally regarded as safe materials (GRAS), clinical trials could proceed without concern of biocompatibility, which hamper progress of poly(ethylene glycol)-*b*-p(L-AA). Furthermore, this research may help to separate the effects of sustained release of AmB and drug deaggregation for poly(ethylene glycol)-*b*-p(L-AA) micelles.

We have incorporated AmB in poly(ethylene glycol)-phospholipid micelles by a method that involves solvent evaporation and reconstitution of a solid film by water. Poly(ethylene glycol)-phospholipid micelles deaggregate AmB based on analysis of the self-aggregation state of the drug by absorption spectroscopy, indicating that there is a reduction in the toxicity of AmB at a membrane level. Size-exclusion chromatography has confirmed the formation of micelles and drug loading, and it suggests that monomeric AmB inside poly(ethylene glycol)-phospholipid micelles is quite stable, staying inside for some time without loss.

Materials and Methods

AmB was obtained from Chem-Impex (Wood Dale, IL). 1,2-distearoyl-*SN*-glycero-3-phosphoethanolamine-N-[methoxy(polyethylene glycol)](PEG-DSPE) (M_n = 5,800 g/mol) was purchased from Avanti Polar Lipids (Alabaster, AL). A clinical bloodstream isolate of *Candida albicans* (K-1) was used for the *in vivo* efficacy study. Sabourand dextrose agar (SDA) slants were obtained from Difco Laboratories (Detroit, MI). Fungizone® was obtained from Bristol-Meyers Squibb (Princeton, NJ) and stored at -70°C. All other chemicals were reagent grade and above and were used without additional purification.

Incorporation of AmB in PEG-DSPE micelles

AmB (654 µg) and PEG-DSPE (2.21, 3.33, 4.00, 5.25, 7.83 and 11.25 mg) were added to 5.0 mL of methanol and sonicated for 5 min to produce a clear solution. The yellow solution was transferred to a round bottom flask and the methanol removed by rotoevaporation at 40°C to produce a thin film of PEG-DSPE and AmB. The solid film was dissolved with 5.0 mL of distilled water, incubated at 40°C for 10 min and vortexed for 30 sec. Unincorporated AmB was removed by filtration (0.22 µm), and the aqueous solution was freeze-dried.

Spectroscopic analysis of AmB

Freeze-dried samples were reconstituted with distilled water and filtered (0.22 μm). Samples were diluted with an equal volume of *N,N'*-dimethyl-formamide (DMF) and the content of AmB assayed by measuring absorbance at 412 nm. In order to analyze the aggregation state of AmB, samples were reconstituted with distilled water, and spectra were acquired from 300 to 450 nm at a rate of 405 nm/min at a scan step of 0.1 nm.

Size-exclusion chromatography

Aqueous samples of AmB incorporated PEG-DSPE micelles were incubated at 37°C in 10 mM PBS, pH=7.4. At selected time intervals, 100 μL samples were injected in a Shodex PROTEIN-KW 804 SEC column. Samples were eluted using 50 mM phosphate buffer at 0.75 mL/min. The elution of AmB and PEG-DSPE micelles were monitored by UV and refractive index detection, respectively.

In vivo antifungal activity

A neutropenic murine model of disseminated candidiasis was used to test the efficacy of AmB (*15*). *Candida albicans* (K-1) was maintained, grown and quantified on SDA slants. Six week old ICR/Swiss specific pathogen free female mice weighing 23 to 27 g were rendered neutropenic by the intra-peritoneal injection of cyclophosphamide over several days. Disseminated infection was induced by tail vein injection of 0.1 mL of inoculum (10^6 colony forming units (CFUs)). AmB in PEG-DSPE micelles at varied doses was injected intravenously 2 h post infection. Over 24 h, mice were sacrificed, and kidneys were removed, homogenized and plated on SDA slants. After 24 h, plates were inspected for determination of CFUs. Results were expressed as mean CFU/kidney for two mice.

Results and Discussion

The direct dissolution of AmB as Fungizone® (2:1 molar ratio of sodium deoxycholate to AmB) in water produces some monomeric drug and various species of soluble aggregates of the drug combined with sodium deoxycholate. Similarly, AmB dissolved in DMF and diluted with excess water produces some soluble monomer and various species of soluble aggregates. The proportion of these species of AmB depends on the method of dissolution, temperature and concentration. In both cases, AmB precipitates given sufficient time. In both cases, the absorption spectrum of AmB is characterized by a broad band centered at 328 nm, corresponding to the soluble aggregated species of AmB (A form). In contrast, AmB is entirely monomeric in methanol or DMF, and it has a characteristic spectrum with four separated bands at 348, 365, 385 and 409 nm (B form).

We were unable to directly dissolve AmB and PEG-DSPE in water and obtain complete dissolution of drug. Instead, we dissolved AmB and PEG-DSPE in methanol, removed the solvent by rotoevaporation to make a solid film of drug and polymer and added water. As a result, PEG-DSPE micelles solubilized AmB (Table I). The yield of AmB (level of incorporated drug/initial level of drug) ranged from 57 to 93%, increasing with the level of PEG-DSPE until about 90%. The ratio of PEG-DSPE to AmB ranged from 0.90 to 3.2 mol:mol. The level of AmB reached 240 µg/mL after reconstitution in water, a level that permits an adequate dose for the treatment of systemic fungal diseases.

Table I. Solubilization of AmB by PEG-DSPE micelles

PEG-DSPE (mg)	AmB (µg)	Solubilized AmB (µg)	PEG-DSPE: AmB (mol:mol)	Yield (%)	I/IV ratio
2.21	650	370	0.90	57	2.97
3.33	650	520	1.0	80	1.68
4.00	650	580	1.1	90	1.30
5.25	650	580	1.4	90	0.75
7.83	650	600	2.0	93	0.50
11.25	650	550	3.2	84	0.35

The self-aggregation state of AmB varied with the ratio of PEG-DSPE to AmB from 0.90 to 3.2 (Figure 3). A broad band at 328 nm that is characteristic of aggregated species of AmB was predominant at a molar ratio of 0.90. The intensity of the band at 328 nm decreased relative to the other bands at higher wavelengths that are associated with monomeric drug at a PEG-DSPE: AmB ratio of 1.0. With an increase in PEG-DSPE content, sharp bands at 368, 388, and 417 nm increased in intensity and were predominant. The ratio of the intensity at 328 nm to 417 nm, i.e., I/IV ratio, a measure of the degree of aggregation of AmB (16), decreased from 2.97 to 0.34 with an increase in PEG-DSPE content, consistent with drug deaggregation.

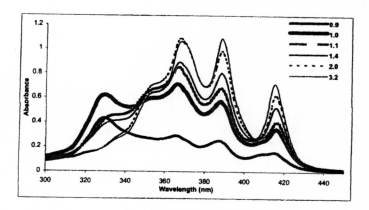

Figure 3. Absorption spectra of AmB solubilized by PEG-DSPE micelle at varied mole ratios of PEG-DSPE to AmB (14 µg/mL).
(See page 1 of color inserts.)

SEC results reveal that the deaggregation of AmB occurs as a result of its incorporation in PEG-DSPE micelles (Figure 4a). Empty PEG-DSPE micelles eluted at 7.68 mL (data not shown). Similarly, PEG-DSPE micelles with AmB eluted at 7.70 mL at a ratio of polymer to drug at 4.0, according to refractive index and UV-VIS detection, and there was no appearance of free drug during an incubation in PBS over several hours at 37°C. In contrast, aggregated AmB at a ratio of 0.50 eluted as three species, one at 6.05 mL and a second at 6.91 mL, both corresponding to aggregated species of AmB, and a third at 7.93 mL, corresponding to PEG-DSPE micelles with incorporated AmB in a monomeric state (Figure 4b). At a ratio of 0.5, a relatively high level of AmB, it appears that the capacity of the PEG-DSPE micelles for the drug was exceeded, and as a result AmB assembles into aggregates, which are larger than the PEG-DSPE micelles that contain drug.

The deaggregation of AmB in the cores of PEG-DSPE micelles likely occurs by a strong binding of AmB and distearoyl chains. A slight bathochromic shift and a difference in the intensity of bands associated with monomeric AmB in cores relative to the drug in methanol was an indication of this interaction. In particular, the intensity of the band at 417 nm was less than the intensity of bands at 368 and 388 nm for monomeric AmB in PEG-DSPE micelles. For AmB in methanol, the opposite was observed, and bands, II, III, and IV, were present at 365, 385, and 409 nm.

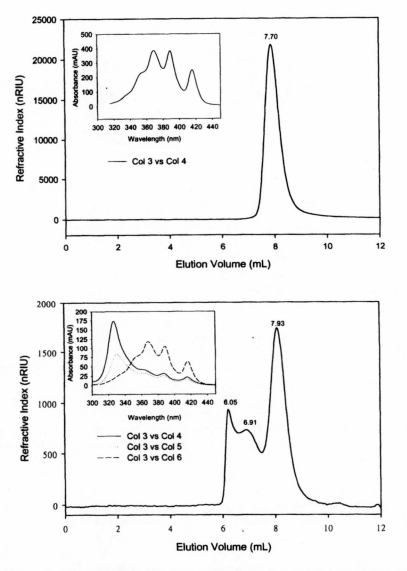

Figure 4. SEC-HPLC of AmB (40 μg/mL) after incorporation in PEG-DSPE micelles at (a) 4.0 (b) 0.5 mole ratio of PEG-DSPE to AmB. Absorption spectra of eluting AmB were obtained by a photo diode array detector (inset).

Several strategies have been studied to deaggregate and thus lower the toxicity of polyene antibiotics (*13*). While nontoxic Pluronics® are unable to deaggregate AmB, they were able to deaggregate nystatin, a structurally related polyene antibiotic, and thus reduce its toxicity in terms of hemolysis (*17*). Gruda and coworkers added excess surfactant such as sodium deoxycholate or lauryl sucrose to AmB and obtained reduced self-aggregation and acute toxicity in mice (*18*). Tassett and coworkers added excess Myrj 59, a PEG derivative of stearic acid, to AmB and obtained a reduction in self-aggregation (*19*). In both cases, however, the high levels of surfactant (mol ratio >50) needed to deaggregate AmB are likely too toxic for parenteral use in humans. Much less PEG-DSPE with a distearoyl chain was required than Myrj 59 with a single stearoyl chain for the deaggregation of AmB. Similarly, poly(ethylene glycol)-*b*-poly(L-aspartate) with 17 stearoyl side chains forms micelles that readily deaggregate AmB, reducing its toxicity, as reflected in hemolysis experiments (*20, 21*). AmB incorporated in these ABC micelles expresses potent *in vivo* antifungal activity (*15*). While promising, extensive toxicological studies will have to validate the safety of this block copolymer. In contrast, PEG-DSPE has been used safely in humans after intravenous injection. Toxicity studies on AmB formulated by PEG-DSPE micelles are underway in mice.

AmB encapsulated in PEG-DSPE micelles has potent antifungal activity in a neutropenic murine model of disseminated candidiasis in a preliminary study (Figure 5). At the start of therapy, the $\log_{10}$CFU at the kidneys of the mice was 3.47 ± 0.16. The $\log_{10}$CFU at kidneys for control mice (saline) increased to 7.01 ± 0.57 after 24 h. The $\log_{10}$CFU at kidneys for mice treated with AmB had lower values over time, and it depended on the dose of AmB from 0.088 to 1.44 mg/kg. At the highest dose of AmB, the reduction in colony count relative to the initial fungal burden was 1.5. The antifungal activity of AmB in PEG-DSPE micelles was not much different than AmB as Fungizone® (data not shown).

Conclusions

In summary, we have incorporated AmB in PEG-DSPE micelles by a procedure based on solvent evaporation. The level of solubilized AmB in water provides adequate doses for evaluation in rodent models and clinical evaluation. PEG-DSPE micelles deaggregate AmB, and encapsulated AmB expresses potent antifungal activity. Additional *in vivo* studies will assess the toxicity of AmB in rodents, focusing on acute toxicity and nephrotoxicity as well further evaluation of antifungal activity with the goal of clinical evaluation in the next few years.

Acknowledgements

This work was funded by NIH grant AI-43346-02.

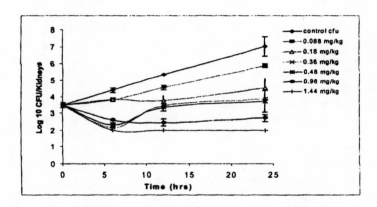

Figure 5. In vivo time-kill plot for AmB in PEG-DSPE micelles (1.34 mol/mol) as a function of dose in a neutropenic murine model of candidiasis. (See page 1 of color inserts.)

References

1. Kataoka, K.; Kwon, G. S.; Yokoyama, M.; Okano, T.; Sakurai, Y. Block copolymer micelles as vehicles for drug delivery. *J. Controlled Release* **1993**, *24*, 119.

2. Torchilin, V. P. Structure and design of polymeric surfactant-based drug delivery systems. *J. Controlled Release* **2001**, *73*, 137.

3. Kwon, G. S. Polymeric micelles for delivery of poorly water-soluble compounds. *Crit. Rev. Ther. Drug Carrier Syst.* **2003**, *20*, 357.

4. Jonkman-de Vries, J. D.; Flora, K. P.; Bult, A.; Beijnen, J. H. Pharmaceutical development of (investigational) anticancer agents for parenteral use. A review. *Drug Dev. Ind. Pharm.* **1996**, *22*, 475.

5. Gelderblom, H.; Verweij, J.; Nooter, K.; Sparreboom, A. Cremophor EL: The drawbacks and advantages of vehicle selection for drug formulation. *Eur. J. Cancer* **2001**, *37*, 1590.

6. Torchilin, V. P.; Lukyanov, A. N.; Gao, Z.; Papahadjopoulos-Sternberg, B. Immunomicelles: Targeted pharmaceutical carriers for poorly soluble drugs. *Proceed. Natl. Acad. Sci. USA* **2003**, *100*, 6039.

7. Liu, J.; Xiao, Y.; Allen, C. J. Polymer-drug compatibility: A guide to the development of delivery systems for the anticancer agent, ellipticine. *J. Pharm. Sci.* **2004**, *93*, 132.

8. Zhang, L.; Eisenberg, A. Multiple morphologies of "crew-cut" aggregates of polystyrene-b-poly(acrylic acid) block copolymers. *Science* **1995**, *268*, 1728.

9. Discher, B.; Won, Y.-Y.; Ege, D. S.; Lee, J. C.-M.; Bates, F. S.; Discher, D. E.; Hammer, D.A. Polymersomes: Tough vesicles made from diblock copolymers. *Science* **1999**, *284*, 1143.

10. Bae, Y.; Fukushima, S.; Harada, A.; Kataoka, K. Design of environment-sensitive supramolecular assemblies for intracellular drug delivery: Polymeric micelles that are responsive to intracellular pH change. *Angew. Chem. Int. Ed.* **2003**, *42*, 4640.

11. Nishiyama, N.; Okazaki, S.; Cabral, H.; Miyamoto, M.; Kato, Y.; Sugiyama, Y.; Nishio, K.; Matsumura, Y.; Kataoka, K. Novel cisplatin-incorporated polymeric micelles can eradicate solid tumors in mice. *Cancer Res.* **2003**, *63*, 8977.

12. Duncan, R. The dawning era of polymer therapeutics. *Nature Rev. Drug Discov.* **2003**, *2*, 347.

13. Brajtburg, J.; Bolard, J. Carrier effects on biological activity of amphotericin B. *Clin. Microbiol. Rev.* **1996**, *9*, 512.

14. Fridkin, S. K.; Jarvis, W. R. Epidemiology of nosocomial fungal infections. *Clin. Microbiol. Rev.* **1996**, *9*, 499.

15. Adams, M. L.; Andes, D. R.; Kwon, G. S. Amphotericin B encapsulated in micelles based on poly(ethylene oxide)-block-poly(L-amino acid)

derivatives exerts reduced in vitro hemolysis but maintains potent in vivo antifungal activity. *Biomacromolecules* **2003**, *4*, 750.

16. Gruda, I.; Dussault, N. Effect of the aggregation state of amphotericin B on its interaction with ergosterol. *Biochem. Cell Biol.* **1988**, *66*, 177.

17. Croy, S. R.; Kwon, G. S. The effects of Pluronic block copolymers on the aggregation state of nystatin. *J. Controlled Release* **2004**, *95*, 161.

18. Barwicz, J.; Christian, S.; Gruda, I. Effects of the aggregation state of amphotericin B on its toxicity to mice. *Antimicrob. Agents Chemother.* **1992**, *36*, 2310.

19. Tasset, C.; Preat, V.; Roland, M. The influence of Myrj 59 on the solubility, toxicity and activity of amphotericin B. *J. Pharm. Pharmacol.* **1991**, *43*, 297.

20. Lavasanifar, A.; Samuel, J.; Sattari, S.; Kwon, G. S. Block copolymer micelles for the encapsulation and delivery of amphotericin B. *Pharm. Res.* **2002**, *19*, 418.

21. Adams, M. L.; Kwon, G. S. Relative aggregation state and hemolytic activity of amphotericin B encapsulated by poly(ethylene oxide)-block–poly(N-hexyl-L-aspartamide)-acyl conjugate micelles: effects of acyl chain length. *J. Controlled Release* **2003**, *87*, 23.

Chapter 3

Polymeric Micelle Drug Carriers for Tumor Targeting

Masayuki Yokoyama

Kanagawa Academy of Science and Technology, KSP East 404,
Sakado 3–2–1, Takatsu-ku, Kawasaki-shi, Kanagawa-ken 213–0012, Japan
(email: masajun@ksp.or.jp)

The methodology of targeting of anti-cancer drugs to solid tumors using polymeric micelle drug carriers is described. Polymeric micelles inherently possess several strong advantages owing to their physico-chemical properties for tumor targeting by a passive targeting mechanism called Enhanced Permeability and Retention (EPR) effect. Several examples for the tumor targeting are discussed.

28

Introduction

Targeting of Anti-cancer Drugs to Solid Tumors

Drug targeting is defined as selective drug delivery to specific sites, organs, tissues or cells, where drug activities are required. By increasing delivery to the therapeutic sites and reducing delivery to unwanted sites, an improved therapeutic index can be obtained with enhanced drug action at the therapeutic site. Drug targeting can be classified into two methods: active and passive targeting (1,2). Active targeting aims at an increase in the delivery of drugs to the target by utilizing biologically specific interactions such as antigen–antibody binding or by utilizing locally applied signals such as heating and sonication. Carriers classified in this method include specific antibodies, transferrin, and thermo-responsive liposomes and polymeric micelles. Passive targeting is defined as a method whereby the physical and chemical properties of carrier systems increase the target/non-target ratio of a quantity of a delivered drug. Which method, active or passive targeting, is superior and more applicable to tumor targeting? Both methods are important for drug targeting, however, it should be emphasized that factors governing passive targeting are also important for active targeting systems for the following reasons:

(1) The vast majority of a living body comprises non-target sites. Even the liver, one of the largest targets of drugs, only occupies approximately 2% of the weight of the entire body. That is, 98% of the body can be considered to be non-target sites, while active targeting can only be achieved within the remaining small body fraction. Minimizing non-specific capture at non-target sites by passive targeting may therefore be important to maximize the amount delivered to active targeting sites.

(2) Passive targeting precedes active targeting in most cases. Exceptions are cases for intravascular targets such as lymphocytes and vascular endothelial cells. Most targets are located in the extravascular space. To reach these targets via the bloodstream, the first step must be extravasation through the vascular endothelia, followed by permeation through the interstitial space to the extravascular targets.

Targeting of anti-cancer drugs to solid tumors using polymeric micelles has been studied mostly with respect to passive targeting, and therefore, only passive targeting to solid tumors is discussed in this chapter.

Passive targeting of polymeric micelles to solid tumors can be achieved through the Enhanced Permeability and Retention (EPR) effect, presented by Maeda and Matsumura in 1986 (3,4). As illustrated in Figure 1, the vascular permeability of tumor tissues is enhanced by secreted factors such as kinin. As a result of this increased vascular permeability, macromolecules increase their transport from blood vessels to tumor tissues, while small molecules do not change their transport. Furthermore, the lymphatic drainage system does not operate effectively in tumor tissues, and therefore, macromolecules are selectively retained for a prolonged time period in the tumor interstitium. For utilizing the EPR effect, specific targeting moieties such as antibodies are not necessary. As a result of this effect, macromolecules including polymeric micelles can selectively accumulate at solid tumor sites.

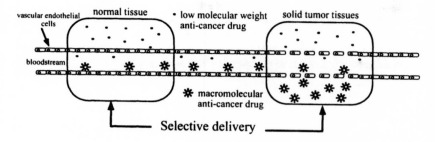

Figure 1. Enhanced Permeability and Retention (EPR) effect of macromolecules at solid tumor tissues.

However, carrier polymers must fulfill two requirements to avoid non-specific capture at non-tumor sites:

(1) Drug carriers must possess an appropriate size or molecular weight. The diameter of a carrier must be smaller than ca. 200 nm if the reticuloendothelial system's uptake is to be evaded (5). Additionally, molecular weights larger than the critical value of approximately 40,000 Da are favorable for evading renal filtration.

(2) Drug carriers must not strongly interact or even being taken up by normal organs, especially the reticuloendothelial systems. This behavior is typically observed for cationic and hydrophobic polymers (6). Therefore, carrier polymers should be hydrophilic with neutral or weak negative overall charge, and they should not contain chemical structur elements that would be biologically recognizable to normal tissues.

Because polymeric micelles are formed in a diameter range from 10 to 100 nm, the size requirement for the EPR effect is inherently fulfilled. Furthermore, the charge requirement can be easily fulfilled by the use of hydrophilic and neutral or weakly negative-charged polymers as building block for the outer shell.

Results and Discussion

Advantages of Polymeric Micelles for TumorTargeting

A polymeric micelle is a macromolecular assembly that forms from block or graft copolymers, has a spherical inner core and an outer shell (7). As shown in Figure 2 in which an AB-type block copolymer is being used, a micellar structure forms if one segment of the block copolymer can provide enough interchain cohesive interactions in a solvent. Most studies of polymeric micelles, both in basic and applied research, have been done with AB or ABA-type block copolymers because the close relationship between micelle-forming behavior and structure of the polymers can be evaluated more easily with these block copolymers than with graft or multi-segmented block copolymers.

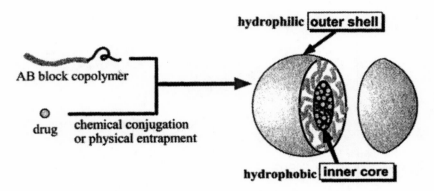

Figure 2. Formation of polymeric micelles as drug carriers.

The cohesive interactions in the inner core utilized as the driving force for micelle formation include hydrophobic, electrostatic, and $\pi-\pi$ interactions as well as hydrogen bonding. Because most anti-cancer drug molecules possess a hydrophobic character, hydrophobic interactions are most commonly used for tumor targeting (*8-20*). Drugs can be incorporated into polymeric micelles both by chemical conjugation and physical entrapment. Polymeric micelles possess strong and unique advantages for tumor targeting, which are summarized in Table 1.

Table I. Advantages of polymeric micelles for tumor targeting

Very small diameter (10 – 100 nm)
High structural stability
High water solubility
Low toxicity
Separated functionality

First, polymeric micelles are formed typically in a diameter range from 10 to 100 nm, with very narrow size distribution. This size range is considered ideal for the attainment of tumor targeting by the EPR effect because it evades the reticuloendothelial system's uptake and renal excretion. Second, polymeric micelles possess high structural stability, provided by the entanglement of polymer chains in the inner core. This stability has static and dynamic aspects (21-24). Static stability is described by a critical micelle concentration (cmc). Generally, polymeric micelles show very low cmc values in a range form 1 to 10 μg/mL. These values are much smaller than typical cmc values of micelles forming from low molecular weight surfactants. The dynamic stability is described by low dissociation rates of micelles, and this aspect might be more important for in-vivo drug delivery in physiological environments, which are in non-equilibrium conditions. The high structural stability of polymeric micelles is an important key to drug delivery through micelles and eliminates the possible contribution of single polymer chains as drug carriers. The third advantage is the high water solubility of polymeric micelles encapsulating hydrophobic anti-cancer drugs. In conventional polymeric drug carrier systems, a loss of water solubility of the polymeric carrier resulting from the interaction with a hydrophobic drug creates a serious problem. Drug conjugation to a homo-polymer easily leads to precipitation because of the high, localized concentration of hydrophobic drug molecules bound along the polymer chain. Several research groups reported this problem during synthesis of drug–homopolymer conjugates (25-27) and during their intravenous injections (28). Therefore, conventional drug–polymer conjugates must be designed with considerably low drug content to avoid or reduce the risk of precipitation. In contrast, polymeric micelles maintain their water solubility because the hydrophilic outer shell works as a barrier against intermicellular aggregation of the hydrophobic cores. This results in much larger hydrophobic drug contents for these carriers than carriers based on conventional polymers. For example, the maximum loading of anticancer drugs adriamycin or daunomycin (an adriamycin derivative) was reported to range from 10 to 35 wt% for conventional polymer-drug conjugates (25,26,29,30), while a polymeric micelle system contained 60 wt% of adriamycin (31). This advantage is especially important in cancer chemotherapy because most of recently developed anti-cancer drugs are strongly hydrophobic.

The beneficial character of low toxicity may be described as the forth advantage. In general, polymeric surfactants are known to be less toxic than low molecular weight surfactants such as sodium dodecyl sulfate, as exemplarily shown for Pluronic (32). Furthermore, polymeric micelles are considered very safe with respect to chronic toxicity. Possessing a much larger size than the critical filtration values in the kidney, polymeric micelles can evade renal filtration, even if the molecular weight of the constituting block copolymer is lower than the critical molecular weight for renal filtration. In addition, all constituent polymer chains of a polymeric micelle can be released as single strands from the micelle during a long time period because these strands are not chemically bound to each other. This phenomenon results in complete excretion of the block copolymers from the renal route if the polymer chains are designed with a lower molecular weight than the critical value for the renal filtration. This phenomenon is a huge advantage of polymeric micelles over conventional (non-micelle forming) and non-biodegradable polymeric drug carriers. The fifth advantage is separated functionality. Polymeric micelles are composed of two phases, the inner core and the outer shell. Various functions required for drug delivery systems can be shared by these structurally separated phases. For example, the outer shell is responsible for interactions with biocomponents such as proteins and cells. These interactions determine where drug carriers go in the living body; and therefore, the shell controls the in-vivo delivery of drugs. The inner core is responsible for pharmacological activities through drug loading and release. The properties of both phases are independently controlled through the selection of the respective constituent block of the polymer strand and makes this heterogeneous structure more favorable for the construction of highly functionalized carrier systems than conventional (non micelle-forming) polymeric carriers.

Examples of Anti-cancer Drug Targeting to Solid Tumors

1. Adriamycin (Doxorubicin)

Yokoyama, Kwon, Okano, and Kataoka et al. succeeded in targeting the anticancer drug adriamycin (ADR) (= doxorubicin) to solid tumors, using a polymeric micelle system (33-36). Adriamycin was chemically conjugated to aspartic acid residues of poly(ethylene glycol)–poly(aspartic acid) block copolymers, PEG-P(Asp) by amide bond formation. The PEG segment was hydrophilic, whereas the ADR-substituted P(Asp) chain was hydrophobic. Therefore, the obtained drug–block copolymer conjugate PEG–P(Asp)-ADR formed micellar structures owing to its amphiphilic character. On the second step, ADR was incorporated into the hydrophobic inner core by physical entrapment. As a result, polymeric micelles containing both the chemically conjugated and the physically entrapped ADR were obtained with the PEG outer shell.

As shown in Figure 3, the physically entrapped ADR was present at much higher concentrations for a long time-period (Fig. 3a) and was delivered to the solid tumor site at much higher concentrations than free ADR (Fig. 3b) (37). The observed time profile with a peak concentration at 24 hours post intravenous injection and an extended retention of this high concentration post injection matched well with passive delivery by the EPR effect (3). On the other hand, accumulation of the physically entrapped ADR at normal organs and tissues was the same or lower than observed for free ADR.

In accordance with this highly selective delivery to solid tumor sites, a dramatic enhancement of anti-tumor activity was observed (37). Figure 4 shows in-vivo anti-tumor activity against murine colon adenocarcinoma 26. For free ADR, only the maximum tolerated dose (10 mg/kg body weight) provided considerable inhibition effects on tumor growth; however, a decrease in tumor volume was never seen from the day of the first injection. For the polymeric micelles, the tumor completely disappeared in two doses (20 and 10 mg physically entrapped ADR/kg of body weight). These results clearly demonstrate the successful passive targeting of an anti-cancer drug using a polymeric micelle carrier system to a tumor. This adriamycin-containing system has passed a Phase I clinical trial and entered the Phase II trial in autumn of 2003 at the National Cancer Hospital in Japan.

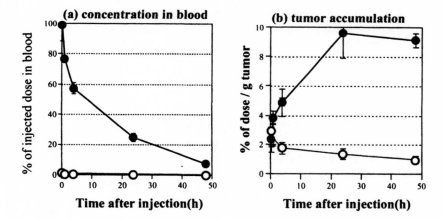

Figure 3. Concentration in blood (a) and tumor accumulation (b) after intravenous injection to tumor-bearing mice. Filled plot: polymeric micelle, vacant plot: free adriamycin. (Adapted with permission from reference 37. Copyright 1999 Taylor & Francis.)

34

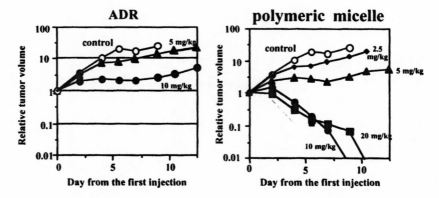

Figure 4. Anti-tumor activity against murine C26 tumor. Doses are expressed in the physically entrapped adriamycin for polymeric micelles. The highest doses are the maximum tolerated doses both for ADR and polymeric micelles. (Adapted with permission from reference 37. Copyright 1999 Taylor & Francis.)

2. KRN-5500

Although adriamycin shows some hydrophobic properties, it is still water-soluble. In the field of cancer chemotherapy, a strong demand for the solubilization of water-insoluble drugs has been voiced particularly since many newly developed and very potent anticancer drugs such as camptothecin and taxol are water-insoluble or poorly water-soluble. In this perspective, the incorporation of a water-insoluble anticancer agent, KRN-5500 (KRN), into polymeric micelles was studied (*38-40*). By choosing an appropriate structure of a block copolymer, the water-insoluble KRN was successfully incorporated into polymeric micelles (*38*). This polymeric micelle showed higher anti-tumor activity than free KRN in a conventional formulation. This polymeric micelle formulation did not exhibit the severe vascular and pulmonary toxic side effects that surfaced in the conventional formulation of KRN, a result that is due to the toxicity of the organic solvents and the surfactants used to dissolve KRN (*39,40*). This fact indicates that polymeric micelle carrier systems are not only a strong strategy for drug targeting but also very potent in dissolving water-insoluble drugs for safe intravenous injection.

3. Cisplatin

Cisplatin is a platinum chelate with two chloride and two ammonium ligands. This compound is not well water-soluble. Nishiyama and colleagues encapsulated cisplatin into the aspartic acid residues of poly(ethylene glycol)-

poly(aspartic acid) block copolymers by ligand exchange reaction between chloride and carboxylate (*41*). These block copolymers aggregated in water to form polymeric micelles. The micelles were not destroyed after addition of the surfactant sodium dodecyl sulfate (SDS) as observed for micelles incorporating adriamycin. This micellar stability was explained by crosslinking of the poly(aspartic acid) blocks through platinum atoms. In physiological saline, cisplatin could be released from the micelles by ligand exchange reaction between carboxylate and chloride, and, simultaneously, disintegration of the micellar structures. This cisplatin delivery system is now on a development stage for a pre-clinical test using a poly(glutamic acid) block instead of the poly(asaprtic acid) block (*42*).

4. Taxol

Taxol is a water-insoluble anti-cancer drug and is recognized as one of the most potent drugs among the approved anti-cancer drugs in the 1990s. For intravenous injection, the surfactant Cremophor® EL and ethanol are used to solubilize taxol. Because of the considerable toxicity of Cremophor® EL, formulations are being studied that do not contain this surfactant. The Samyoung Company developed poly(ethylene glycol)-*b*-poly(lactic acid) block copolymer-based micelles containing taxol (*43*). The water solubility of this formulation was significantly increased compared to the conventional Cremophor® EL containing formulation, and the toxicity was significantly reduced. This system is now in a clinical trials. This system, however, does not show targeting effects because all of the incorporated drug is released within several minutes after intra-venous injection. Therefore, another micelle carrier system with targeting ability is studied, using a poly(ethylene glycol)-b-poly(aspartate) block copolymer (*44*).

5. Methotrexate

Methotrexate (MTX) is a widely used anti-cancer drug and recently has also been used to treat rheumatoid arthritis. Kwon and associates conjugated MTX to a block copolymer and obtained polymeric micelles. In this system, the MTX-conjugated hydrophobic block served as the driving force for the micelle formation (*45*). The micelle stability was successfully controlled by controlling the amount of conjugated MTX. The dynamic micelle stability, i.e., the equilibrium between micelle and single polymer chains, is correlated with the drug release rate because hydrolysis of MTX from the polymer strands can proceed more rapidly from a single polymer chain due to easier access to both water and hydrolysing small molecules. This is the first example of release control of a chemically conjugated drug.

Conclusions

Polymeric micelles are powerful carrier systems for the delivery of drugs utilizing passive targeting via the EPR effect. The advantages of polymeric micelles formed by hydrophilic-*b*-hydrophobic block copolymers have been discussed and substantiated by several examples, including micelle formulations of the anti-cancer drugs adriamycin, KRN-5500, cisplatin, taxol, and methotrexate.

References

1. Yokoyama M. and Okano T. Targetable drug carriers: Present status and a future perspective. *Advanced Drug Delivery Reviews* **1996**, *21*, 77-80.
2. Sugiyama Y. Importance of pharmacokinetic considerations in the development of drug delivery systems. *Adv Drug Delivery Reviews* **1996**, *19*, 333-334.
3. Matsumura Y. and Maeda H. A new concept for macromolecular therapeutics in cancer chemotherapy: Mechanism of tumoritropic accumulation of proteins and the antitumor agent smancs. *Cancer Res.* **1986**, *46*, 6387-6392.
4. Maeda H., Seymour L.-W., and Miyamoto Y. Conjugates of anticancer agents and polymers: Advantages of macromolecular therapeutics in vivo. *Bioconjugate Chem* **1992**, *3*, 351-361.
5. Litzinger D.C., Buiting A.-M.-J., van Rooijen N., and Huang L. Effect of liposome size on the circulation time and intraorgan distribution of amphipathic poly(ethylene glycol)-containing liposomes. *Biochim. Biophys. Acta* **1994**, *1190*, 99-107.
6. Illum L., Davis S.S., Miller R.-H., Mak E., and West P. The organ distribution and circulation time of intravenously injected colloidal carriers sterically stabilized with a block copolymer – Poloxamine 908. *Life Sci.* **1987**, *40*, 367-374.
7. Tuzar Z. and Kratochvil P. Block and graft copolymer micelles in solution. *Advances in Colloid and Interface Science* **1976**, *6*, 201-232.
8. Yokoyama M., Inoue S., Kataoka K., Yui N., Okano T., and Sakurai Y. Molecular design for missile drug: Synthesis of adriamycin conjugated with IgG using poly(ethylene glycol)-poly(aspartic acid) block copolymer as intermediate carrier. *Die Makromolekulare Chemie* **1989**, *190*, 2041–2054.
9. Kabanov A.V., Chekhonin V.P., Alakhov V.Y., Batrakova E.V., Lebedev A.S., Melik-Nubarov N.S., Arzhakov S.A., Levashov A.V., Morozov G.V., Severin E.S., and Kabanov V.A. The neuroleptic activity of haloperidol increases after its solubilization in surfactant micelles: Micelles as microcontainers for drug targeting. *FEBS Lett.***1989**, *258*, 343-345 .
10. Yokoyama M., Fukushima S., Uehara R., Okamoto K., Kataoka K., Sakurai Y., and Okano T. Characterization of physical entrapment and chemical conjugation of adriamycin in polymeric micelles and their design for in vivo delivery to a solid tumor. *J Controlled Release* **1998**, *50*, 79–92.

11. Yokoyama M., Satoh A., Sakurai Y., Okano T., Matsumura Y., Kakizoe T., and Kataoka K. Incorporation of water-insoluble anticancer drug into polymeric micelles and control of their particle size. *J Controlled Release* **1998**, *55*, 219–229.

12. Li Y. and Kwon G.S. Methotrexate esters of poly(ethylene oxide)-block-poly(2-hydroxyethyl-L-aspartamide). Part 1: Effects of the level of methotrexate conjugation on the stability of micelles and on drug release. *Pharm. Res.* **2000**, *17*, 607-611.

13. Lavasanifar A., Samuel J., and Kwon G.S. The effect of fatty acid substitution on the in vitro release of amphotericin B from micelles compoased of poly(ethylene oxide)-block-poly(N-hexyl stearate-L-aspartamide). *J Controlled Release* **2002**, *79*, 165-172.

14. Allen C., Han J., Yu Y., Maysinger D., and Eisenberg A. Polycaprolactone-b-poly(ethylene oxide) copolymer micelles as a delivery vehicle for dihydrotestosterone. J Controlled Release **2000**, *51*, 275-286.

15. Trubetskoy V.S. and Torchilin V.P. Polyethyleneglycol based micelles as carriers of the therapeutic and diagnostic agents. *S.T.P. Parma Science* **1996**, *6*, 79-86.

16. Zhang X., Burt H. M., Von Hoff D., Dexter D., Mangold G., Degen D., Oktaba A. M., and Hunter W. L. An investigation of the antitumor activity and biodistribution of polymeric micellar paclitaxel. *Cancer Chemother. Pharmacol.* **1997**, *40*, 81-86.

17. Rolland A., O'Mullane J., Goddard P., Brookman L., and Petrak K. New macromolecular carriers for drugs. I. Preparation and characterization of poly(oxyethylene-b-isoprene-b-oxyethylene) block copolymer aggregates. *J Appl. Polym. Sci.* **1992**, 1195-1203.

18. Inoue T., Chen G., Nakamae K., and Hoffman A.S. An AB block copolymer of oligo(methyl methacrylate) and poly(acrylic acid) for micellar delivery of hydrophobic drugs. *J Controlled Release* **1998**, *55*, 221-229.

19. Rapoport N.Y., Herron J.N., Pitt W.G., and Pitina L. Micellar delivery of doxorubicin and its paramagnetic analog, ruboxyl, to HL-60 cells: Effect of micelle structure and ultrasound on the intracellular drug uptake. *J Controlled Release* **1999**, *58*, 153-162.

20. Benahmed A., Ranger M., and Leroux J-C. Novel polymeric micelles based on the amphiphilic diblock copolymer poly(N-vinyl-2-pyrrolidone)-block-poly(D,L-lactide). *Pharm. Res.* **2001**, *18*, 323-328.

21. Calderara F., Hruska Z., Hurtrez G., Lerch J-P., Nugay T., and Riess G. Investigation of olystyrene-poly(ethylene oxide)block copolymer micelle formation in organic and aqueous solutions by nonradiative energy transfer experiments. *Macromolecules* **1994**, *27*, 1210–1215.

22. Wang Y., Kausch C. M., Chun M., Quirk R. P., and Mattice W. L. Exchange of chains between micelles of labeled polystyrene-block-poly(oxyethylene) as monitored by nonradiative singlet energy transfer. *Macromolecules* **1995**, *28*, 904–11.

23. Wilhelm M., Zhao C-L., Wang Y., Xu R. and Winnik R. A. Poly(styrene-ethylene oxide) block copolymer micelle formation in water: A fluorescence probe study. *Macromolecules* **1991**, *24*, 1033–40.

24. Desjardins A. and Eisenberg A. Colloidal properties of block ionomers. I. Characterization of reverse micelles of styrene-b-metal methacrylate diblocks by size-exclusion chromatography. *Macromolecules* **1991**, *24*, 5779–90.

25. Hoes C. J. T., Potman W., van Heeswijk W. A. R. , Mud J., de Grooth B. G., Grave J., and Feijen J. Optimization of macromolecular prodrugs of the antitumor antibiotic adriamycin. *J Controlled Release* **1985**, *2*, 205–213.

26. Duncan R., Kopeckova-Rejmanova P,, Strohalm J., Hume I., Cable H.C., Pohl J., Lloyd J. B., and Kopecek J. Anticancer agents coupled to N-(2-hydroxypropyl)methacrylamide copolymers I. Evaluation of daunomycin and puromycin conjugates in vitro. *Br. J Cancer* **1987**, *55*, 165–174.

27. Endo N., Umemoto N., Kato Y., Takeda Y., and Hara T. A novel covalent modification of antibodies at their amino groups with retention of antigen-binding activity. *J. Immunol. Methods* **1987**, *104*, 253–258.

28. Zunino F., Pratesi G., and Micheloni A. Poly(carboxylic acid) polymers as carriers for anthracyclines. *J Controlled Release* **1989**, *10*, 65–73.

29. Hirano T., Ohashi S., Morimoto S., Tsukada K., Kobayashi T., and Tsukagoshi S. Synthesis of antitumor-active conjugates of adriamycin or daunomycin with the copolymer of divinyl ether maleic anhydride. *Makromol. Chem.* **1986**, *187*, 2815–2824 .

30. Tsukada Y., Kato Y., Umemoto N., Takeda Y., Hara T., and Hirai H. An anti-α-fetoprotein antibody-daunorubicin conjugate with a novel poly-L-glutamic acid derivative as intermediate drug carrier. *J. Natl. Cancer Inst.* **1984**, *73*, 721-729 .

31. Yokoyama M., Kwon G. S., Okano T., Sakurai Y., Seto T., and Kataoka K. Preparation of micelle-forming polymer-drug conjugates. *Bioconjugate Chemistry* **1992**, *3*, 295–301.

32. Kabanov A.V. and Alakhov, V.Y. Micelles of amphiphilic block copolymers as vehicles for drug delivery. In Alexandridis P. and Lindmay B. eds. Amplhiphilic Block Copolymers: Self Assembly and Applications, Elsevier, Netherlands, 1997, pp. 1-31

33. Kwon G.S., Naito M., Kataoka K., Yokoyama M., Sakurai Y., and Okano T. Block copolymer micelles as vehicles for hydrophobic drugs. *Colloids and Surfaces, B: Biointerfaces* **1994**, *2*, 429-434.

34. Yokoyama M., Okano T., Sakurai Y., and Kataoka K. Improved synthesis of adriamycin-conjugated poly(ethylene oxide)-poly(aspartic acid) block copolymer and formation of unimodal micellar structure with controlled amount of physically entrapped adriamycin. *J Controlled Release* **1994**, *32*, 269-277.

35. Kwon G.S., Naito M., Yokoyama M., Okano T., Sakurai Y., and Kataoka K., Physical entrapment of adriamycin in AB block copolymer micelles, *Pharm. Res.* **1995**, *12*, 192-195.

36. Yokoyama M., Fukushima S., Uehara R., Okamoto K., Kataoka K., Sakurai Y., and Okano T. Characterization of physical entrapment and chemical conjugation of adriamycin in polymeric micelles and their design for in vivo delivery to a solid tumor. *J Controlled Release* **1998**, *50*, 79–92.

37. Yokoyama M., Okano T., Sakurai Y., Fukushima S., Okamoto K., and Kataoka K. Selective delivery of adriamycin to a solid tumor using a polymeric micelle carrier system. *J Drug Targeting* **1999**, *7*, 171–186

38. Yokoyama M., Satoh A., Sakurai Y., Okano T., Matsumura Y., Kakizoe T., Kataoka K. Incorporation of water-insoluble anticancer drug into polymeric micelles and control of their particle size. *J Controlled Release* **1998**, *55*, 219–229.

39. Matsumura Y., Yokoyama M., Kataoka K., Okano T., Sakurai Y., Kawaguchi T., and Kakizoe T. Reduction of the adverse effects of an antitumor agent, KRN 5500 by incorporation of the drug into polymeric micelles. *Jap J Cancer Research* **1999**, *90*, 122–128.

40. Mizumura Y., Matsumura Y., Yokoyama M., Okano T., Kawaguchi T., Moriyasu F., and Kakizoe T. Incorporation of the anticancer agent KRN 5500 into polymeric micelles diminishes the pulmonary toxicity. *Jap J Cancer Research* **2002**, *93*, 1237–1243.

41. Nishiyama N., Yokoyama M., Aoyagi T., Okano T., Sakurai Y., and Kataoka K. Preparation and characterization of self-assembled polymer-metal complex micelle from cis-dichlorodiammineplatinum (II) and poly(ethylene glycol)-poly(aspartic acid) block copolymer in an aqueous medium. *Langmuir* **1999**, *15*, 377–383.

42. Nishiyama N., Okazaki S., Cabral H., Miyamoto M., Kato Y., Sugiyama Y., Nishio K., Matsumura Y., and Kataoka K. Novel Cisplatin-Incorporated Polymeric Micelles Can Eradicate Solid Tumors in Mice, *Cancer Res* **2003**, *63*, 8977–8983.

43. Zhang X., Burt H.M., von Hoff D., Dexter D., Mangold G., Degen D., Oktaba A.M., and Hunter W.L. An investigation of the antitumor activity and biodistributiion ofg polymeric micllear pactitaxel, *Cancer Chemether. Pharmacol.* **1997**, *40*, 81–81.

44. Hamaguchi T., Matsumura Y., Suzuki M., Shimizu K., Goda R., Nakamura I., Nakatomi I., Yokoyama M., Kataoka, and Kakizoe T. NK105, a paclitaxel-incorporating micellar nanoparticle formulation, can extend in vivo antitumour activity and reduce the neurotoxicity of paclitaxel. *Br. J. Cancer*, in press.

45. Li Y. and Kwon G.S. Methotrexate ester of poly(ethylene oxide)-block-poly(2-hydroxyethyl-L-aspartamide). Part 1: Effects of the level of methotrexate conjugation on the stabliity of micelles and on drug release. *Pharm. Res.* **2000**, *157*, 607-611.

Chapter 4

Polymeric Micelles with Transient Stability: A Novel Delivery Concept

Cornelus F. van Nostrum[*], Dragana Neradovic, Osamu Soga, and Wim E. Hennink

Department of Pharmaceutics, Utrecht Institute for Pharmaceutical Sciences (UIPS), Utrecht University, P.O. Box 80082, 3508 TB Utrecht, The Netherlands
*Corresponding author: C.F.vanNostrum@pharm.uu.n

The conversion of polymers from a hydrophobic to a more hydrophilic state by a chemical or enzymatic process such as hydrolysis or oxidation, is presented as a way to destabilize self-assembled micelles or vesicles. Amphiphilic block copolymers of poly(ethylene glycol) (PEG) and a thermo-sensitive block containing N-isopropylacrylamide (NIPAAm) and 2-hydroxypropylmethacrylamide lactate (HPMAm-lactate) form micelles in aqueous solution above the tunable lower critical solution temperature (LCST). The micelles are destabilized upon hydrolysis of the lactate esters in the side chains. This process causes swelling and ultimately disso-lution of the micelles due to increasing hydrophilicity of the core, and provides a unique mechanism to control the release of encapsulated drugs in a responsive way.

Introduction

Colloidal carriers are frequently used to transport and deliver drugs through the body for the reason of protecting the drug against degradation and/or excretion, to prevent adverse side effects of toxic drugs, or to accomplish targeted drug delivery. Examples of such carriers are micro/nanospheres, polymer-drug conjugates, liposomes, and (polymeric) micelles or vesicles. Polymeric micelles have been emerging as a convenient carrier system during the past two decades. Some recently published papers provide excellent reviews on the use of polymeric micelles as drug carriers in general (*1-3*).

Polymeric micelles are formed in aqueous solution from amphiphilic block or graft copolymers. They contain hydrophobic segments which form the core of the micelles, while the soluble segments form the corona. There are a number of reasons why polymeric micelles are interesting as drug carriers. They are highly versatile in terms of preparation methods, composition and physical/chemical properties. As solubilizing agent for hydrophobic drugs, they have a distinct advantage over low molecular surfactants in view of the higher stability of the micelles. This higher stability is reflected in terms of the usually very low critical micelle concentration (cmc) of polymeric surfactants (*3*). This means that polymeric micelles are resistant to dilution effects upon, for example, i.v. administration of the drug formulation. Another important characteristic of micelles when compared withparticles such as microspheres or many liposomal formulations, is their small and uniform particle size. In theory, particle sizes can go down to the order of 10 nanometers for non-loaded polymeric micelles. This size is still large enough to accomplish passive targeting to *e.g.* tumors and inflamed tissues by the so-called enhanced permeation and retention (EPR) effect (*4*). The hydrophilic corona of the micelles may prevent interaction with blood components. This characteristic and their small size will prevent recognition by proteins and macrophages, and thus long circulation times in the blood stream may be achieved (*5*). Finally, active targeting is possible by modifying the peripheral chain ends of the polymers with targeting ligands (*6-8*).

For drug delivery purposes, large variations in the composition of the core have been reported, *e.g.* polyesters, poly(amino acids), poly(meth)acrylates, and poly(acrylamides). However, the corona has almost exclusively been constituted from poly(ethylene glycol) (PEG), because it is a highly bio-compatible polymer which shows little or no undesirable interactions with proteins and cells. PEG is frequently used to 'shield' colloidal drug carriers from its environment in order to extend the residence time in the blood circulation (*9*). Polymeric micelles have been used to carry hydrophobic drugs, which are physically entrapped in and/or covalently bound to the hydrophobic core. Usually, physical entrapment is achieved by electrostatic interactions between drug and polymer (the resulting particles are called polyion complex (PIC) micelles(*10*)), by dialysis from an organic solvent, or by oil-in-water emulsion procedures. High drug loadings can be accomplished even up to the level of 25 weight% (*11*).

In most cases the drugs have to be released from the micelles *in vivo* to become active. Normally, drugs are released from the micelles by passive diffusion after administration. Ideally, drug-loaded micelles should be stable during circulation in the blood stream and not release the drug before they have reached the target site. One of the methods that have been proposed in literature to actively release the drugs is by using polymers that respond to temperature and/or pH changes. Introducing pH sensitivity would be a valuable approach, since it is known that for example tumors and inflamed tissues exhibit a decreased extracellular pH (*12*). Moreover, after cellular uptake, the carrier may end up in cellular compartments such as endosomes/lysosomes that exhibit an acidic pH. As a consequence, the polymer polarity and structure may change upon protonation, causing destabilization of the endosomal membranes and/or release of the drug (*13-16*).

Temperature sensitive block polymers have been used to prepare micelles. For example, poly(*N*-isopropylacrylamide) (pNIPAAm) displays temperature-dependent solubility in water. Below the cloud point (CP, *i.e.,* 32 °C for pNIPAAm), the polymer is soluble but it precipitates above that critical temperature. When attached to a hydrophilic block, micelles will be formed above the CP of the thermosensitive block (*17-19*). For drug delivery applications, it would be advantageous to have such a polymer with a CP just below body temperature. Drug loading is quite easy because the polymer is soluble in water at room temperature and can be mixed with the drug. The mixture will form micelles upon heating to 37 °C, and release of the drug after administration of the material could take place, for example, by local hypothermia (cooling of relevant body sections). However, local hypothermia is only limited applicable, and therefore, the use of additional triggers to control drug release is required, *e.g.* a chemical conversion of the thermosensitive polymers.

The concept of drug release from polymeric micelles that is controlled by a chemical conversion has been proposed by us and a few other groups in the past few years, and is schematically illustrated in Figure 1. The amphiphilic block copolymers are designed such that functional groups present in the hydrophobic blocks are converted through a chemical or an enzymatic process, e.g. oxidation, reduction or hydrolysis. This conversion causes a gradual decrease of the hydrophobicity of the core blocks and, consequently, a destabilization of the micelles as soon as the hydrophilic-lipophilic balance becomes inadequate to keep the micelles together at a certain concentration and temperature. In other words, such micelles display transient stability. This chapter focuses on the

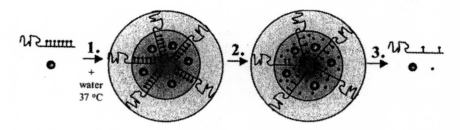

Figure 1. A schematic representation of the concept of thermosensitive polymeric micelles with transient stability. (1) Self assembly and drug loading of polymeric micelles above the critical micelle temperature. (2) Conversion and hydrophilization of the core. (3) Destabilization and the release of the drug.

work that has been done along this line in our laboratory, and will end with a short overview of some new developments by other groups.

Results and Discussion

1. NIPAAm Copolymers

Micelle Formation and Dissolution Kinetics

The CP of pNIPAAm depends on several factors such as concentration or ionic strength and, importantly, can be influenced by copolymerization: the CP decreases when hydrophobic comonomers are used, while hydrophilic comonomers cause an increase of the CP. Copolymers of NIPAAm and a monomer containing one or more hydrophobic ester groups have been described, which upon hydrolysis of the ester bonds convert into a more hydrophilic polymer with increasing CP from below to above 37 °C (*20-22*). As a consequence, these polymers gradually dissolve when incubated at physiological pH and temperature. For example, Figure 2 shows the change of CP when copolymers of 2-hydroxypropylacrylamide lactate (HPMAm-lactate) and NIPAAm were hydrolysed (Scheme I, $n_{average}$ = 3). When the HPMAm-lactate content was high enough (≥35 mol%), the CP indeed passed 37 °C during hydrolysis and eventually the copolymers dissolved at that temperature (*22*).

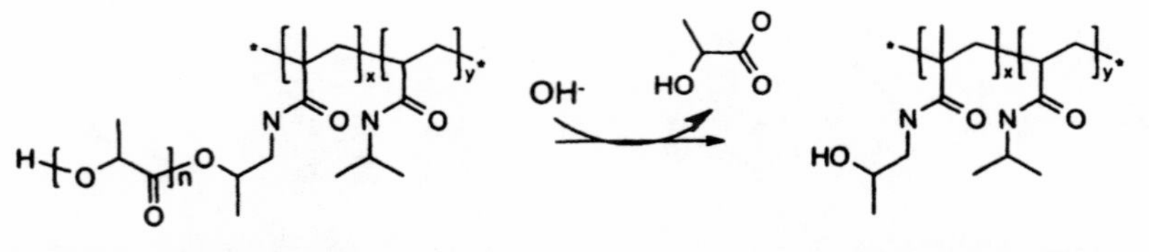

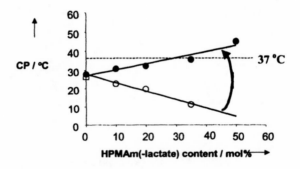

poly(NIPAAm-co-HPMAm-lactate) (n=1 or 2)   *poly(NIPAAm-co-HPMAm)*

Scheme I. Hydrolysis of poly(NIPAAm-co-HPMAm-lactate), resulting in an increasing hydrophilicity.

Figure 2. Effect of polymer composition on the cloud point of poly(NIPAAm-co-HPMAm(-lactate)) copolymers; when the hydrophobic copolymers (O, with HPMAm-lactate,) are transformed to hydrophilic ones (●, with HPMAm) by hydrolysis, as indicated by the arrow, the cloud point is gradually increased. At the indicated temperature, the polymers start to dissolve. (Adapted from Reference 22. Copyright 2001 American Chemical Society.)

Diblock copolymers of the above mentioned thermosensitive copolymer and PEG 5000 as the hydrophilic block were prepared by free radical polymerization using an azo macroinitiator containing two PEG chains (PEG$_2$-ABCPA, Scheme II) (22). Nanoparticles of approximately 70 to 800 nm in diameter (depending on the medium and the heating rate, as will be shown in the next section) were formed upon heating a block copolymer solution above the cloud point. Critical aggregation concentrations (cac) at 37 °C were determined with the aid of an environment-sensitive fluorescent probe (pyrene) and are listed in Table I for a selection of block copolymers. Cac values were of the same order of magnitude as commonly observed for polymeric micelles, and low enough to allow intravenous injection without dissolving the micelles by dillution *(23)*. Interestingly, the particles were very stable at pH —5 (were hydrolysis is minimized), but at higher pH the particles started to dissolve after a while during hydrolytic conversion of the lactic acid side chains, owing to the increasing CP of the thermosensitive block (further on referred to as the critical micelle temperature, cmt). It was shown for block copolymers containing NIPAAm and HPMAm- lactate in a molar ratio of 65:3 5 that it took almost one day at 37 $^{\circ}$C and pH 8.5 to convert polymers to such degree that the cmt increased to the incubation temperature and the micelles dissolved *(22)*. Because ester hydrolysis is known to be first order in hydroxyl ion concentration, it was anticipated that it would take approximately one week to dissolve the micelles at physiological pH (7.4) by this process.

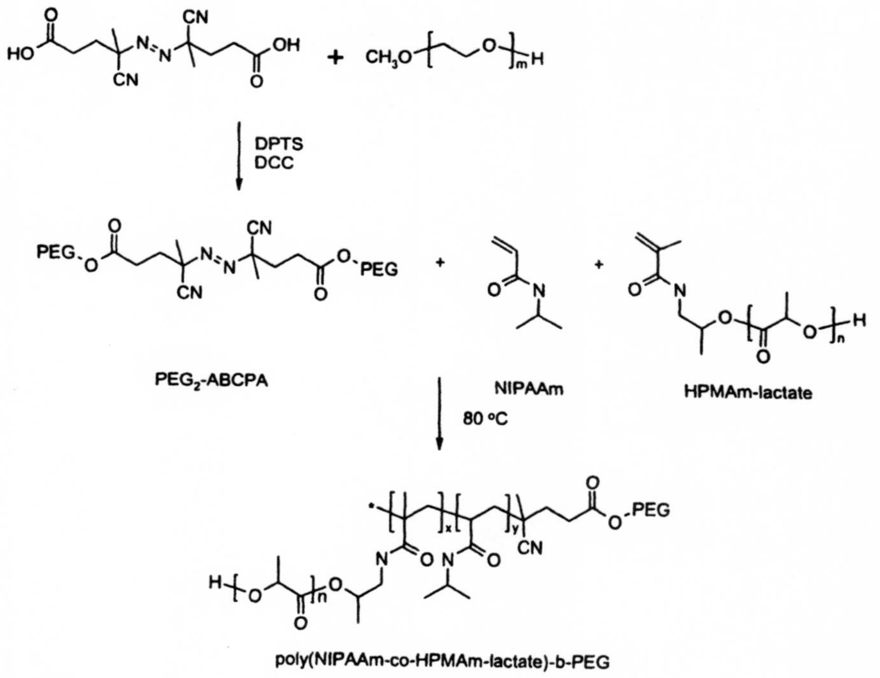

poly(NIPAAm-co-HPMAm-lactate)-b-PEG

Scheme II. Synthesis of the block copolymers. Typically, PEG with M_n = 5000 Da (m ≈ 113) and HPMAm-dilactate (n = 2) are used unless stated otherwise.

Table I. Critical Aggregation Concentrations (cac's) and Critical Micelle Temperatures (cmt's) of Poly(NIPAAm-*co*-HPMAm-dilactate)-*b*-PEG in Water

M_n PEG (Da)	M_n thermosensitive block (Da)[a]	ratio NIPAAm/ HPMAm-dilactate (mol/mol)[a]	cac (mg L⁻¹)[b]	cmt (°C)
5 000	10 800	73:27	n.d.	24.0
5 000	23 800	68:32	8.5 ± 0.5	23.8
10 000	30 600	66:34	6 ± 1	23.2

a Determined by ¹H NMR. *b* At 37 °C.

SOURCE: cac: reproduced with permission from Reference 23, Copyright 2003 D. Neradovic; cmt: reproduced with permission from Reference 24, Copyright 2004 Elsevier Ltd.

We investigated the dissolution of the micelles in more detail by dynamic light scattering at accelerated degradation conditions (pH 10, 37 °C) (*25*). It was shown (see Figure 3) that particle formation during the first 10 minutes at pH 10 was followed by a significant increase in scattering intensity and particle size (swelling phase).

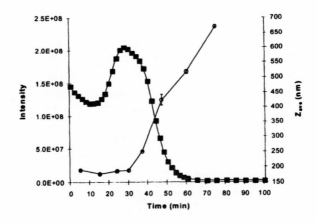

Figure 3. Incubation of poly(NIPAAm-co-HPMAm-dilactate)-b-PEG 5000 in 0.15 M NaHCO₃ buffer at pH 10 and 37 °C, as measured by static (■, intensity) and dynamic (○, average particle size, Z_{ave}) light scattering. (Reproduced from ref. 25. Copyright 2003 American Chemical Society.)

However, a dramatic decrease in scattering intensity was observed after 30 minutes until finally a clear solution was obtained with only a few very large particles left (dissolution phase). We attributed this swelling and dissolution phenomenon to the increasing hydrophilicity and water uptake of the micellar core due to the gradual removal of the lactic acid side chains.

In order to get an idea of the mechanism of the hydrolysis process, we studied the hydrolysis kinetics of the lactic acid-containing monomers (HPMAm-monolactate and dilactate used to prepare the copolymers) at physiological temperature and pH (37 °C, pH 7.5) (25). As shown in Scheme III, the dilactate contains two ester bonds, which can hydrolyze independently. The reaction rate constant (being first order in monomer concentration) of the ester bond between the two lactic acid units was $k_1 = 32 \times 10^{-3}$ h^{-1}, while the hydrolysis rate of the other ester bond was slightly slower, with $k_2 = 14 \times 10^{-3}$ h^{-1} and $k_3 = 8 \times 10^{-3}$ h^{-1} for the dilactate and monolactate, respectively. The difference between k_2 and k_3 can be explained by the fact that the hydrolysis of the dilactate was accelerated by a so-called backbiting mechanism, i.e., attack of the ester bond by the terminal hydroxy group with the formation of an intermediate six-membered ring structure (26). Likewise, the degradation kinetics of the NIPAAm copolymer containing 35 mol% of HPMAm-mono/ dilactate was determined indirectly by measuring the amount of lactic acid and lactoyl lactate (the linear dimer of lactic acid) formed during hydrolysis. It was concluded that in the dissolved state (i.e., below the CP of the copolymers) the reaction rate constants were close to those of the monomers, while above the CP the hydrolysis was slowed down by approximately one to two orders of magnitude. From a detailed kinetic analysis, it was estimated that at 37 °C and pH 10 approximately 75% of the HPMAm-lactate units were completely hydrolyzed after 30 min., i.e., the time point at which the micelles started to dissolve.

Particle Size

The size of the particles formed during heating of the above mentioned solutions of block copolymers of PEG 5000 and NIPAAm-*co*-HPMAm-dilactate was highly dependent on the polymer concentration, ionic strength of the medium and, interestingly, the heating rate (Table II) (24). Particles with a hydrodynamic diameter of 212 nm and quite large polydispersity were formed during slow heating of a 1 gL^{-1} solution in pure water. The size dramatically increased in phosphate buffered saline (PBS, ionic strength 0.3), probably due to a salting out effect. With lower concentrations, the size and polydispersity decreased to acceptable values. However, the size of the particles were still larger than 200 nm, which suggests the formation of aggregates or nanoparticles rather than micelles with a core-shell morphology.

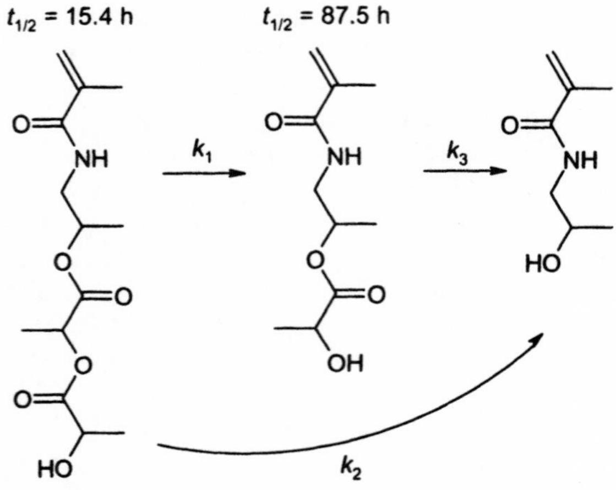

$t_{1/2}$ = 15.4 h $t_{1/2}$ = 87.5 h

Scheme III. Half life times ($t_{1/2}$) of HPMAm-dilactate and monolactate during hydrolysis. Values of the hydrolysis rate constants k_1, k_2 and k_3 are given in the text.

Table II. Size (nm) (and Polydispersity) of Poly(NIPAAm-*co*-HPMAm-dilactate)-*b*-PEG at 37 °C at various conditions.

M_n of Thermo-sensitive Block[a]	0.1 M PBS (pH 7.2, μ=0.3) Slow Heating			Water			
				Slow Heating	Fast Heating		Heat Shock
	0.01 $g\,L^{-1}$	0.1 $g\,L^{-1}$	1 $g\,L^{-1}$	1 $g\,L^{-1}$	0.1 $g\,L^{-1}$	1 $g\,L^{-1}$	0.1 $g\,L^{-1}$
10 800	263 (0.087)	626 (0.097)	789 (0.347)	n.d.	n.d.	n.d.	n.d.
25 800	268 (0.051)	319 (0.054)	770 (0.212)	212 (0.217)	106 (0.020)	141 (0.067)	68 (0.096)

[a] As determined by ^{1}H NMR (M_n of PEG was 5 000).

SOURCE: Adapted with permission from Reference 24. Copyright 2004 Elsevier Ltd.

A further and significant reduction in size was obtained by a new "heat shock" procedure, whereby the cold aqueous solutions were rapidly diluted in hot water. In this way, particles were obtained with a diameter of approximately 70 nm at 0.1 mg mL^{-1}, probably micelles.

2. Poly(HPMAm-lactate)

Above we summarized our results obtained with block copolymer micelles containing NIPAAm and HPMAm-lactate in the core-forming blocks. These monomers were chosen in order to combine thermosensitivity (provided by the NIPAAm) and hydrolytical sensitivity (provided by the lactate esters) as a mechanism to destabilize the thermosensitive core during incubation at physiological conditions. Recently, we also prepared some polymers from HPMAm-monolactate and/or dilactate and, surprisingly, they appeared to be thermosensitive as well (27).

Upon heating aqueous solutions of homopolymers of HPMAm-monolactate or HPMAm-dilactate, a thermoreversible cloud point was observed at 65 °C and 10 °C, respectively. This difference in CP is related to the difference in hydrophobicity of the polymers. For the corresponding copolymers, the CP increased linearly from 10 to 65 °C with increasing amount of the most hydrophilic monomer HPMAm-monolactate (Figure 4). This observation can be exploited for controlled release materials, e.g. block copolymer micelles, because during hydrolysis of poly(HPMAm-dilactate) the amount of monolactate (and eventually HPMAm) will increase in time causing a transition from the insoluble state to the soluble state (see Figure 4). It was shown that it took 5 days to dissolve the polymers at physiological temperature and pH. The time required for dissolution can be shortened because it will depend on the initial ratio of monolactate to dilactate in the copolymer, which can be tailored and provides a tool to adjust the dissolution rate and, consequently, drug release.

The mechanism of swelling and dissolution (as discussed in the previous section) was confirmed by the spectroscopic characteristics of a fluorescent probe (pyrene) that was entrapped in the micelles. The fluorescence spectrum indicated a biphasic change of the local environment of pyrene, i.e., the polarity of the core slightly increased during the swelling, but it increased more rapidly during the dissolution phase because the probe was released into the aqueous phase.

PEG-b-(pHPMAm-dilactate) micelles were prepared, whose particle size (typically 45 – 80 nm) can be controlled by the polymer concentration, heating rate (as for the NIPAAm containing block copolymers shown above) and by the molecular weight of the thermosensitive block (28). Up to 22 weight% of the anticancer drug paclitaxel (PTX) was loaded into these micelles by adding a concentrated solution of PTX in ethanol to the aqueous polymer solution at 0 °C, and then quickly heating to above the cloud point. The PEG-b-(pHPMAm-dilactate) micelles indeed showed a PTX release behavior which could be attributed to the unique dissolution properties of the micelles (29).

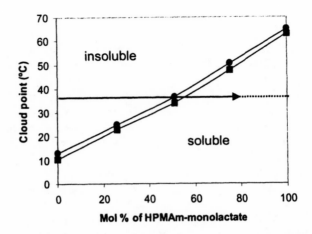

Figure 4. Cloud point of poly(HPMAm-monolactate-co-HPMAm-dilactate) as a function of the mol% HPMAm-monolactate in the copolymer. ●, 1 gL⁻¹ solution in water; ■, 1 gL⁻¹ solution in isotonic 120 mM ammonium acetate buffer (pH = 5.0). The arrow indicates the change in polymer composition during hydrolysis at 37 °C. (Reproduced from ref. 27. Copyright 2004 American Chemical Society.)

3. Recent Developments

To the best of our knowledge, changing the hydrophilicity of a polymer by a chemical process to control its dissolution behavior has been reported by only a few groups so far. Since the work described in the paper of Pitt *et al.* in 1997 (*20*) and the research that has been initiated in our laboratories at about the same time (*21*), there are only two groups that reported results based on a similar concept.

Katayama *et al.* (*30*) synthesized copolymers of NIPAAm, containing PEG and oligopeptide side chains. The peptide side chains are a substrate for protein kinase A (PKA), an enzyme which phosphorylates the peptide and thereby increases the hydrophilicity (and thus the LCST) of the polymer. The PKA-induced degradation of the micelles was not characterized by a delayed swelling and dissolution profile as observed with the system described above, but by immediate and gradual *decrease* in hydrodynamic radius and aggregation number of the particles. No report on the application of these kinase-responsive micelles as a controlled release device have been published yet.

Another system, recently described by Hubbell and coworkers (*31*) relies on the oxidation of poly(propylene sulphide) (PPS). This polymer was used as the hydrophobic B segment in ABA block copolymers with PEG, which form polymeric vesicles and can be oxidized to the much more hydrophilic poly(propylene sulphoxide) and ,ultimately, poly(propylene sulphone). Upon exposure to low concentrations of H_2O_2 (0.03 – 10 vol%), the vesicles were gradually converted into worm-like and spherical micelles and finally dissolved completely. In another report, the group reported encapsulation of glucose oxidase inside the vesicles, which generated H_2O_2, causing vesicle destabilization in response to applied glucose levels (*32*).

Conclusions

In summary, the controlled destabilization of polymeric vesicles and micelles by increasing the hydrophilicity of the polymers upon hydrolytic, oxidative or enzymatic conversion, as described in this chapter, opens the way to novel materials. These materials, in principle, can be loaded with a drug and release their contents in a responsive way or after a tailored induction time post administration. Such unique behavior is unattainable with other, more commonly used biodegradable micelles such as those prepared from PEG-poly(lactic acid)-based block copolymers.

References

1. Torchilin, V. P. Structure and design of polymeric surfactant-based drug delivery systems. *J. Controlled Release* **2001**, *73*, 137-172.
2. Kataoka, K.; Harada, A.; Nagasaki, Y. Block copolymer micelles for drug delivery: design, characterizatioin and biological significance. *Adv. Drug Deliv. Rev.* **2001**, *47*, 113-131.
3. Adams, M. L.; Lavasanifar, A.; Kwon, G. S. Amphiphilic block copolymers for drug delivery. *J. Pharm. Sci.* **2003**, *92*, 1343-1355.
4. Maeda, H.; Wu, J.; Sawa, T.; Matsumura, Y.; Hori, K. Tumor vascular permeability and the EPR effect in macromolecular therapeutics: a review. *J. Controlled Release* **2000**, *65*, 271-284.
5. Kwon G.; Suwa, S.; Yokoyama, M.; Okano, T.; Sakurai, Y.; Kataoka, K. Enhanced tumor accumulation and prolonged circulation times of micelle-forming poly(ethylene oxide-aspartate) block copolymer-adriamycin conjugates. *J. Controlled Release* **1994**, *29*, 17-23.

6. Vinogradov, S.; Batrakova, E.; Li, S.; Kabanov, A. Polyion complex micelles with protein-modified corona for receptor-mediated delivery of oligonucleotides into cells. *Bioconjugate Chem.* **1999**, *10*, 851-860.

7. Yasugi, K.; Nakamura, T.; Nagasaki, Y.; Kato, M.; Kataoka, K. Sugar-installed polymer micelles: Synthesis and micellization of poly(ethylene glycol)-poly(D,L-lactide) block copolymers having sugar groups at PEG chain end. *Macromolecules* **1999**, *32*, 8024-8032.

8. Torchilin, V. P.; Lukyanov, A. N.; Gao, Z.; Papahadjopoulos-Sternberg, B. Immunomicelles: targeted pharmaceutical carriers for poorly soluble drugs. *Proc. Natl. Acad. Sci. USA* **2003**, *100*, 6039-6044.

9. Storm, G.; Belliot, S. O.; Daemen, T.; Lasic, D. D. Surface modification of nanoparticles to oppose uptake by the mononuclear phagocyte system. *Adv. Drug Delivery Rev.* **1995**, *17*, 31-48.

10. Kakizawa, Y.; Kataoka, K. Block copolymer micelles for delivery of gene and related compounds. *Adv. Drug Delivery Rev.* **2002**, *54*, 203-222.

11. Liggins, R. T.; Burt, H. M. Polyether-polyester diblock copolymers for the preparation of paclitaxel loaded polymeric micelle formulations. *Adv. Drug. Deliv. Rev.* **2002**, *54*, 191-202.

12. Stubbs, M.; McSheehy, P. M.; Griffiths, J. R.; Bashford, C. L. Causes and consequences of tumor acidity an implications for treatment. *Mol. Med. Today* **2000**, *6*, 15-19.

13. Meyer, O.; Papahadjopoulos, D.; Leroux, J.-C. Copolymers of N-isopropylacrylamide can trigger pH sensitivity to stable liposomes. *FEBS Lett.* **1998**, *42*, 61-64.

14. Chung, J. E.; Yokoyama, M.; Yamato, M.; Aoyagi, T.; Sakurai, Y.; Okano, T. Thermo-responsive drug delivery from polymeric micelles constructed using block copolymers of poly(N-isopropylacrylamide) and poly(butylmethacrylate). *J. Controlled Release* **1999**, *62*, 115-127.

15. Taillefer, J.; Jones, M.-C.; Brasseur, N.; van Lier, J. E.; Leroux, J.-C. Preparation and characterization of pH-responsive polymeric micelles for the delivery of photosensitizing anticancer drugs. *J. Pharm. Sci.* **2000**, *89*, 52-62.

16. Zhang, G.-D.; Nishiyama, N.; Harada, A.; Jiang, D.-L.; Aida, T.; Kataoka, K. pH-sensitive assembly of light-harvesting dendrimer zinc porphyrin bearing peripheral groups of primary amine with poly(ethylene glycol)-b-poly(aspartic acid) in aqueous solution. *Macromolecules* **2003**, *36*, 1304-1309.

17. Topp, M. D. C.; Dijkstra, P. J.; Talsma, H.; Feijen, J. Thermosensitive micelle-forming block copolymers of poly(ethylene glycol) and poly(N-isopropylacrylamide). *Macromolecules* **1997**, *30*, 8518-8520.

18. Virtanen, J.; Holappa, S.; Lemmetyinen, H.; Tenhu, H. Aggregation in aqueous poly(N-isopropylacrylamide)-block-poly(ethylene oxide) solutions studied by fluorescence spectroscopy and light scattering. *Macromolecules* **2002**, *35*, 4763-4769.
19. Lin, H. H.; Cheng, Y. L. In-situ thermoreversible gelation of block and star copolymers of poly(ethylene glycol) and poly(N-isopropylacrylamide) of varying architectures. *Macromolecules* **2001**, *34*, 3710-3715.
20. Shah, S. S.; Wertheim, J.; Wang, C.T.; Pitt, C. G. Polymer-drug conjugates: manipulating drug delivery kinetics using model LCST systems. *J. Controlled Release* **1997**, *45*, 95-101.
21. Neradovic D.; Hinrichs W. L. J.; Kettenes-van den Bosch J. J.; Hennink W. E. Poly(N-isopropylacrylamide) with hydrolyzable lactic acid ester side groups: a new type of thermosensitive polymer. *Macromol. Rapid Commun.* **1999**, *20*, 577-581.
22. Neradovic, D.; van Nostrum C. F.; Hennink W. E. Thermoresponsive polymeric micelles with controlled instability based on hydrolytically sensitive N-isopropylacrylamide copolymers. *Macromolecules* **2001**, *34*, 7589-7591.
23. Neradovic, D. Thermoresponsive polymeric nanoparticles with controlled instability based on hydrolytically sensitive N-isopropylacrylamide copolymers, Ph.D. thesis, Utrecht University, Utrecht, The Netherlands, 2003.
24. Neradovic, D.; Soga, O.; van Nostrum, C. F.; Hennink, W. E. The effect of processing and formulation parameters on the size of nanoparticles based on block copolymers of poly(ethylene glycol) and poly(N-isopropylacrylamide) with and without hydrolytically sensitive groups. *Biomaterials* **2004**, *25*, 2409-2418.
25. Neradovic, D.; van Steenbergen, M. J.; Vansteelant, L.; Meijer, Y. J.; van Nostrum, C. F.; Hennink, W. E. Degradation mechanism and kinetics of thermosensitive polyacrylamides containing lactic acid side chains. *Macromolecules* **2003**, *36*, 7491-7498.
26. De Jong, S. J.; Arias, E. R.; Rijkers, D. T. S.; van Nostrum, C. F.; Kettenes-van den Bosch, J. J.; Hennink, W. E. New insights into the hydrolytic degradation of poly(lactic acid): participation of the alcohol terminus, *Polymer* **2001**, *42*, 2795-2802.
27. Soga, O.; van Nostrum, C. F.; Hennink, W. E. Poly(N-(2-hydroxypropyl) methacrylamide mono/di-lactate): A new class of biodegradable polymers with tuneable thermosensitivity. *Biomacromolecules* **2004**, *5*, 818-821.
28. Soga, O.; van Nostrum, C. F.; Ramzi, A.; Visser, T.; Soulimani, F.; Frederik, P. M.; Bomans, P. H. H.; Hennink, W. E. Physicochemical

characterization of degradable thermosensitive polymeric micelles. *Langmuir* **2004**, *20*, 9388-9395.

29. Soga, O.; van Nostrum, C. F.; Fens, M.; Rijcken, C. J. F.; Schiffelers, R. M.; Storm, G.; Hennink, W. E. Thermosensitive and biodegradable polymeric micelles for paclitaxel delivery. *J. Controlled Release*, in press.

30. Katayama, Y.; Sonoda, T.; Maeda, M. A polymer micelle responding to the protein kinase A signal. *Macromolecules* **2001**, *34*, 8569-8573.

31. Napoli, A.; Valentini, M.; Tirelli, N.; Müller, M.; Hubbell, J. A. Oxidation-responsive polymeric vesicles. *Nature Mater.* **2004**, *3*, 183-189.

32. Napoli, A.; Boerakker, M. J.; Tirelli, N.; Nolte, R. J. M.; Sommerdijk, N. A. J. M.; Hubbell, J. A. Glucose-oxidase based self-destructing polymeric vesicles. *Langmuir* **2004**, *20*, 3487-3491.

Chapter 5

Synthesis and Evaluation of Hydrophobically-Modified Polysaccharides as Oral Delivery Vehicles for Poorly Water-Soluble Drugs

Mira F. Francis[1], Mariella Piredda[1], Mariana Cristea[1,3], and Françoise M. Winnik[1,2,*]

[1]Faculty of Pharmacy, University of Montreal, C.P. 6128 Succ. Centre-ville, Montreal, Quebec H3C 3J7, Canada
[2]Department of Chemistry, University of Montreal, C.P. 6128 Succ. Centre-ville, Montreal, Quebec H3C 3J7, Canada
[3]"Petru Poni" Institute of Molecular Chemistry, Iasi 6600, Romania
*Corresponding author: francoise.winnik@umontreal.ca

Novel polysaccharide-based micelles were prepared to exploit their potential towards solubilizing poorly water-soluble drugs in order to improve their oral bioavailability. Hydrophobically-modified (HM) dextran (DEX) and hydroxypropylcellulose (HPC) copolymers were synthesized. Cyclosporine A (CsA) was selected as a model drug. In aqueous solution, HM-DEX and HM-HPC form polymeric micelles with low onset of micellization. The CsA incorporation in HM-polysaccharide micelles was significantly higher than in corresponding un-modified polysaccharides. The polymeric micelles exhibited no significant cytotoxicity towards Caco-2 cells at concentrations up to 10 g/L. The apical to basal permeability of CsA across Caco-2 cells increased significantly, when loaded in polymeric micelles, compared to free CsA.

Introduction

Oral drug treatment is by far the preferred administration route as it is convenient to patients, reduces administration costs, and facilitates the use of more chronic treatment regimens. However, low oral bioavailability, which is related to the rate and extent to which a drug is absorbed into the systemic circulation, has limited the development of treatments by the oral route. Important factors affecting the oral bioavailability of drugs are their structural instability in the gastrointestinal fluids, low aqueous solubility, sluggish dissolution, affinity of the drug for the intestinal and liver cytochrome P450 metabolizing enzymes, and the multidrug efflux pump P-glycoprotein, which serves to protect the body from xenotoxins. (1)

Among various strategies investigated for oral delivery of poorly water-soluble drugs, such as nanosuspensions (2), microemulsions (3), and liposomes (4), the use of polymers as active agents in drug formulations has gained much attention. Polymers have long been a part of drug formulations as passive ingredients, but it is only recently that they have been endowed with specific functions in order to facilitate or target the delivery of drugs. This approach exploits the high diversity of polymers in terms of structure and functionalities that enable conjugation of various pilot molecules. (5) Among the different polymer-based drug delivery systems, "polymeric micelles" represent a promising delivery vehicle for poorly water-soluble pharmaceutical active ingredients. (6) Polymeric micelles form spontaneously when amphiphilic polymers, containing both hydrophilic and hydrophobic fragments, are dissolved in water. (7,8) They consist of a hydrophobic core, created upon assembly of the hydrophobic residues, stabilized by a corona of highly hydrated hydrophilic polymeric chains. (9) Unlike surfactant micelles, polymeric micelles are thermodynamically stable and form even under extremely dilute conditions.

Hydrophobically-modified (HM) polysaccharides are known to form polymeric micelles in water. Their size, stability, and colloidal properties depend on their chemical composition, the number of saccharide units, and the architecture of the polymer. (10,11) While a number of fundamental studies of HM-polysaccharides have been reported, their use as nanometric carriers of poorly water-soluble drugs has been largely overlooked. We recently initiated a study of polysaccharide-based micelles as oral drug delivery vehicles and review here the key features of this investigation. (12,13) Two polysaccharides, dextran (DEX) and hydroxypropylcellulose (HPC), were selected as starting materials, since they are known to present no toxicity upon oral ingestion. (14) Dextran (DEX), a mostly linear polymer of glucose with $\alpha(1\text{-}6)$-glycosidic linkages has been used in medicine as a plasma substitute. (15) Hydroxypropylcellulose (HPC) is an excipient in many oral solid dosage forms, in which it acts as a disintegrant and a binder in granulation. (16) In vivo, HPC tends to undergo intimate contact with the absorbing intestinal membrane, thus facilitating drug transport into the systemic circulation. (*17*)

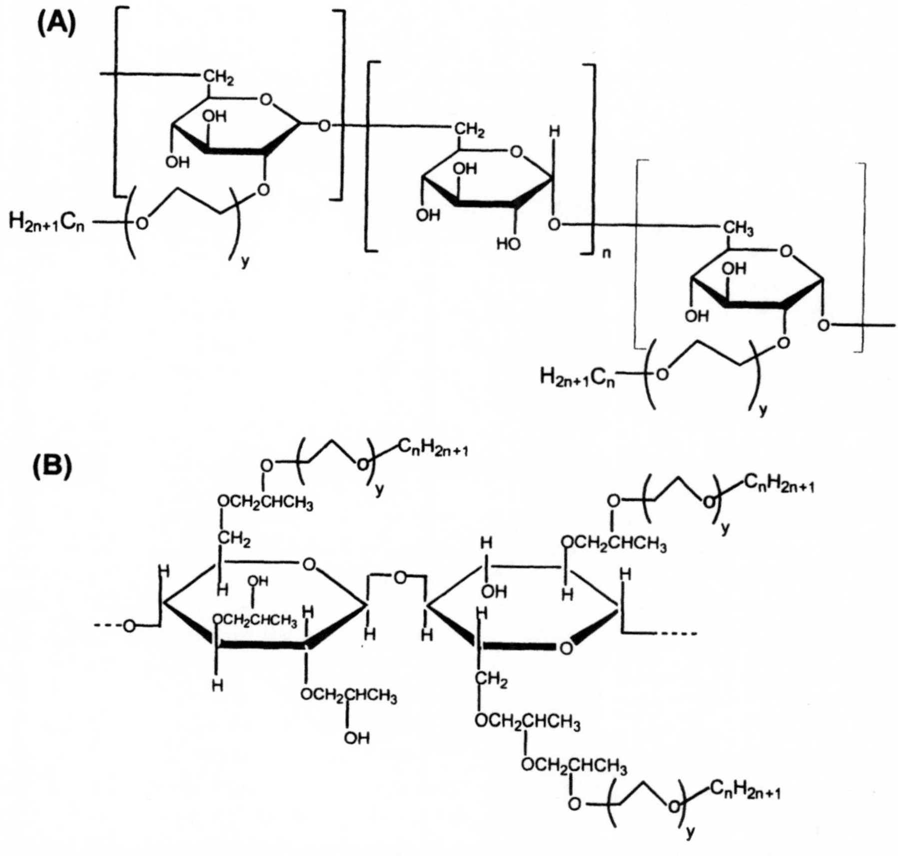

Figure 1. Chemical structures of (A) DEX-g-POE$_y$-C$_n$ copolymer where y = 10, n = 16 or 18; and (B) HPC-g-POE$_y$-C$_n$ copolymer where y = 10 or 20, n = 16 or 18.

Cyclosporine A (CsA) was selected as a model drug. It is given to patients as immunosuppressive agent to prevent graft rejection in various organ transplantations. It is a neutral cyclic undecapeptide with seven *N*-methylated amide linkages. Four intramolecular hydrogen bonds contribute to the extremely low water solubility (23 μg/mL at 20°C) of cyclosporine A. The absorption of CsA through the intestinal mucosa is plagued with many inherent difficulties, not only as a consequence of the minute CsA solubility in water, but also due to the presence of several metabolizing enzymes, such as cytochrome P-450 and of the multidrug transporter P-glycoprotein (P-gp) present in the small intestine. (*18*)

We present here the micellar properties of HM-DEX and HM-HPC (Figure 1) in water, either alone or in the presence of CsA. We assess the permeability of CsA trapped within polymeric micelles, through model intestinal cells monolayers in order to gain insight into the absorption of the micelle-entrapped drug from the gastrointestinal tract into the blood stream.

Materials and Methods

Materials

HM-polysaccharides were prepared as described previously. (*19,20*) The level of POE-C_n grafting was determined by ^{1}H-NMR spectroscopy (Bruker ARX-400 400 MHz spectrometer) measurements carried out with solutions of the polymers in DMSO-d_6.

Micellar Properties of HM-DEX and HM-HPC in Water

The critical association concentration (CAC) of the polymeric micelles in water was estimated by a steady-state fluorescence spectroscopy assay (SPEX Industries Fluorolog 212 spectrometer equipped with a GRAMS/32 data analysis system), using pyrene as a probe. Pyrene partitions preferentially into the hydrophobic core of the micelles. It undergoes spectral changes as a result of changes in the micropolarity it experiences upon diffusion from bulk water (hydrophilic environment) into the micelle core (hydrophobic environment). (*21*) To carry out the measurements, polymer solutions of increasing concentration in pyrene-saturated water ([Py] ~7 x 10^{-7} M) were prepared and equilibrated overnight prior to spectroscopic analysis. The CAC was obtained from plots of the changes in the vibronic fine structure of the pyrene emission by monitoring the changes in the ratio of the intensities I_1 and I_3 of the [0,0] and [0,2] bands, respectively, as a function of polymer concentration (Figure 4). (*22*) The hydrodynamic diameter of the micelles was determined by dynamic laser light scattering (DLS) at 25°C, at a scattering angle of 90°.

Physical Loading of CsA in HM-DEX and HM-HPC Polymeric Micelles

The polymer was dissolved in deionized water while CsA was dissolved in ethanol. Subsequently, different mixtures of polymer with varying CsA initial concentrations (2.5–40% *w/w*) were prepared by mixing the two solutions. Following 48 h of dialysis, each solution was filtered to remove free CsA and the filtrate was freeze-dried yielding a white powder which easily redispersed in water forming a clear solution (Figure 2).

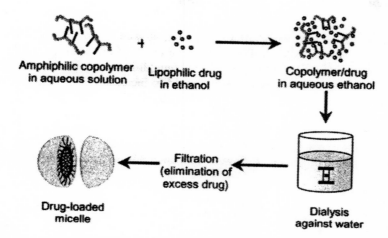

Figure 2. Schematic representation of drug loading in polymeric micelles using the dialysis method.

High Performance Liquid Chromatography (HPLC) Analysis

CsA was extracted from freeze-dried micelles with acetonitrile (ACN). This treatment yielded suspensions which were sonicated for 10 min then agitated for 8 h. They were then filtered and the filtrate was assayed by HPLC using a symmetry® octadecylsilane C_{18} column. *(23)* Drug loading (DL) was calculated using Eq. 1:

$$DL (\%) = 100 (W_c/W_M) \qquad (1)$$

where W_c is the weight of CsA loaded in micelles and W_M is the weight of micelles before extraction.

Cell Culture

The human colon adenocarcinoma cells, Caco-2, were routinely maintained in Dulbecco's modified Eagle medium, supplemented with 10% (*v/v*) heat-inactivated fetal bovine serum, 1% (*v/v*) non-essential amino acids and 1% (*v/v*) penicillin-streptomycin antibiotics solution. Cells were allowed to grow in a monolayer culture at 37 °C, 5% CO_2 and 90% relative humidity.

Cytotoxicity Assay

Caco-2 cells were seeded in triplicate in 96-well culture plates at a density of approximately 5×10^4 cells in 100 μL of cell culture medium per well. The cells were cultured for 48 h at 37 °C in a humidified atmosphere of 5% CO_2 in air. Thereafter, Caco-2 cells were exposed to increasing concentrations (0–10 g/L) of DEX, HPC, POE-C_n, DEX-g-POE-C_n or HPC-g-POE-C_n for 24 h. Cell viability was then evaluated using the MTT colorimetric assay. The assay is based on the reduction of MTT by mitochondria in viable cells to water insoluble formazan crystals. The absorbance was measured with a multiwell-scanning spectrophotometer (PowerWave; Biotek Instruments) at 570 nm.

CsA Transport Study

CsA transport across Caco-2 cells was evaluated. Briefly, cells were grown in Transwell dishes until a tight monolayer was formed as measured by transepithelial electrical resistance (TEER). The integrity of the monolayers following the transport experiments was similarly evaluated. Free or micelle-loaded CsA was added at a final concentration of 1 μM to the apical compartments and aliquots were withdrawn from the basal chamber at predetermined time points. After sample withdrawal, an equivalent volume of the transport medium (Hank's buffered salt solution) was added to the receiving compartment to keep the receiver fluid volume constant. Pluronic P85® was added as a P-glycoprotein inhibitor to the apical compartment. Caco-2 monolayers were then solubilized in 1% Triton X-100, and aliquots were taken for determining protein content using the Pierce BCA method.

Results and Discussion

Synthesis of the Polymers

The HM-polysaccharides were synthesized *via* ether formation between a tosylated POE-C_n (commercially available under the Brij® trade name) and hydroxyl groups of the corresponding polysaccharide. As the polymers and POE-C_n have similar solubility characteristics, the coupling can be carried out in homogeneous solution. Under these conditions, high levels of hydrophobic modification can be achieved and the distribution of alkyl chains along the polymer chain tends to be random rather than "blocky". A series of polymers was obtained, by varying parameters, such as (i) the composition and molecular weight of the hydrophilic chain, *i.e.* DEX of different molecular weights (10,000 and 40,000 Da or approx. 62 and 247 glucose units per chain, for DEX10 and DEX40, respectively) or HPC (ca 80,000 Da); (ii) the level of grafting, *i.e.* the number of hydrophobic substituents linked to the chain; and (iii) the size of the hydrophobic group (hexadecyl or octadecyl) (Table 1). (*12,13*)

Table I. Characteristics of DEX-g-POE$_y$-C$_n$ and HPC-g-POE$_y$-C$_n$ copolymers with various compositions.

Polymer composition	Grafted POE-C$_n$[a] (mol %)	CAC[b] (mg/L)	Mean diameter[c] (nm ± SD)
DEX10-g-(POE)$_{10}$-C$_{16}$	3.0	7.5 ± 3.4	18 ± 2
DEX10-g-(POE)$_{10}$-C$_{18}$	3.9	12.5 ± 2.5	21 ± 1
DEX40-g-(POE)$_{10}$-C$_{16}$	3.5	18.0 ± 2.0	30 ± 1
DEX10-g-(POE)$_{10}$-C$_{16}$	7.0	6.5 ± 3.5	9 ± 3
HPC-g-(POE)$_{20}$-C$_{16}$	3.9	15.0 ± 5.0	78 ± 1
HPC-g-(POE)$_{20}$-C$_{18}$	3.1	22.0 ± 6.0	83 ± 2
HPC-g-(POE)$_{10}$-C$_{16}$	4.7	17.0 ± 3.0	80 ± 1

[a]Determined by ^{1}H-NMR measurement; [b]Determined by change in I_1/I_3 ratio of pyrene fluorescence with log polymer concentration (25 °C, water); [c]Determined by DLS measurements with a scattering angle of 90° (25 °C, water).

The composition of the polymers was established from the ^{1}H-NMR spectra of polymer solutions in DMSO-d_6 based, in the case of DEX-g-POE-C$_n$, on the integral of the signal due to the terminal methyl protons of the POE-C$_n$ groups (δ ~0.85 ppm) and the integral of the signal due to the anomeric protons of dextran (δ ~4.7 ppm) (Figure 3).

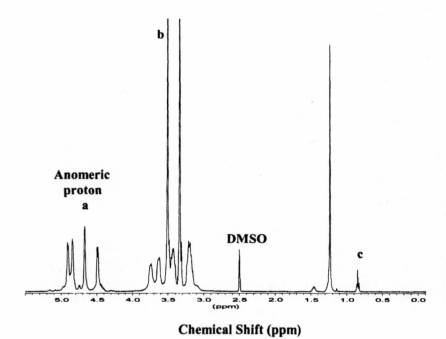

Figure 3. ^1H NMR spectrum of DEX10-g-POE$_{10}$-C$_{16}$ copolymer solution in DMSO-d$_6$.

Micellar Properties of the Polymer

In aqueous solution, HM-DEX and HM-HPC form polymeric micelles. A fluorescence assay using pyrene as a hydrophobic probe was used to determine the polymer concentration at which micellization first takes place, monitoring the changes in the intensity ratio I_1/I_3 as a function of polymer concentration. A sharp decrease of the ratio indicates the formation of polymeric micelles (Figure 4). The CAC values (Table 1) range from approx. 6 to 22 mg/L. They depend mostly on the nature of the polysaccharide main chain and, to some extend, on the POE-C$_n$ residue, *i.e.* decreasing with increasing POE-C$_n$ molar content. The hydrodynamic diameters of the polymeric micelles, as determined by DLS, are below 100 nm (Table 1), with a unimodal distribution indicative of the absence of free polymer chains and of large aggregates. We note that copolymers with longer polysaccharide chain form micelles with larger mean diameter, and that micelles formed by HM-HPC copolymers are larger than those formed by HM-DEX copolymers.

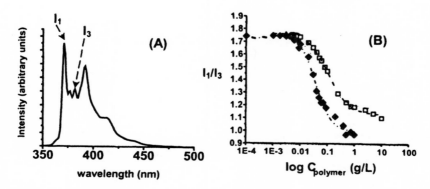

Figure 4. (A) Emission spectrum of pyrene in water showing I_1 and I_3 intensities. (B) Change in the I_1/I_3 ratio of pyrene fluorescence intensity in presence of DEX-g-POE-C_{16} (□) 3 mol% and 15 mol% (♦) copolymers.

Cytotoxicity of HM DEX and HPC Copolymers

From previous reports we know that both HPC and DEX possess no toxicity *(14,15)*; however, free POE-C_n affects the integrity of cell membranes and inhibits cell growth due to its surfactant properties. *(24)* Therefore, it was important to assess if linking POE-C_n chains to a polysaccharide framework would alleviate their toxicity. Thus, we carried out toxicity tests of DEX and HPC, the POE-C_n copolymers, and POE-C_n towards human intestinal epithelial cells (Caco-2 cells), which are widely used to investigate intestinal absorption mechanisms of drugs. *(25)* As expected, DEX and HPC show no toxicity at concentrations as high as 10 g/L, whereas POE-C_{16} inhibits cell growth when added to cells at concentrations as low as 0.5 g/L. Like DEX and HPC, the modified polysaccharides DEX-g-POE-C_{16} and HPC-g-POE-C_{16} exhibit no significant cytotoxicity at a concentration as high as 10 g/L, independently of the grafting level. *(12,13)* These results confirm (i) that POE-C_{16} looses its cytotoxicity upon linking to a polymer chain, and (ii) that the polymer purification method efficiently removed any free POE-C_{16} from the polymer.

Incorporation of CsA in Polymeric Micelles

The level of CsA incorporation within polymeric micelles, expressed in % w/w (CsA/polymer), ranged from approx. 3-4% in case of HM-DEX to approx. 5-7% for HM-HPC. The loading efficiency varied, depending on the initial

CsA/polymer ratio, the degree of grafting, and the chemical composition of the polymer. In Figure 5, we present the CsA loading in polymeric micelles as a function of the initial CsA/copolymer ratio using two DEX-g-POE-C$_{16}$ copolymers and two HPC-g-POE-C$_{16}$ copolymers as well as unmodified DEX and HPC.

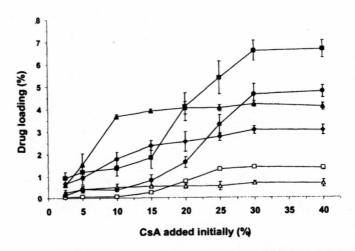

Figure 5. CsA final loading (% w/w) in micelles of (▲) DEX10-g-POE-C$_{16}$ (3 mol%), (♦) DEX10-g-POE-C$_{18}$ (3.9 mol%), (■) HPC-g-POE-C$_{16}$ (3.9 mol%) and (●) HPC-g-POE-C$_{18}$ (3.1 mol%) copolymers at 2.5 – 40 (% w/w) of initially added CsA. For comparison, CsA was incorporated in (△) unmodified dextran T10 and (□) unmodified HPC polymers. Mean ± SD (n = 3).

Note that HPC-g-POE-C$_n$ micelles incorporate higher amounts of CsA than DEX-g-POE-C$_n$ micelles with the same molar content of POE-C$_n$, and that CsA displays a higher affinity for HPC than for DEX. This trend is consistent with the inherent hydrophobicity of HPC due to the presence of isopropoxy substituents. We observe also that, at low CsA initial concentrations (<15% w/w), HPC-g-(POE)$_y$-C$_n$ polymeric micelles exhibit a solubilizing/loading trend different from that displayed by DEX-based micelles. While the degree of CsA incorporation within DEX-based micelles increases steadily with increasing initial CsA concentration, the level of CsA incorporation in polymeric micelles remains low and nearly constant with initial CsA concentrations of <15%, then increases rapidly as the initial CsA concentration exceeds 15%. The origin of this trend is currently under investigation.

Permeability Studies

Intestinal permeability is a key factor in determining overall absorption of orally administered drugs. To model the *in vivo* situation, we investigated the permeability of CsA entrapped in polymeric micelles across Caco-2 cells, since a strong correlation was observed between *in vivo* human absorption and *in vitro* permeability across Caco-2 cells for a variety of compounds. (*26*) The transport of CsA encapsulated in polymeric micelles across Caco-2 monolayers is significantly enhanced when it is incorporated in polymeric micelles, as shown in Figure 6. The apical to basal permeability of CsA across Caco-2 cells increased by factors of 1.5 and 3 for CsA incorporated in DEX-*g*-POE-C$_{16}$ and HPC-*g*-POE-C$_{16}$ micelles, respectively, compared to that of free CsA, after an incubation time of 240 min.

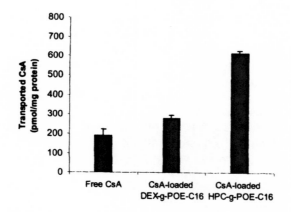

Figure 6. Cumulative CsA (pmol/mg protein) transported across Caco-2 monolayers after a 4 h-incubation at 37°C/5%CO$_2$/90% relative humidity. Mean ± SD.

Conclusions

The drug permeation-enhancing effect of polysaccharide-based polymeric micelles across model intestinal membranes was demonstrated for CsA. Initial results indiquate that this effect is more pronounced in the case of HPC, a polymer endowed with inherent bioadhesive properties, compared to DEX. The thermodynamic stability, very low onset of micellization, small size, and low cytotoxicity of polysaccharide-based polymeric micelles contribute to render them promising novel polymeric drug carrier for the oral delivery of poorly water-soluble drugs.

Acknowledgements

This work was financially supported by the Natural Sciences and Engineering Research Council of Canada under its strategic grants program.

66

M.F. Francis acknowledges a scholarship from the Rx&D Health Research Foundation (HRF)/Canadian Institutes of Health Research (CIHR). We thank L. Lavoie for his help in developing some of the synthetic procedures.

References

1. Bardelmeijer, H. A., van Tellingen, O., Schellens, J. H. and Beijnen, J. H. The oral route for the administration of cytotoxic drugs: Strategies to increase the efficiency and consistency of drug delivery. *Invest. New Drugs* **2000**, *18*, 231-241.
2. Muller, R. H., Jacobs, C. and Kayser, O. Nanosuspensions as particulate drug formulations in therapy. Rationale for development and what we can expect for the future. *Adv. Drug Deliv. Rev.* **2001**, *47*, 3-19.
3. Itoh, K., Matsui, S., Tozuka, Y., Oguchi, T. and Yamamoto, K. Improvement of physicochemical properties of N-4472. Part II: characterization of N-4472 microemulsion and the enhanced oral absorption. *Int. J. Pharm.* **2002**, *246*, 75-83.
4. Minato, S., Iwanaga, K., Kakemi, M., Yamashita, S. and Oku, N. Application of polyethyleneglycol (PEG)-modified liposomes for oral vaccine: effect of lipid dose on systemic and mucosal immunity. *J. Controlled Release* **2003**, *89*, 189-197.
5. Sakuma, S., Hayashi, M. and Akashi, M. Design of nanoparticles composed of graft copolymers for oral peptide delivery. *Adv. Drug Deliv. Rev.* **2001**, *47*, 21-37.
6. Kataoka, K., Harada, A. and Nagasaki, Y. Block copolymer micelles for drug delivery: design, characterization and biological significance. *Adv. Drug Deliv. Rev.* **2001**, *47*, 113-131.
7. Antonietti, M. and Goltner, C. Superstructures of functional colloids: chemistry on the nanometer scale. *Angew. Chem. Int. Ed. Engl.* **1997**, *36*, 911-928.
8. Otsuka, H., Nagasaki, Y. and Kataoka, K. Self-assembly of poly(ethylene glycol)-based block copolymers for biomedical applications. *Curr. Opinion Colloid Interface Science* **2001**, *6*, 3-10.
9. Allen, C., Maysinger, D. and Eisenberg, A. Nano-engineering block copolymer aggregates for drug delivery. *Colloids and Surfaces B: Biointerfaces* **1999**, *16*, 3-27.
10. Akiyoshi, K., Kang, E. C., Kurumada, S. and Sunamoto, J. Controlled association of amphiphilic polymers in water: thermosensitive nanoparticles formed by self-assembly of hydrophobically modified pullulans and poly(*N*-isopropylacrylamides). *Macromolecules* **2000**, *33*, 3244-3249.
11. Pelletier, S., Hubert, P., Payan, E., Marchal, P., Choplin, L. and Dellacherie, E. Amphiphilic derivatives of sodium alginate and hyaluronate for cartilage repair: rheological properties. *J. Biomed. Mater. Res.* **2001**, *54*, 102-108.

12. Francis, M. F., Piredda, M. and Winnik, F. M. Solubilization of poorly water soluble drugs in micelles of hydrophobically modified hydroxypropylcellulose copolymers. *J. Controlled Release* **2003**, *93*, 59-68.

13. Francis, M. F., Lavoie, L., Winnik, F. M. and Leroux, J. C. Solubilization of cyclosporin A in dextran-*g*-polyethyleneglycolalkyl ether polymeric micelles. *Eur. J. Pharm. Biopharm.* **2003**, *56*, 337-346.

14. Obara, S., Muto, H., Kokubo, H., Ichikawa, N., Kawanabe, M. and Tanaka, O. Primary dermal and eye irritability tests of hydrophobically modified hydroxypropyl methylcellulose in rabbits. *J. Toxicol. Sci.* **1992**, *17*, 21-29.

15. Couch, N. P. The clinical status of low molecular weight dextran: a critical review. *Clin. Pharmacol. Ther.* **1965**, *6*, 656-665.

16. Skinner, G. W., Harcum, W. W., Barnum, P. E. and Guo, J. H. The evaluation of fine particle hydroxypropylcellulose as a roller compaction binder in pharmaceutical applications. *Drug. Dev. Ind. Pharm.* **1999**, *25*, 1121-1128.

17. Eiamtrakarn, S., Itoh, Y., Kishimoto, J., Yoshikawa, Y., Shibata, N., Murakami, M. and Takada, K. Gastrointestinal mucoadhesive patch system (GI-MAPS) for oral administration of G-CSF, a model protein. *Biomaterials* **2002**, *23*, 145-152.

18. Tjia, J. F., Webber, I. R. and Back, D. J. Cyclosporin metabolism by the gastrointestinal mucosa. *Br. J. Clin. Pharmacol.* **1991**, *31*, 344-346.

19. Cristea, M. and Winnik, F. M. Synthesis of hydrophobically-modified dextrans. *Macromolecules* **2003**, *submitted.*

20. Piredda, M., Francis, M. F. and Winnik, F. M. Hydrophobically-modified hydroxypropyl celluloses: synthesis and self-assembly in water. *Biomacromolecules* **2003**.

21. Zhao, C. L., Winnik, M. A., Riess, G. and Croucher, M. D. Fluorescence probe techniques used to study micelle formation in water-soluble block copolymers. *Langmuir* **1990**, *6*, 514-516.

22. Kalyanasundaram, K. and Thomas, J. K. Environmental effects on vibronic band intensities in pyrene monomer fluorescence and their application in studies of micellar systems. *J. Am. Chem. Soc.* **1977**, *99*, 2039-2044.

23. Francis, M. F., Cristea, M., Winnik, F. M. and Leroux, J. C. Dextran-*g*-Polyethyleneglycolcetyl Ether Polymeric Micelles For Oral Delivery of Cyclosporin A. *Proceed. Intern. Symp. Control. Rel. Bioact. Mater.* **2003**, *30*, 68-69.

24. Dimitrijevic, D., Shaw, A. J. and Florence, A. T. Effects of some non-ionic surfactants on transepithelial permeability in Caco-2 cells. *J. Pharm. Pharmacol.* **2000**, *52*, 157-162.

25. krishna, G., Chen, K. J., Lin, C. C. and Nomeir, A. A. Permeability of lipophilic compounds in drug discovery using in-vitro human absorption model, Caco-2. *Int. J. Pharm.* **2001**, *222*, 77-89.

26. Yee, S. In vitro permeability across caco-2 cells (colonic) can predict in vivo (small intestinal) absorption in man: fact or myth. *Pharm. Res.* **1997**, *14*, 763-766.

Chapter 6

Oral Delivery of Macromolecular Drugs

Pingwah Tang[*], NaiFang Wang, and Steven M. Dinh

Emisphere Technologies, Inc., 765 Old Saw Mill River Road,
Tarrytown, NY 10591
[*]Corresponding author: ptang@emisphere.com

Recent and rapid progress in the field of biotechnology has resulted in an increasing number of novel macromolecular drugs with great clinical promise. However, the delivery of these macromolecular drugs by routes other than the parenteral route is difficult. The pipeline of macromolecular drugs derived from biotechnology presents a challenging opportunity to develop practical dosage forms that could be dosed via the oral route. Given this, the successful oral delivery of macromolecular drugs presents an enormous opportunity. Emisphere has developed specific molecules of low molecular weight (named "delivery agents") to facilitate the gastrointestinal (GI) absorption, and the subsequent delivery of macromolecules. Herein, we provide an overview of this innovative oral delivery technology that has demonstrated success in human testing.

Introduction

Parenteral delivery has been the primary method for delivering macromolecular drugs including proteins, peptides and carbohydrates. However, injection can cause pain and inconvenience, and thereby can reduce patient compliance. In some cases, parenteral delivery may not provide the optimal therapeutic effect, but is used due to a lack of alternative mode of administration. In the case of insulin, for instance, the sustained release of insulin from subcutaneous injection can cause hypoglycemic episodes. A number of non-invasive routes to administer macromolecular drugs, such as transdermal, transmucosal, nasal, pulmonary, ocular, buccal and oral routes have been pursued. The advantage includes the reduction of frequency of injections. In order to successfully apply these technologies to administer macromolecular drugs, one must overcome the enzymatic and/or metabolic degradation, and the limited permeability of the biological barriers at the site of absorption (*1,2*).

Non-parenteral Routes of Delivery for Macromolecular Drugs

1. Dermal and Transdermal (2,3)

Dermal delivery uses the skin as the site for drug absorption for local and systemic applications. A key advantage of dermal delivery is the avoidance of hepatic first-pass drug metabolism, thereby optimizing drug utilization. Chemical and physical enhancements are important for this route. Another advantage includes the lower demand of specificity for absorption enhancers. However, the drawback is that the penetration rate of macromolecular drugs across skin is usually very low. In addition, the extent of local and systemic tolerability may need to be considered as part of product development. Given these issues, the dermal delivery of macromolecules has met with limited success.

2. Transmucosal(2)

The transmucosal routes of administration include buccal, sublingual, nasal, pulmonary, rectal and vaginal routes. Transmucosal delivery uses the mucous membranes as the sites for drug absorption for local and systemic applications. Like transdermal, the advantage of the transmucosal route is the elimination of drug first-pass metabolism in the liver. In particular, oral mucosa has greater perfusion than the skin while the oral cavity provides an environment almost free from acidity and protease activity. However, histology and physiology of mucosae degrade protein and peptide drugs and retard macromolecular drug transport.

3. Oral(4-9)

The oral route is the most convenient and preferred method of drug delivery. In the case of insulin, oral administration would have the potential to provide a convenient means of dosing , and also would mimic the physiological approach that regulates blood glucose via the portal absorption in a non-diabetic. However, the oral delivery of naturally occurring macromolecular drugs such as insulin, along with well-known synthetic and newly discovered macromolecular drugs, is technically difficult (4-9).

The delivery to the systemic circulation by oral administration has often been severely limited by biological, chemical and physical barriers, such as low pH in the stomach, powerful digestive enzymes, and the low permeability of the GI epithelium. Macromolecular drugs are easily rendered ineffective or destroyed in the stomach, by acid hydrolysis and enzymatic degradation. The rapid and extensive degradation of the drug, coupled with its poor absorption, render the oral route to deliver macromolecular drugs a challenging technical problem. The low permeability of macromolecular drugs is primarily due to their polar functional groups and their high molecular weights. There are numerous pathways for cellular transport. In the case of passive diffusion, a macromolecule needs to have a certain degree of lipophilicity to partition into the cell membrane. Heparin, for example, by virtue of its anionic charge, high molecular weight, and hydrophilic nature, is not readily absorbed transcellularly in the GI tract.

Attempts of Oral Delivery of Macromolecular Drugs

In an effort to overcome the problems of degradation of macromolecular drug molecules in the GI system and to increase the absorption of macromolecules, pharmaceutical and biotechnology companies have devoted years of research to developing oral delivery technologies (10-12). Many attempts have been reported in the literature including, but not limited to the following:

1. Covalent Modification (Drug Modification)

(a) Prodrug Approach (13-15). The addition of beneficial functional groups to the macromolecular drug molecule to improve the physicochemical properties such as better resistance to enzymatic and/or low pH degradation, greater lipophilicity, and less hydrogen bonding than the parent molecule.

(b) Molecular Recognition (16-18). The drug molecule is attached with a ligand that is recognized by an endogenous cellular transport system. For example, a cell membrane transporter or a receptor that can trigger endocytosis.

2. Non-covalent Modification (Formulation Approach)

(a) Incorporation) (19). Methods comprise loading or incorporating the macromolecular drug into lipid-based vehicles (e.g. liposome) or polymeric particles in order to protect it from the hostile environment in GI (low pH in stomach and from digestive enzymes). The drug will be released in a more favorable environment for absorption.

(b) Co-administration (20, 21). Methods are based on co-administration of protein drugs with protease inhibitors to increase bioavailability. Improvement of bioavailability has been observed when a combination of enzyme inhibitors and absorption enhancers are used.

(c) Absorption enhancers (22-26). A variety of GI absorption enhancers (citric acid, salicylates, surfactants, such as sodium lauryl sulfate, fatty acids and their derivatives, chelating agents) are employed to increase the permeability of macromolecular drugs into epithelial cell membranes. For some enhancers, the increase of permeability is mainly through the disruption of the enterocyte membrane, but also by inducing the opening of tight junctions between epithelial cells. Although the absorption of drugs is enhanced, the absorption of toxic substances present in the GI tract can also increase. Therefore, it may not be safe for long-term use (27-33).

Carrier-aided Delivery of Macromolecular Drugs by Oral Route

Although the above identified methods and other technologies for oral routes could present potentially useful ways to deliver the macromolecule orally, they have met limited success. An alternative strategy to effectively deliver macromolecular drugs orally is the technology of carrier-aided delivery of macromolecular drugs. Without covalently modifying the drugs, Emisphere Technologies, Inc. pioneered a novel technology (trademarked as the *eligen*® technology), and perhaps the most convenient and practical approach, by using specific proprietary molecules, referred to as EMISPHERE® delivery agents to effectively deliver macromolecular drugs by oral administration (34).

The mechanism of oral macromolecule absorption based on the co-administration of a proprietary EMISPHERE® delivery agent was studied using intestinal epithelial cells. The enhanced drug permeation across the intestinal membranes is neither due to the alteration in membrane structure (mucosal damage) nor results of direct inhibition of physiological mechanisms of degradation. Instead, the results indicated that the delivery agents interact with the macromolecular drugs to transiently render them more lipophilic (44). This weak interaction enables drug transport across the intestinal mucosa (9, 35-43).

72

Studies conducted by Robinson *et al.* *(44,45)* using recombinant human growth hormone (rhGH) also demonstrated the interaction of an EMISPHERE® delivery agent with the protein drug to facilitate transcellular drug transport in a Caco-2 cell model. Confocal microscopy studies conducted at Emisphere confirmed the transcellular absorption of protein drugs using the proprietary carriers without disruption of the cell membrane and tight junctions *(46, 47)*. Robinson *et al.* have shown the site-dependence of the carrier-aided absorption of macromolecular drugs *(44,45)*. Recent studies *(48)* of the mechanism of enhanced absorption of insulin mediated by the delivery agent, 8-[(2-hydroxy-benzoyl)amino]octanoic acid in rat model reveal that the intensity of absorption increases as the dose of the delivery agent increases. The same studies also disclose that the magnitude of such absorption is GI site-dependent. Other studies have shown the site-dependence of the carrier-aided absorption of insulin *(26)*. Chemistry of the carriers and their synthesis are described next.

Results and Discussion

1. Carrier Chemistry and Preparation

The key components for the success of oral delivery of macromolecular drugs are EMISPHERE® proprietary delivery agents. They can consist of compounds of low molecular weight, or they can be PEGlylated materials. These delivery agents are orally administered in combination with macro-molecular drugs to facilitate the GI absorption, and the subsequent delivery of the drugs. For illustration proposes, the following are the examples of carriers and their syntheses *(49-54)*.

(a) Examples of delivery agents

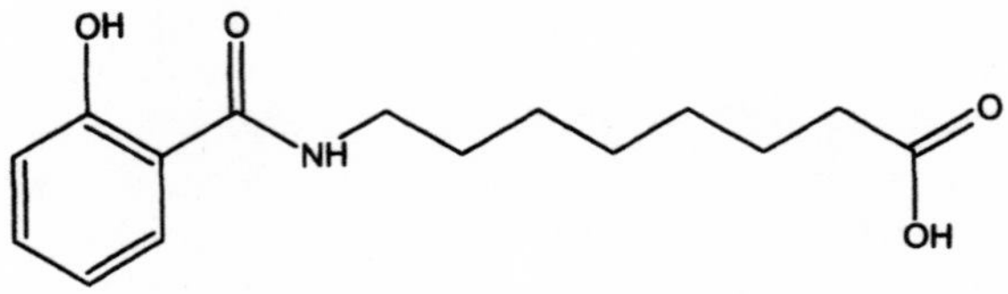

1

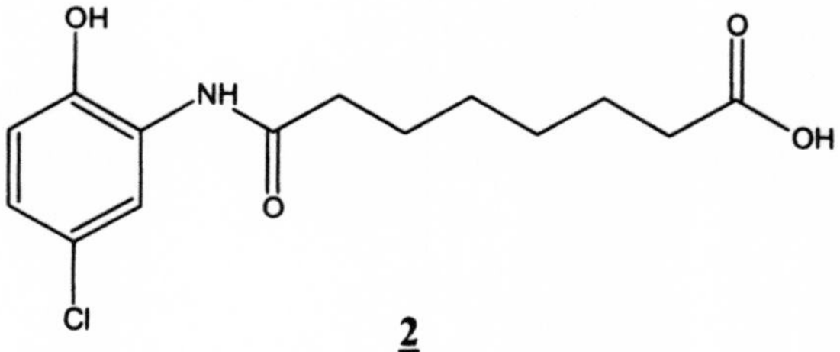
OH
NH
O
O
OH
Cl

2

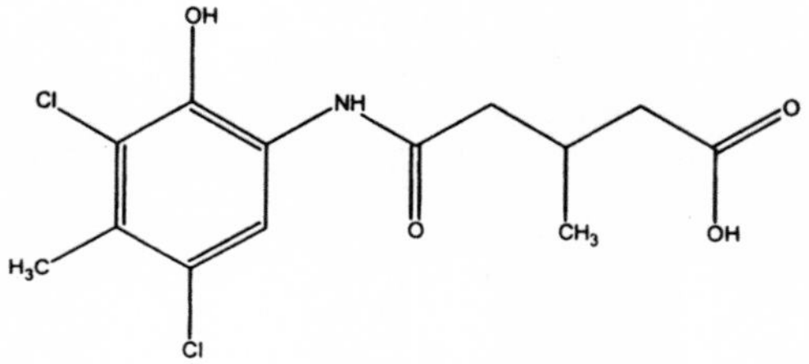
OH
Cl
NH
O
CH_3
O
OH
H_3C
Cl

3

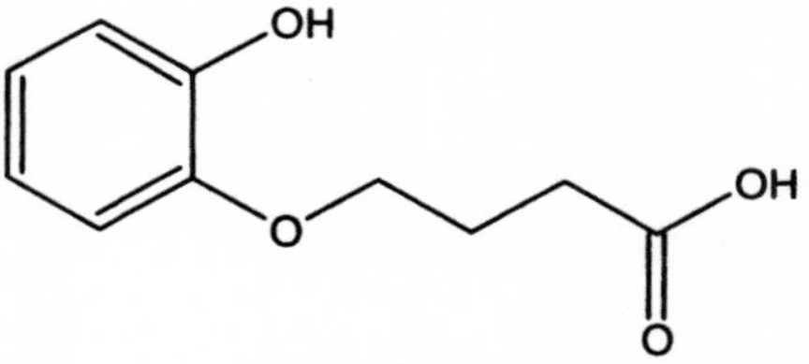
OH
O
OH
O

4

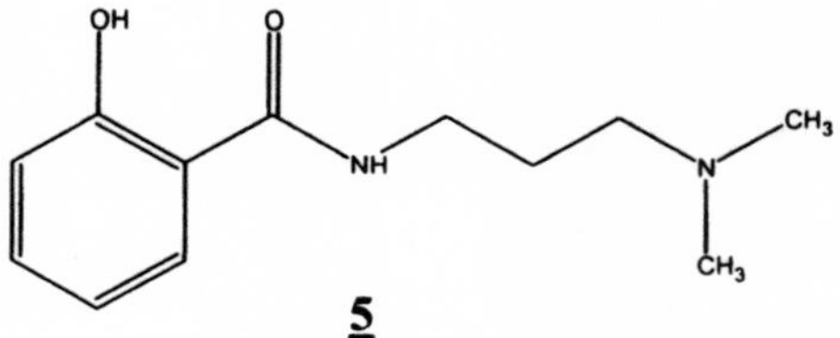
OH
O
NH
CH_3
N
CH_3

5

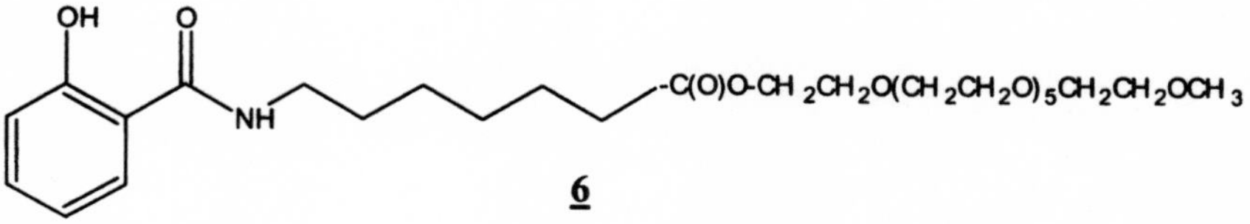
OH
O
NH
-C(O)O-CH $_2$CH $_2$O(CH $_2$CH $_2$O) $_5$CH $_2$CH $_2$OCH $_3$

6

(b) Examples of synthesis

The carriers **2**, **3**, and carriers of similar structures can be conveniently synthesized using boric acid-mediated amidation as illustrated by the following scheme. Table 1 depicts further examples synthesized by this method. *(52-54).*

2. Selected Clinical Data

Emisphere's oral delivery technology has been demonstrated with many medicinally important macromolecular drugs in humans. Examples from clinical trial results using heparin, insulin, and human growth hormone (rhGH) are highlighted below.

Heparin

The oral delivery of unfractionated heparin in combination with an Emisphere delivery agent was demonstrated in healthy human volunteers. The results from two solid formulations (one in tablet form, and the second in a capsule form) are shown in Figure 1. In both cases, heparin absorption and its pharmacological effect, as determined by the change in activated partial thromboplastin time (APTT), were noted.

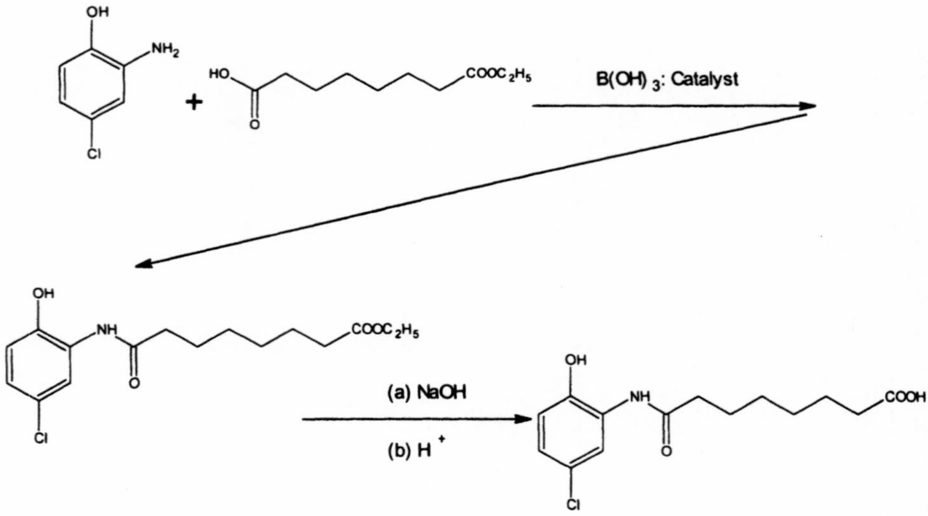

Table I. Examples of EMISPHERE® proprietary delivery agents.

Entry	Carboxylic acid	Amine	Carboxamide 3	Time (h)	Yield % of 3
1	EtO$_2$C-C$_6$H$_{12}$ (acid)	2-aminophenol	HOOCC$_6$H$_{12}$ (amide)	16	75
2	EtOOC-C$_4$H$_8$ (acid)	aminocresol	HOOCC$_4$H$_8$ (amide)	4	77
3	EtOOC-C$_4$H$_8$ (acid)	aminocresol	HOOCC$_4$H$_8$ (amide)	4	78
4	EtO$_2$C-C$_6$H$_{12}$ (acid)	chloroaminophenol	HOOCC$_6$H$_{12}$ (amide)	16	57
5	MeO$_2$C- (acid)	dichloro-methyl aminophenol	HOOC- (amide)	16	68
6	MeO$_2$C- (acid)	methyl aminophenol	HOOC- (amide)	16	70

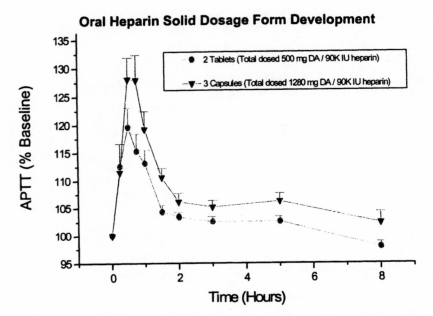

Figure 1. Oral Heparin solid dosage: APTT vs. Time (mean ± SEM, n=10), (DA denotes delivery agents). (Data presented at the Annual Meeting of the American Society of Hematology, 2002.)

77

Insulin

The oral delivery of insulin in combination with an Emisphere delivery agent was demonstrated in healthy human volunteers. The formulation was administered in capsules containing insulin and the delivery agent. Figure 2 shows the plasma concentration of insulin from oral delivery. We also display the plasma concentration from subcutaneous administration.

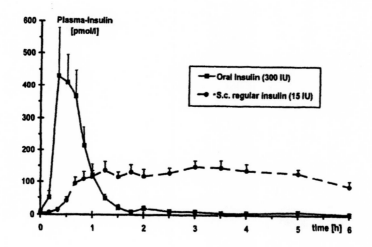

Figure 2. Insulin PK results in patients with Type 2 Diabetes (mean ± SEM, n=10). (Data presented at the Annual Meeting of the American Diabetes Association (ADA), 2003.)

Growth Hormone

The oral delivery of rhGH in combination with an Emisphere delivery agent was demonstrated in healthy human volunteers. The formulation was administrered in capsules containing rhGH and the delivery agent. As shown in Figure 3, the plasma concentration of rhGH from oral delivery is significantly higher than the physiological baseline.

Conclusions

Emisphere Technologies, Inc. has demonstrated that the carrier-aided delivery of macromolecular drugs by the oral route is feasible. The technology uses a delivery agent to alter transiently the physico-chemical properties of the drug for absorption. The pharmacological effect of the drug is unaffected as

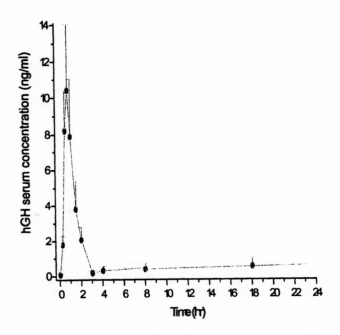

Figure 3. hGH PK profiles after oral administration of hGH/DA capsules in humans (mean ± SEM, n=9). (Data presented at the Controlled Release Society, 2004.)

demonstrated from clinical studies with diverse drug molecules. Additional challenges to develop a commercially viable formulation still remain. However, we are confident that this breakthrough technology will enable important drug molecules to reach their full potential of improved therapy.

Acknowledgements

We are indebted to the staff of Emisphere Technologies, Inc. for all the technical support and for helpful discussion. We thank Dr. Puchun Liu of Emisphere Technologies, Inc. for many precious suggestions during the preparation of this manuscript. Thanks are due to our academic collaborators for scientific support and valuable advice. We would like to express our appreciation to the industrial partners for their technical and financial supports.

References

1. Lee, H.J. Protein Drug Oral Delivery: The Recent Progress, *Arch Pham. Res.* **2002**, *25*, 572.
2. Patki, V.P.; Jagasia, S.H. Progress Made in Non-Invasive Insulin Delivery, *Ind. J. Pharmacol.*, **1996**, *28*, 143.
3. Ghosh, T.K.; Pfisher, W.R.; Yum, S. I. In " Transdermal and Topical Drug Delivery Systems" Interpharm Press, Inc. Buffalo Grove, Il. 1997.
4. Shah, R.B.; Ahsan, F.; Khan, M.A. Oral Delivery of Proteins: Progress and Prognostication, *Crit. Rev. ther. Drug Carrier Syst.* **2002**, *19*, 135.
5. Sood, A.; Panchagnula, R. Peroral Route: An Opportunity for Protein and Peptide Delivery, *Chem. Rev.* **2001**, *101*, 3275.
6. Amidon, G.L.; Lee, H.J. Absorption of Peptide and Peptidomimetric Drugs, *Ann. Rev. Pharmacol. Toxicol.* **1994**, *34*, 321.
7. Wearly, L.L. Recent Progress in Protein and Peptide Delivery by Non-invasive Routes, *Crit. Rev. ther. Drug Carrier Syst.* **1991**, *8*, 331.
8. Bemkop-Schnurch, A.; Walker, G. Multifunctional Matrices for Oral Delivery, *Crit. Rev. ther. Drug Carrier Syst.* **2001**, *18*, 459.
9. Mahato, R.; Narang, A.S.; Thoma, L.; Miller, D. Emerging Trends in Oral Delivery of Peptides and Protein Drugs, *Crit. Rev. Ther. Drug Carrier Syst.* **2003**, *20*, 153.
10. Gomez-Orellana, I.; Paton, D.R. Advances in the Oral Delivery of Proteins: Review, *Exp. Opin. Ther. Patents*, **1998**, *8*, 223.
11. Gomez-Orellana, I.; Paton, D.R. Advances in the Oral Delivery of Proteins: Update, *Exp. Opin. Ther. Patents*, **1999**, *9*, 247.
12. Hwang, S.H.; Maitani, Y.; Qi, X.-R.; Takayama, K.; Nagai, T. Remote loading of diclofenac, insulin and fluorescein isothiocyanate labeled insulin

into liposomes by pH and acetate gradient methods, *Int. J. Pharm.* **1999**, *179*, 85.

13. Asada, H.; Douen, M.; Waki, M.; Adaci, S.; Fujita, T.; Yamamoyo, A. Absorption Characteristics of Chemically Modified-Insulin Derivatives with Various Fatty Acids in the Small and Large Intestine, *J. Pharm. Sci.* **1995**, *84*, 682.

14. Lee, Y.; Nam, J.H.; Shin, H.C.; Byun, Y. Conjugation of Low-Molecular-Weight Heparin and Deoxycholic Acid for the Development of a New Oral Anticoagulant Agent, *Circulation* **2001**, *104*, 3116.

15. Barnett, A.H.;Owen, D.R. Insulin Analogues, *Lancet* **1997**, *349*, 47.

16. Rubio-Aliaga, I.; Daniel, H. Mammalian Peptide Transporters as Targets for Drug Delivery, *Trends Pharmacol. Sci.* **2001**, *23*, 434.

17. Swaan, P.W. Recent Advances in Intestinal Macromolecular Drug Delivery via Receptor-Mediated Transport Pathways, *Pharm. Res.* **1998**, *15*, 826.

18. Oh, D.M.; Han, H.K.; Amidon, G.L. Drug Transport and Targeting Intestinal Tansport, *Pharm. Biotech.* **1999**, *12*, 59.

19. Patel, H.M.; Ryman, B.E. Oral Administration of Insulin by Encapsulation within Liposomes, *Febs Lett.* **1976**, *62,* 60.

20. Bernkop-Schnurch, A. The Use of Inhibitory Agents o Overcome the Enzymatic Barrier to Perorally Administered Therapeutic Peptided and Proteins, *J. Contr. Release*, **1998**, *52*, 1.

21. Bernkop-Schnurch, A.; Walker, G. Multifunctional Matrices for Oral Peptide Delivery, *Crit. Rev. Ther. Drug Carrier Syst.* **2001**, *18*, 459.

22. Sinko, P.J.; Lee, Y.H. Markhey, V.; Leesman, G.D.; Sutyak, J.P.; Yu, H. . Biopharmaceutical Approaches for Development and Assessing Oral Peptide Delivery Strategies and Systems: In-vitro Permeability and Invo Oral Absorption of Calcitonin (sCT), *Pharm. Res.* **1999**, *16*, 527.

23. Swenson, E.S.; Curatolo, W.J. Intestinal Permeability Enhancement for Proteins, Peptides, and Other Polar Drugs: Mechanisms and Potential Toxicity, *Adv. Drug Delivery Rev.* **1992**, *8*, 39.

24. Madara, J.L. Regulation of the Movement of Solutes Across tight Junctions, *Annu. Rev. Physiol.* **1998**, *60*, 143.

25. Cho, S.Y.; Sim, J.-S.; Kang, S.S.; Jeong, C.-S.; Linhardt, R.J.; Kim, Y.S. Enhancement of Heparin and Heparin Disaccharide Absorption by the Phytolacca Americana Saponins, *Arch. Pharm. Res.* **2003**, *26*, 1102.

26. Morishita, M.; Morishita, I.; Takayama, K.; Machida, Y.; Nagai, T. Site Dependent Effect of Aprotinin Sodium caprate, Na_2EDTA and sodium glycocholate on intestinal Absorption of Insulin, *Biol. Pharm. Bull.* **1993**, *16*, 68.

27. Aungt, B.J. Intestinal Permeation Enhancers, *J. Pharm. Sci.* **2000**, *89,* 429.

28. Lee, V.; Yamamoto A.; Kompella, U. Mucosal Penetration Enhancers for Facilitation of Peptided and Protein Drug Absorption, *Crit. Rev. Ther. Drug Carrier Syst.* **1991,** *8,* 91.

29. Muranishi, S. Absorption Enhancers, *Crit. Rev. Ther. Drug Carrier Syst.* **1990,** *7,* 1.

30. Fix, J. Strategies for Delivery of Peptides Utilizing Absorption Enhancing Agents, *J. Pharm. Sci.* **1996,** *85,* 1282.

31. Anderberg, E.; Nystrom, C.; Artursson, P. Epithelial Transport of Drugs in Cell Culture. VII. Effects of Pharmaceutical Surfacetant Excipients and Bile Acids on Transepithelial Permeability in Monolayers of Human Intestinal Epithelial (CaCo2) Cells, *J. Pharm. Sci.* **1992,** *81,* 879.

32. Kajii, H.; Hore, T.; Hayashi, M.; Awazu, S. Fluorescence Study on the Interaction of Salicylate with Rat Small Intestinal Epithelial Cells: Possible Mechanism for the Promotion Effects of Salicylate of Drug Absorption In-Vivo, *Life Sci.* **1985,** *37,* 523.

33. Yeh. P.-Y.; Smith. P.L.; Ellens, H. Effect of Medium Chain Glycerides on Physiological Properties of Rabbit Intestinal Epithelium In-Vitro, *Pharm. Res.* **1994,** *11,* 1148.

34. Golberg, M.; Gomez-Orellana, I. Challenges for the Oral Delivery of Macromolecules, *Nature Rev. Drug Discovery,* **2003,** *2,* 289.

35. Leone-Bay, A.; Paton, D.R.; Weidner, J.J. The Development of Delivery Agents that Facilitate the Oral Absorption of Macromolecular Drugs, *Med. Res. Rev.* **2000,** *20,* 169.

36. Leone-Bay, A.; Paton, D.R.; Variano, B.; Leopold, H.; Rivera, T.; Miura-Fraboni, J.; Baughman, R.A.; Santiago, N. Acylated Non-α–Amino Acids as Novel Agents for the Oral Delivery of Heparin Sodium USP, *J. Contr. Release 1998, 50,* 41.

37. Leone-Bay, A.; Ho, K.K.: Agarwal, R.; Baughman, R.A.; Chaudhary, K.; DeMorin, F.; Genoble, L.; McInnes, C.; Lercara, C.; Milstein, S.; O'Toole, D.; Sarubbi D.;Variano, B.; Paton, D.R. 4-{4-{(2-Hydroxyl benzoyl) amino}phenyl Butyric Acid as a Novel Oral Delivery Agent for Recombinant Human Growth Hormone, *J. Med. Chem.* **1996,** 39, 2571.

38. Leone-Bay, A.; Santiago, N.; Achan, D.; Chaudhary, K.; DeMorin, F.; Falzarano, L.; Haas, S.; Kalbag, S.; Kaplan, D.; Leipold, H.; Lercara, C.; O'Toole, D.; Rivera, T.; Rosado, C.; Sarubbi D.; Vuocolo, E.; Wang, N.-F.; Baughman, R.A. N-Acylated Alpha-Amino Acids as Novel Oral Delivery Agents for Proteins, *J. Med. Chem.* **1995,** 38, 4263.

39. Milstein, S.J.; Leopold, H.; Sarubbi D.; Leone-Bay, A.; Mlynek, G.M.; Robinson, J.R.; Kasimova, M.; Freire, E. Partially Unfolded Protein Efficiently Penetrate Cell Membranes-Implication for Oral Drug Delivery, *J. Contr. Release,* **1998,** *53,* 259

40. Nishihata, T. Tomida, H.; Frederick, G.;Rytting, J.; Higuchi, T. Comparison of the Effect of Sodium Salicylate, Disodium Ethylenediaminetetraacetic Acid, and Polyoxyethylene-23-lauryl Ether as Adjuvants for the Rectal Absorption of Sodium Cefoxitin, *J. Pharmacy Pharmacology* **1985**, *37*, 159.

41. Stoll. B.R.; Leipold, H.R.; Milstein, S.; Edwards, D.A. A Mechanistic Analysis of Carrier-Mediated Oral Delivery of Protein Therapeutics, *J. Contr. Release* **2000**, *64*, 217.

42. Mlynek, G.M.; Calvo, L.J.; Robinson, J.R. Carrier-Enhanced Human Growth Hormone Absorption Across Isolated Rabbit Intestinal Tissue, *Int. J. Pharma*, **2000**, *197*, 13.

43. Said, H.M.; Redha, R. A Carrier-mediated System for Transport of Biotin in Rat Intestine In-Vitro, *Am. J. Physiol.* **1987**, *252*, G52.

44. Wu, S.-J.; Robinson, J.R. Transcellular and Lipophilic Complex-enhanced Intestinal Absorption of Human Growth Hormone, *Pharm. Res.*, **1999**, *16*, 1266.

45. Wu, S.-J.; Robinson, J.R. Transport of Human Growth Hormone Across Caco-2 Cells with Novel Delivery Agents: Evidence for P-glycoprotein Involvement, *J. Contr. Release*, *1999*, *62*, 171.

46. Malkov, D.; Wang, H.-W.; Dinh, S.;Gomez-Orellana, I. Pathway of Oral Absorption of Heparin with Sodium N-[8-(2-Hydroxybenzoyl) amino]caprylate, *Pharm. Res.* **2002**, *19*, 1180.

47. Malkov, D.; Wang, H.-W.; Angelo, R.; Tang, H.; Dinh, S.; Gomez-Orellana, I. Oral protein absorption. Mechanistic studies using intestinal epithelial cells. Controlled Release Society 30th Annual Meeting, Glasgow, UK, July 19-23, **2003**. Poster # 711.

48. Rong, Q.; Ping, Q.-N.; Zao, W. Studies on Gastrointestinal Absorption Enhancement Effect and Mechanism of Sodium N-[8-(2-hydroxybenzoyl)amino]caprylate to Insulin Solution, *Acta Pharma. Sinica,* **2003**, *38*, 953.

49. Leone-Bay, A.; Leipold, H.; Paton, D.R.; Milstein, S.J.; Baughman, R.A. Oral Delivery of rhGH: Preliminary Mechanistic Considerations, *DN&P* **1996**, *9*, 586.

50. Leone-Bay, A.; McInnes, C.; Wang, N.F.; DeMorin, F.; Achan, D.; Lercara, C.; Sarubbi, D.; Haas, S.; Press, J.; Barantsevich, E.; O'Broin, B.; Milstein, S. J.; Paton, D.R. Microsphere Formation in a Series of Derivatized Alpha-Amino Acids: Properties, Molecular Modeling and Oral Delivery of Salmon Calcitonin, *J. Med. Chem.* **1995**, 38, 4257.

51. Tang, P.; Ye, F. Boron-Mediated Amidation of Carboxylic Acid, U.S. Patent No. 6,384,278, Feb. 05, 2001.

52. Tang, P.; Leone-Bay, A.; Gschneidner, D. Compounds and Compositions for Delivery of Active Agents, U.S. Patent No. 6,646,162, Nov. 11, 2003.

53. Tang, P. In Advanced in Controlled Drug Delivery, ACS Symposium Series 846, Chapter 8, Editors: Dinh, S. M.; Liu, P. **2003**, 103.
54. Milstein, S.J.; Barantsevitch, E.N.; Wang, N.F.; Liao, J.; Smart, J.E.; Conticello, R.D.; Ottenbrite, R.M. Polymeric Delivery Agents and Delivery Agents Compounds, U.S. Patent No. 6,627,228. Sept. 30, 2003.

Chapter 7

Immunoliposomes Directed against Colon Adenocarcinoma Cells: Delivery of 5-Fluorodeoxyuridine to Colon Cancer Cells and Targeting to Liver Metastases

Jan A. A. M. Kamps[1,*], Henriëtte W. M. Morselt[1], Gerrit L. Scherphof[2], and Gerben A. Koning[3]

Departments of [1]Pathology and Laboratory Medicine, Medical Biology Section and [2]Cell Biology, Section Liposome Research, Groningen University Institute for Drug Exploration (GUIDE), Hanzeplein 1, 9713 GZ Groningen, The Netherlands
[3]Department of Radiochemistry, Delft University of Technology, Delft, The Netherlands
[*]Corresponding author: j.a.a.m.kamps@med.rug.nl

Tumor cell-specific immunoliposomes have been applied to deliver a dipalmitoyl derivative of 5-fluorodeoxyuridine (FUdR-dP) to CC531 adenocarcinoma cells. Immunoliposomes containing FUdR-dP caused stronger inhibition of CC531 *in vitro* cell growth than FUdR-dP in non-targeted liposomes. The immunoliposomes were not internalized but FUdR-dP was hydrolyzed intracellularly to FUdR within 24 h. Uptake and processing of FUdR-dP involves selective transfer from liposomes to the cell and subsequent digestion. *In vivo*, immunoliposomes enhanced tumor accumulation two-fold compared to non-immunoliposomes. The immunoliposomes appeared to be associated with tumor-associated macrophages rather than tumor cells. Both Fc receptors and scavenger receptors on macrophages are involved in the cell uptake.

Introduction

Liposomes or liposome-like particles have been shown over the last 30 years to be potentially useful carriers for a wide variety of drugs, including drugs with poor water solubility *(1)*. The development of long circulating liposome formulations such as those containing poly(ethylene glycol) (PEG-liposomes) has improved the utility of liposomes for tumor targeting because of increased levels of extravasation compared to conventional liposomes. Coupling of a tumor cell-specific antibody to PEG-liposomes is expected to further increase specificity and efficacy *(2,3)*.

The occurrence of liver metastases after surgical treatment of colon cancer is a serious clinical problem because chemotherapeutic treatment of hepatic metastases has limited efficacy, and the prognosis of patients with liver metastases is poor *(4)*. Antibody-directed PEG-liposomes selective for metastatic tumor cells, may provide an alternative treatment for non-resectable liver metastases.

In our studies over the last 6 years, we applied several types of immunoliposomes to deliver a lipophilic, dipalmitoyl derivative of 5-fluorodeoxyuridine (FUdR-dP) to colon adenocarcinoma cells. The dipalmitoyl ester of FUdR was synthesized because of a lack of liposomal retention of native FUdR in serum *(5)*. On the other hand, FUdR-dp does not exchange with serum components such as (lipo)proteins and remains liposome associated in *in vivo* conditions *(6)*. FUdR-dP-containing immunoliposomes were prepared using a tumor-specific antibody, CC52, which is directed against a surface antigen on CC531 colon adenocarcinoma cells *(7)*. CC52 was coupled to the liposomes by applying different coupling techniques, affecting the orientation and exposition of the antigen-binding portion of the antibody. *In vitro*, the interaction of these differently designed immunoliposomes was studied using CC531 colon cancer cells and primary cultures of rat liver macrophages (Kupffer cells). The *in vivo* targeting potential of the CC52-immunoliposomes was tested in an established syngeneic rat liver CC531 metastasis model *(8)*.

Materials and Methods

Liposomes

Liposomes were composed of egg yolk phosphatidylcholine (PC)/ cholesterol (Chol)/ methoxypoly(ethylene glycol)$_{2000}$-distearoylphosphatidyl-ethanolamine (PEG-DSPE) in a molar ratio of 23:16:1. Liposomes were prepared and characterized as described earlier *(9,10)*. FUdR-dP and [^{3}H]FUdR-dP were synthesized and incorporated in liposomes as reported before

(6,11). Hydrazide-PEG-DSPE was synthesized as described *(12).* When required, the liposomes were labeled with trace amounts of [³H]cholesteryloleyl ether or with 1,1'-di-octadecyl-3,3,3',3'-tetramethyl indocarbocyanine per-chlorate C18 (DiI) as a fluorescent lipid bilayer marker. Liposomes containing colloidal gold were prepared as described before *(13).* The monoclonal antibody CC52, recognizing a surface antigen on CC531 colon adenocarcinoma cells, was coupled to maleimido-4-(p-phenylbutyryl)-phosphatidylethanolamine (MPB-PE)-containing liposomes by a sulfhydrylmaleimide coupling method *(14)* (Figure 1a,b). PEG-liposomes with the antibody attached in an oriented way were prepared by coupling CC52 to Hz-PEG-DSPE-containing liposomes via a hydrazone linkage between the hydrazide moiety at the distal end of the PEG chain and oxidized carbohydrates in the Fc region of the antibody as described previously *(15,16)* (Figure 1c).

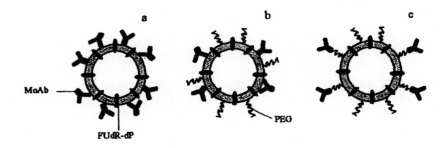

Figure 1. FUdR-dP containing immunoliposomes: (a) conventional immunoliposome with the antibody attached in a random orientation; (b) PEG-liposomes with the antibody attached in a random orientation; and (c) PEG-liposomes with the antibody attached to the distal end of the PEG-chain in a Fab- oriented way.

In Vitro Studies

Effects of FUdR and FUdR-dP-containing liposomes on the proliferation of CC531 cells were determined by a [methyl-³H]-thymidine incorporation assay *(10).* Association of liposomes with CC531 cells was determined after a 3 h incubation of the cells with labeled liposomes. Depending on the type of experiment, liposomes were labeled with [³H]cholesteryloleyl ether or DiI as lipophilic markers, and with encapsulated colloidal gold for electron microscopy studies *(9,11).* [³H]FUdR-dP was synthesized to follow the fate of the liposomal drug. Cells were incubated with liposomes containing [³H]FUdR-dP followed by a liposome free incubation of 24 h. Extracellular FUdR was determined in the culture medium, surface–bound FUdR after trypsin treatment and centrifugation of the cells in the supernatant, and internalized drug in the cell pellet. To distinguish between hydrophobic FUdR-dP and hydrophilic FUdR or metabolites, chloroform/methanol extractions were performed *(11)*.

Kupffer cell isolation and culturing was established as described before *(17)*. Interaction of liposomes with Kupffer cells was determined after 3 h incubation of the cells with labeled liposomes as indicated.

In Vivo Studies

Liver metastases were induced by inoculation of CC531 cells into the superior mesenteric vein of WAG/Rij rats as described before *(8)*. CC531 cells are syngeneic with WAG/Rij rats. Twenty-two days after inoculation of CC531 cells, radioactive or fluorescently labeled liposomes were injected intravenously into rats and allowed to circulate for 24 h. After this time period, tissues were removed from the rat and further processed for measurement of radioactivity or for fluorescence microscopy *(9)*.

Results and Discussion

Immunoliposomes containing FUdR-dP caused a much stronger inhibition of CC531 cell growth than FUdR-dP in non-targeted liposomes (Table I). The anti-proliferative effect was shown not to be influenced by liposome size and only marginally influenced by the presence of PEG.

Table I. IC_{50} Values of Inhibition of Proliferation of CC531 Colon Adenocarcinoma Cells by Liposomal FUdR-dP (type a and b) or Free FUdR

Liposome	Treatment (h)	IC_{50} FUdR (ng/ml)
FUdR-dP-liposomes	72	170
CC52-FUdR-dP-liposomes	72	5
CC52-FUdR-dP-PEG-liposome	72	24
Free FUdR	72	1

To establish the mechanism of interaction of the CC52 immunoliposomes with CC531 colon cancer cells, we studied both the fate of the liposomal carrier and the FUdR-dP. The interaction of [³H]cholesteryloleyl ether-labeled CC52 immunoliposomes with CC531 cells was saturable and could be blocked completely by co-incubation with unlabeled CC52 antibody. Human IgG, mouse IgG or bovine serum albumin had no effect on the association of CC52 immunoliposomes with the tumor cells, demonstrating that CC52-immuno-liposomes specifically bind to CC531 cells. Inhibitors of endocytosis such as monensin, colchicine and cytochalasin B and D did not significantly affect the interaction of the immunoliposomes with the cells, suggesting that the antibody-mediated binding to the cells is not followed by internalization. This was confirmed by transmission electron microscopy of the interaction of colloidal-gold labeled immunoliposomes with CC531 cells after incubation for 3 h at 37 °C. Figure 2 shows that the immunoliposomes are not internalized after binding to the cell surface.

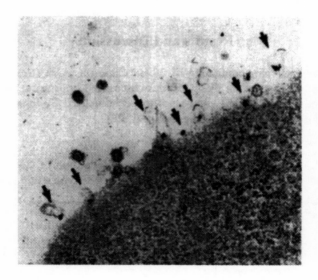

Figure 2. Transmission electron micrograph of a thin section of cultured CC531 cells incubated for 3 h with immunoliposomes containing colloidal gold (arrows). Liposomes were only observed bound to the cell surface.

Since FUdR-dP-containing immunoliposomes were, despite the lack of internalization of immunoliposomes by the tumor cells, able to efficiently inhibit CC531 cell proliferation, we also investigated the fate of the FUdR-dP.

Therefore we used FUdR-dP labeled with ^{3}H in the uracil moiety of the drug to monitor the metabolic fate of the FUdR-dP. The FUdR-dP from CC52 immunoliposomes that was allowed to associate with CC531 cells during a 3 h incubation was rapidly hydrolyzed upon further (liposome free) incubation with the cells, as was demonstrated by the release of the ^{3}H in the medium. The cellular localization of the remaining cell-associated label was determined by separating the trypsin-removable (membrane bound) label and non-removable label (internalized). Further analysis of the nature of the cell-associated label and the label released in the medium was assessed by fractionating these samples into water-soluble and chloroform-soluble compounds to be able to distinguish between the chloroform-soluble FUdR-dP and the water soluble intralysosomally hydrolyzed prodrug (FUdR or metabolites thereof). FUdR-dP initially associated with the cells, was partly released into the medium principally in the form of water-soluble FUdR or FUdR metabolites. The cell- associated remainder of the prodrug was present as intact prodrug at the cell surface or intrcellularly as a mixture of metabolized and non-metabolized prodrug as shown in Table II.

Table II. Chemical Form of the Drug after Incubating CC531 Cells for 3 h with [^{3}H]FUdR-dP Containing Immunoliposomes Followed by a Liposome-Free Incubation of 24 h.

Location	FUdR-dP (%)	FUdR (%)
Surface-bound	89±4	11±4
Intracellular	42±3	58±3
Extracellular	17±2	83±2

Given these results, the lack of uptake of the immunoliposomes and the efficacy of the incorporated FUdR-dP in cultured CC531 cells, we propose that the mechanism of uptake and processing of the FUdR-dP involves selective transfer of the lipophilic prodrug from the liposome bilayer to the cell membrane and subsequent internalization and lysosomal digestion of the prodrug. The FUdR or FUdR metabolites diffuse into the cytosol either to remain intracellularly or to be released extracellularly.

To study the in vivo behavior, including the targeting potential of the CC52-immunoliposomes, we used an established rat metastasis model. PEG-liposomes were synthesized to which the tumor-specific CC52 antibody was coupled either in a random orientation via a lipid anchor or uniformly oriented at the distal end of the PEG chains exposing their Fab portion. In addition to exposure more distant from the liposome surface, a diminished Fc-exposition was expected to be achieved with the latter method, as this part of the antibody was involved in coupling. Both types of PEGylated immunoliposomes efficiently retained the prodrug FUdR-dP in serum and their long circulating properties upon coupling of the antibody. Both visualization studies and quantitative measurements using radiolabeled CC52 immunoliposomes revealed that the presence of antibodies on the long circulating liposomes enhanced tumor accumulation two-fold compared to non-immunoliposomes (Figure 3).

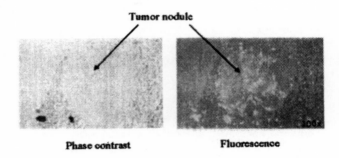

Figure 3. Fluorescently labeled immunoliposomes accumulated in intrahepatic tumor nodules. Micrographs of cryosections of a liver from a tumor bearing rat were taken 24 h after injection of DiI labeled CC52-PEG-liposomes.

Besides the tumor tissue, spleen and (non-tumor) normal liver tissue contributed substantially to the uptake of immunoliposomes from the blood. When an irrelevant murine IgG-1 was coupled to the immunoliposomes, uptake of these control immunoliposomes was comparable or at best only slightly less than the uptake of CC52 immunoliposomes. More detailed visualization studies using colloidal gold-labeled CC52 immunoliposomes, showed that the larger fraction of the tumor-localized immunoliposomes was not associated with tumor cells but rather with tumor associated cells, presumably macrophages. Therefore we performed a series of experiments to directly compare the interaction of CC52 immunoliposomes with CC531 colon cancer cells and with macrophages, isolated from rat liver (Kupffer cells), with the aim to improve target binding to the CC531 cells and to reduce interference by macrophage uptake.

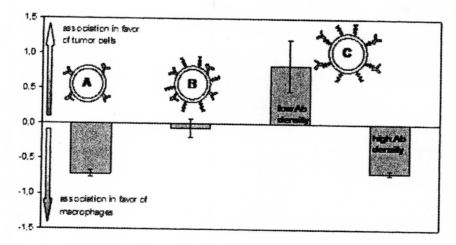

Figure 4. The association of type a,b and c liposomes (see figure1) with tumor cells and macrophages is depicted as the tumor cell / Kupffer cell (TC/KC) ratio. Only when CC52 is coupled at a low density to the distal end of the PEG chains via its Fc moiety, the interaction with the two cell types is in favor of the tumor cells.

With increasing antibody density, both types of PEGylated immuno-liposomes showed increased binding to CC531 cells. For the immunoliposomes with uniformly oriented PEG-distal end coupled antibodies there is apparently an optimal antibody density of ~30 μg/μmol of lipid for optimal tumor cell binding. There is, however, also a positive correlation between antibody density and uptake by macrophages. Immunoliposomes with the antibody coupled via its Fc moiety showed at low antibody density (8 μg/μmol of lipid) the lowest level of macrophage uptake (Figure 4). The role of Fc receptors on macro-phages was elucidated by co-incubation of the immunoliposomes with aggregated IgG, a ligand for these receptors. As expected, the uptake by macrophages of liposomes with randomly surface-coupled CC52 antibodies could be inhibited efficiently by aggregated IgG, demonstrating involvement of Fc receptors. More surprisingly, this was also the case with immunoliposomes to which the antibody was coupled via the Fc moiety. Since the coupling chemistry used (oxidation with periodate) may also influence the charge of the coupled antibody, we also looked at the involvement of macrophage scavenger receptors on the uptake of immunoliposomes with the antibody coupled to the distal end of bilayer-anchored PEG. Co-incubation of macrophages with these immunoliposomes and poly(inosinic acid), an established inhibitor of scavenger receptor mediated uptake significantly reduced the uptake of the liposomes.

Conclusions

We have developed a targeted liposomal preparation that enables intracellular delivery of FUdR in colon carcinoma cells without the need of internalization of the whole carrier. The proposed mechanism of selective transfer of the FUdR-dP prodrug and subsequent activation of the prodrug leads to cytotoxic effects in the tumor cells involved in the uptake of the prodrug. In addition, drug released from the tumor cells will diffuse into neighbouring cells causing a so-called bystander effect. A liposomal drug delivery system, which is independent of internalization of the drug carrier itself may also be more generally applicable for lipophilic anti-cancer drugs. The interference of tumor-associated macrophages with efficient tumor cell targeting is explained by recent experiments in which we demonstrated that immunoliposomes with the antibody coupled to the cell surface in a random orientation are readily taken up by cells expressing Fc receptors, e.g. macrophages. Distal end coupling of antibodies to bilayer-anchored PEG chains by means of the Fc moiety did not effectively prevent recognition by Fc receptors. In addition, it is likely that the coupling conditions in the latter method causes oxidative damage, leading to additional involvement of scavenger receptor systems in the processing of these immunoliposomes *(17)*. To further improve the *in vivo* targetability of these immunoliposomes, new advanced coupling techniques and the use of smaller antigen recognizing antibody-derived entities (e.g. single chain Fv) for coupling to liposomes will be required.

Acknowledgements

We gratefully acknowledge the collaboration with Dr. T. M. Allen, University of Alberta, Canada; Dr. S. Zalipsky, Alza Corporation, USA; and Dr. A. Gorter, University of Leiden, The Netherlands. This work was partly supported by Grant 94-767 from the Dutch Cancer Society.

References

1. Allen, T. M.; Cullis, P.R. Drug Delivery Systems: Entering the Mainstream, *Science* **2004**, *303*(5665), 1818-1822.
2. Bendas, G. Immunoliposomes - A Promising Approach to Targeting Cancer Therapy, *BioDrugs.* **2001**, *15*(4), 215-224.
3. Mastrobattista, E.; Koning, G.A.; Storm, G. Immunoliposomes for the Targeted Delivery of Antitumor Drugs, *Adv.Drug Deliv.Rev.* **1999**, *40*(1-2), 103-127.
4. Vahrmeijer, A.L.; van Dierendonck, J.H.; van de Velde, C.J. Treatment of Colorectal-Cancer Metastases Confined to the Liver, *Eur.J.Cancer* **1995**, *31A*(7-8), 1238-1242.
5. Nishizaw, Y; Casida, J.E. 3',5'-Diesters of 5-Fluoro-2'-deoxyuridine-Synthesis and Biological Activity, *Biochemical Pharmacology* **1965**, *14*(11), 1605-1619.
6. Borssum Waalkes, M.; van Galen, M.; Morselt, H.; Sternberg, B.; Scherphof, G.L. In-vitro Stability and Cytostatic Activity of Liposomal Formulations of 5-Fluoro-2'-deoxyuridine and its Diacylated Derivatives, *Biochim.Biophys.Acta* **1993**, *1148*(1), 161-172.
7. Beun, G.D.M.; van Eendenburg, D.H.; Corver, W.E.; van de Velde, C.J.; Fleuren, G.J. T-Cell Retargeting Using Bispecific Monoclonal Antibodies in a Rat Colon-Carcinoma Model. 1. Significant Bispecific Lysis of Syngeneic Colon-Carcinoma CC531 is Critically Dependent on Prolonged Preactivation of Effector Lymphocytes-T by Immobilized Anti-T-Cell Receptor Antibody, *J. Immunother.* **1992**, *11*(4), 238-248.
8. Thomas, C.; Nijenhuis, A.M.; Timens, W.; Kuppen, P.J.; Daemen, T.; Scherphof, G.L. Liver Metastasis Model of Colon-Cancer in the Rat-Immunohistochemical Characterization, *Invasion & Metastasis* **1993**, *13*(2), 102-112.
9. Kamps, J.A.A.M.; Koning, G.A.; Velinova, M.J.; Morselt, H.W.; Wilkens, M.; Gorter, A.; Donga, J.; Scherphof, G.L. Uptake of Long-circulating Immunoliposomes, Directed Against Colon Adenocarcinoma Cells, by Liver Metastases of Colon Cancer, *J.Drug Target* **2000**, *8*(4), 235-245.
10. Koning, G.A.; Gorter, A.; Scherphof, G.L.; Kamps, J.A.A.M. Antiproliferative Effect of Immunoliposomes Containing 5-Fluoro-deoxyuridine-dipalmitate on Colon Cancer Cells, *Br.J.Cancer* **1999**, *80*(11), 1718-1725.
11. Koning, G.A.; Morselt, H.W.; Velinova, M.J.; Donga, J.; Gorter, A.; Allen, T.M.; Zalipsky, S.; Kamps, J.A.A.M.; Scherphof, G.L. Selective Transfer of a Lipophilic Prodrug of 5-Fluorodeoxyuridine from

Immunoliposomes to Colon Cancer Cells, *Biochim.Biophys.Acta* **1999,** *1420*(1-2), 153-167.

12. Zalipsky, S. Synthesis of an End-group Functionalized Polyethylene glycol-Lipid Conjugate for Preparation of Polymer-grafted Liposomes, *Bioconjug.Chem.* **1993,** *4*(4), 296-299.

13. Hong, K.; Friend, D.S.; Glabe, C.G.; Papahadjopoulos, D. Liposomes Containing Colloidal Gold are a Useful Probe of Liposome-Cell Interactions, *Biochim.Biophys.Acta* **1983,** *732*(1), 320-323.

14. Kamps, J.A.A.M.; Swart, P.J.; Morselt, H.W.; Pauwels, R.; De Bethune, M.P.; De Clercq, E.; Meijer, D.K.; Scherphof, G.L. Preparation and Characterization of Conjugates of (Modified) Human Serum Albumin and Liposomes: Drug Carriers with an Intrinsic Anti-HIV Activity, *Biochim.Biophys.Acta* **1996,** *1278*(2), 183-190.

15. Koning, G.A.; Morselt, H.W.; Gorter, A.; Allen, T.M.; Zalipsky, S.; Kamps, J.A.A.M.; Scherphof, G.L. Pharmacokinetics of Differently Designed Immunoliposome Formulations in Rats with or without Hepatic Colon Cancer Metastases, *Pharm.Res.* **2001,** *18*(9), 1291-1298.

16. Hansen, C.B.; Kao, G.Y.; Moase, E.H.; Zalipsky, S.; Allen, T.M. Attachment of Antibodies to Sterically Stabilized Liposomes-Evaluation, Comparison and Optimization of Coupling Procedures, *Biochim. Biophys. Acta* **1995,** *1239*(2), 133-144.

17. Koning, G.A.; Morselt, H.W.; Gorter, A.; Allen, T.M.; Zalipsky, S.; Scherphof, G.L.; Kamps, J.A.A.M. Interaction of Differently Designed Immunoliposomes with Colon Cancer Cells and Kupffer Cells. An In-vitro Comparison, *Pharm.Res.* **2003,** *20*(8), 1249-1257.

Chapter 8

Polymer-Protected Liposomes: Association of Hydrophobically-Modified PEG with Liposomes

Debra T. Auguste[1], Robert K. Prud'homme[1,*], Patrick L. Ahl[2,4], Paul Meers[2,5], and Joachim Kohn[3]

[1]Department of Chemical Engineering, Princeton University, Princeton, NJ 08544
[2]Elan Pharmaceuticals, One Research Way, Princeton, NJ 08540
[3]Department of Chemistry, Rutgers, the State University of New Jersey, Piscataway, NJ 08854
[4]Current address: BioDelivery Systems International, 4 Bruce Street, Newark, NJ 07103
[5]Current address: Transave, 11 Deer Park Drive, Suite 117, Monmouth Junction, NJ 08852
[*]Corresponding author: prudhomm@Princeton.edu

We demonstrate the impact of cooperative interactions on liposome protection by incorporating multiple hydrophobic sites on poly(ethylene glycol) (PEG) polymers. Our hydrophobically-modified PEGs (HMPEGs) are comb-graft copolymers with precisely alternating, monodisperse PEG blocks (MW= 6, 12, or 35 kDa), bonded to C18 stearylamide hydrophobes. Cooperativity is controlled by varying the extent of oligomerization at a constant ratio of PEG to stearlyamide. The association of polymer with liposomes increases with the degree of oligomerization; equilibrium constants for 6 kDa PEG increases from 6.1 ± 0.8 $(mg/m^2)/(mg/ml)$ for 3 loops to 78.1 ± 12.2 $(mg/m^2)/(mg/ml)$ for 13 loops. In addition, HMPEG6k-DP3 (with three 6 kDa loops) results in superior, irreversible protection from complement protein binding than DSPE-PEG5k after 12 hours.

Introduction

Liposomes, or vesicles, are concentric lipid bilayers separated by aqueous volumes [1]. Because they are biodegradable, nontoxic, and able to non-covalently encapsulate molecules (i.e., chemotherapeutic agents, hemoglobin, imaging agents, drugs, and genetic material) within their 100-200 nm diameter interior, liposomes have advantages over other drug delivery methods [2]. In addition, liposomes can target specific sites for drug delivery [3]. Clearance by macrophages, as a result of immune recognition, can localize liposomes in sites of inflammation. Enhanced permeability of immature tumor vasculature to particles with sizes less than 100-780nm provides localization in tumors [4]. Controlled release of agents from liposomes can be exploited to maintain therapeutic drug concentrations in the bloodstream or in local areas. Other benefits of encapsulation are protection of the drug from degradation and efficient intracellular delivery [5].

Liposomes are typically categorized in one of four categories: (1) conventional liposomes, which are composed of neutral and/or negatively-charged lipids and often cholesterol; (2) long-circulating liposomes, which incorporate PEG covalently bound to a lipid; (3) immunoliposomes, used for targeting where antibodies or antibody fragments are bound to the surface, and (4) cationic liposomes, that are used to condense and deliver DNA [5]. Liposomes are characterized by surface charge, size, composition, and number and fluidity of the lamellae [6]. Certain lipids have been shown to affect the mechanics of liposomes by making them more rigid (i.e., cholesterol) or more fusogenic (i.e., phosphatidylethanolamine and N-dodecanoyl-phosphatidyl-ethanolamine) [7,8]. These parameters have been investigated to increase liposome accumulation in sites of interest, which is typically hindered by immune recognition [3,8-16]. Immune recognition results in liposome accumulation in the liver, kidney, and spleen, which are the organs whose primary responsibility is to remove toxins and foreign agents from the bloodstream [12]. Some scientists have focused on liposomal delivery to these organs based on formulations that cause immune recognition [11,17]. For delivery to tumors, evasion of the immune system is vital. Increased circulation has been shown to correspond to an increase in liposome accumulation in tumors [12,18].

Looking at nature, scientists studied natural cell membranes to devise methods to reduce immune recognition. Glycolipids, receptors, and eventually hydrophilic polymers were covalently attached to lipids and incorporated into liposomes [2,3,19-21]. The most advantageous liposome formulation (the "Stealth" liposome) integrated a PEG covalently bound to a lipid (i.e., DSPE-PEG5k) in the liposome formulation [2]. A range of PEG molecular weights were investigated. Ultimately, 2-5 kDa showed the strongest ability to reduce immune recognition [22]. Blume and Cevc demonstrated that incorporation of at least 2.5 mol% DSPE-PEG5k increased circulation times from 0.47 to 8 hours [23].

It is postulated that the limitation of DSPE-PEG5k for extending the circulation times arises from deprotection of the liposomes by partitioning of the PEG-lipid off of the surface. The PEG layer produces a steric barrier to protein binding. Senior showed a decrease in the rate of protein adsorption onto liposomes incorporating DSPE-PEG5k [16]. In parallel, Chonn et al. showed greater protein binding on bare liposomes, which correlated with shorter circulation times [24]. Evidence that immune recognition was a result of complement protein adsorption was demonstrated by Liu et al. [14]. Complement proteins are known to cause a cascade of events that result in immune recognition. Therefore, we employ an automated in vitro assay that measures the ability of the HMPEG polymers to shield liposomes from complement interactions.

We present multiply attached polymers as a means for constructing polymer protected liposomes. The method exploits current interest in cooperativity, where multiple, relatively weak binding leads to strong overall association. The concept is established by a series of PEG-based comb copolymers with concatenated PEG chains having hydrophobic anchoring groups between the linked PEG chains. These polymers allow the direct comparison of binding in relation to protection with polymers with the same ratio of PEG to hydrophobe but with varied oligomerization and polymers with the same degree of oligomerization but with different ratio of PEG to hydrophobe.

The HMPEG polymers provide substantial benefits over PEGylated lipids. The ability to add the soluble polymer to preformed liposomes permits greater flexibility in manufacturing and tailoring of liposome formulations. Also, PEGylated lipids are limited in the molecular weight of PEG and in the amount able to be incorporated into the liposome. The former occurs due to partitioning and the latter results from phase transitions. With multiple binding interactions, higher molecular weight PEGs can be bound at higher binding affinities. The HMPEG polymers we investigate in this study are comb-graft copolymers with precisely alternating, monodisperse PEG blocks (Mw= 6, 12, or 35 kDa), bound to C18 stearylamide hydrophobes with 3 to 13 hydrophobic anchors per chain. We report the ability of these polymers to associate with fusogenic liposomes and shield from complement binding. We benchmark our results against DSPE-PEG5k, which has been studied previously.

Materials and Methods

Liposome Preparation

N-C12-DOPE:DOPC (7:3, mol:mol) liposomes were prepared as described by Shangguan *et al.*[7]. Lipids solubilized in chloroform at 25 mg/ml were added to a vial at a final molar mass of 80 μmol. Chloroform was removed under reduced pressure using a Büchi RE 111 Rotovapor and 461 water bath (Büchi Labortechnik AG, Flawil, Switzerland), then left under vacuum overnight to remove residual solvent. The lipid bilayer film was hydrated with a TES buffer solution (10 mM TES, 150 mM NaCl, 0.1 mM EDTA, pH 7.4), vortexed, subjected to five cycles of freezing in liquid nitrogen and thawing in a room temperature water bath. The lipid solution was extruded ten times through a 0.2 μm polycarbonate membrane filter at 250 psi using a 10 ml Lipex extruder (Northern Lipids, Inc. Vancouver, BC, Canada). The liposomes were stored at 4°C.

To aid in separation from free polymer in solution, the liposome density was increased by encapsulating a sucrose solution (10 mM TES, 250 mM sucrose, pH 7.4) and a fluorescent marker (DiD) was added for visibility. Unencapsulated sucrose was removed by dialysis at 4°C overnight against the TES buffer using a Slide-A-Lyzer cassette with a 10k MWCO. Sucrose-encapsulating liposomes containing DSPE-PEG5k were prepared similarly however the amount of DSPE-PEG5k (x) added was subtracted from the mol ratio of DOPC (2.9-x), keeping the ratio of DiD and N-C12-DOPE constant. Therefore, the composition of liposomes incorporating DSPE-PEG5k is denoted as N-C12-DOPE:DOPC:DSPE-PEG5k:DiD (7:(2.9-x):x:0.1, mol:mol:mol:mol).

Liposome concentration was determined by the phosphate assay as described by Chen *et al.*[25]. The liposome radii and polydispersity were determined by quasi-elastic light scattering using a NICOMP 270 submicron particle sizer (NICOMP Instruments, Goleta, CA). The number-averaged diameters for sucrose-encapsulating and buffer-encapsulating liposomes were 62.1 ± 16.4 nm and 112.8 ± 28.5 nm, respectively.

Adsorption

HMPEG polymers (for polymer description see Table I) were equilibrated with liposomes at 1.4mM lipid in 0.5 ml TES buffer. Polymer was then added, either solubilized in TES buffer or as dry mass, to achieve the desired concentration. For liposomes with DSPE-PEG5k, four liposome formulations were prepared with increasing amounts of polymer coverage (0.14, 0.28, 0.70, and 2.4 times Γ*, i.e., full surface coverage as calculated from the dimensions of the PEG loops [26]. The DSPE-PEG5k containing liposomes were diluted to 1.4mM lipid in TES buffer as described above. Each sample was vortexed and allowed to equilibrate overnight at room temperature in an Eppendorf 5436 Thermomixer (Brinkman Instruments, Westbury, NY, USA). After 24 hours, an initial fraction of 200 µl was removed. The remaining 0.3 ml was centrifuged at 21000xg in the Eppendorf 5417r Centrifuge (Brinkmann Instruments, Westbury, NY, USA) at 4°C. The supernatant was extracted. Phosphate and PEG assays were completed on the supernatant and initial fractions to detect the amount of PEG per lipid.

Table I. Description of hydrophobically-modified PEG (HMPEG) polymers.

Polymer	N_b	$\bar{\xi}$, Å	MW, kDa	Polymer Area, x 10^{17} m^2	Γ^*, mg/m^2	Γ_{hm}^*, anchor mol%
HMPEG6k-DP3	4	50	42	7.9	0.88	0.89
HMPEG6k-DP13	24	43	112	35.0	0.53	1.21
HMPEG12k-DP2.5	3	75	48	13.3	0.60	0.40
HMPEG12k-DP5	8	65	106	26.5	0.66	0.53
HMPEG35k-DP2.5	3	128	138	38.8	0.59	0.14
DSPE-PEG5k	1	54	5781	2.3	0.42	1.55

To convert Γ^* (mg/m^2) to Γ_{hm}^*, determine the moles polymer per 4.7×10^{-6} moles of lipid then multiply by the number of hydrophobes (DP-1) and 100 to obtain the mol% hydrophobes per lipid, or C_{hm}^*.

The association constant K (see Table II), was determined by the initial slope (first 4 data points) of each adsorption profile, such that

$$K = \frac{d\Gamma}{d[C_p(free)]} \equiv \frac{(mg/m^2)}{(mg/ml)} \qquad (1)$$

PEG Assay

The concentration of PEG was quantified by an assay described by Baleux [27], wherein 25 µl of an iodine-potassium iodide solution (0.04 M I$_2$, 0.12 M KI) was added to 1 ml of a diluted sample. Samples were adjusted to an optimal adsorption range (0.1<AU<1.0). The sample and color reagent were mixed in a disposable semi-micro cuvette with a 1.0 cm path-length. After five minutes, the optical density (OD) was determined at room temperature by a UV-2101PC spectrophotometer (Shimadzu Scientific Instruments, Princeton, NJ, USA) in the

Table II. Association constants of HMPEGs to fusogenic liposomes. Error is given as one standard deviation.

Polymer	Association Constant, K $(mg/m^2)/(mg/ml)$
HMPEG6k-DP3	6.1 ± 0.8
HMPEG6k-DP13	78.1 ± 12.2
HMPEG12k-DP2.5	1.9 ± 0.1
HMPEG12k-DP5	16.4 ± 2.5
HMPEG35k-DP2.5	4.3 ± 0.5
DSPE-PEG5k	0.4 ± 0.1

visible region, λ=500nm. Because lipid in the assay affects the OD, multiple calibration curves were required. Table III details the correlation between HMPEG concentration and OD for 0, 10, 30, and 50 μM lipid. Similarly, calibration curves were conducted for DSPE-PEG5k at different lipid concentrations; however, there was no dependence on lipid concentration. The slope reported is an average of all of the calibration curves taken at different lipid concentrations. The variation in the PEG assay with respect to different polymer architectures is a result of differences in the hindrance to helix formation, the origin of the colored complex.

The square of the Pearson product moment correlation coefficient, R^2, interprets the proportion of the variation in Y attributable to the variation in X. It is given as follows, where a value of one indicates that the estimated value is equal to the actual value [45]:

$$R^2 = \frac{n(\sum XY) - (\sum X)(\sum Y)}{\sqrt{\left[n \sum X^2 - (\sum X)^2\right]\left[n \sum Y^2 - (\sum Y)^2\right]}}$$

In vitro complement depletion assay

The depletion of complement protein from serum has been shown to correlate with *in vivo* immune response [28-30]. We measure the ability of the treated serum to achieve complement-mediated lysis of activated sheep erythrocytes. The complement assay was conducted as described by Ahl *et al.* [28]. Activation of sheep erythrocytes (or red blood cells), was performed by first washing the cells three times in 10 ml GVB^{2+}, centrifuging at 8000xg for 4 minutes, and removing the supernatant. The cells were resuspended at 10^8 cells/ml, determined by hemacytometry, and incubated with hemolysin rabbit anti-sheep erythrocyte stromata serum at 1/500 (v/v). Excess hemolysin was removed by rinsing three times in GVB^{2+} and resuspended at 10^8 cells/ml. Activated cells were stored at 4°C and used within seven days.

Table III. HMPEG correlation values for quantification of PEG (mg/L) as a function of the optical density at increasing concentrations of lipid in µM.

Polymer	0 µM lipid		10 µM lipid	
	$d[OD]/d[C_p]$	R^2	$d[OD]/d[C_p]$	R^2
HMPEG6k-DP3	0.038	0.987	0.041	0.980
HMPEG6k-DP13	0.022	0.971	0.023	0.980
HMPEG12k-DP2.5	0.054	0.996	0.053	0.982
HMPEG12k-DP5	0.046	0.973	0.050	0.982
HMPEG35k-DP2.5	0.049	0.976	0.048	0.978
DSPE-PEG5k	0.046	0.931	-	-

Polymer	30 µM lipid		50 µM lipid	
	$d[OD]/d[C_p]$	R^2	$d[OD]/d[C_p]$	R^2
HMPEG6k-DP3	0.043	0.980	0.046	0.973
HMPEG6k-DP13	0.022	0.998	0.023	0.989
HMPEG12k-DP2.5	0.056	0.982	0.057	0.988
HMPEG12k-DP5	0.048	0.997	0.053	0.992
HMPEG35k-DP2.5	0.052	0.979	0.053	0.992
DSPE-PEG5k	-	-	-	-

Each individual complement assay consisted of six samples prepared in 200 μl volumes: TES buffer (the negative control, no liposomes, 0% complement activation), 8 mM unmodified liposomes in TES buffer (the positive control, 100% complement activation), and four test samples containing 8 mM liposomes in TES buffer with increasing amounts of polymer. The samples were equilibrated overnight at 4°C with gentle shaking. Each sample was incubated at 37°C for thirty minutes with 100 μl reconstituted rat sera, diluted 1:1 (v/v) with GVB^{2+}. Subsequently, 300 μl of GVB^{2+} was added followed by vortexing and centrifugation at 8000xg for 4 minutes. We utilized a Biomek® 2000 Laboratory Automation Workstation (Beckman Coulter, Brea, CA, USA) to perform the pipetting for the remaining steps. A 100 μl aliquot of the supernatant was diluted 1:1 (v/v) preceding eight successive serial dilutions in GVB^{2+}. To each diluted sample of treated serum, 100 μl of activated sheep cells was added and incubated for thirty minutes at 37°C. Hemolysis was quenched by the addition of 300 μl of GVB^{2+}-EDTA. Cells that were intact were sedimented by centrifuging the samples at 8000xg for 4 minutes. A 200 μl aliquot from the supernatant of each sample was placed into a 96 well plate. The optical density of each sample well was determined at 415 nm using a 3550-UV spectrophotometer plate reader (Bio-Rad Laboratories, Hercules, CA, USA). Buffer encapsulating N-C12-DOPE:DOPC:DiD (7:2.9:0.1, mol:mol:mol) and sucrose encapsulating N-C12-DOPE:DOPC:DSPE-PEG5k:DiD (7:(2.9-x):x:0.1, mol:mol:mol:mol) liposomes were used. Sucrose encapsulation enabled pelleting of the liposomes without affecting the complement assay.

The results are plotted as the percent hemolysis versus the log of the inverse of the serum dilution [-Log(1/SD)]. The CH50, commonly utilized in related literature [29,31,32], is the serum dilution necessary to achieve 50% hemolysis and is directly related to the level of active complement in the serum. The CH50 of each hemolysis curve was acquired by a linear fit to a log-log version of the von Krough equation [33]. The surface "protection" mediated by HMPEG adsorption or DSPE-PEG5k addition can be quantitatively described using the following equation:

$$\% protection = \frac{CH50_{PEG-Liposomes} - CH50_{BareLiposomes}}{CH50_{Buffer} - CH50_{BareLiposomes}} \times 100 \qquad (2)$$

Adsorption Theory and Models

Defining surface coverage, Γ*

We define full surface coverage as the mass of polymer required to cover an equivalent area of lipid. The area of lipid in square meters is determined from the lipid concentration. The calculation utilizes the average lipid head area (approximately 70Å^2 [34]) and assumes an equivalent number of lipids on the internal and external liposome bilayer. Since the HMPEGs are post-added, the only available surface to adsorb is the outer shell of the liposome. From previous study, it has been shown that the polymers are unable to penetrate the liposome bilayer. Our study focuses on hydrophobically-modified PEGs (HMPEGs). These polymers have strictly alternating PEG backbones with $C_{18}H_{37}$ hydrophobes. They are described by the size of the PEG spacer (molecular weight of PEG between hydrophobes) and number of loops (or degree of polymerization, denoted DP) [35,36]. The polymers are designated: HMPEG"X"-DP"Y", where X is the molecular weight of the PEG spacers in Daltons and Y is the average number of PEG spacers, or loops.

A schematic of the PEG polymer chain tethered to a two-dimensional lipid bilayer is shown in Figure 1. Previous studies have shown [26] that the polymer can be treated as a series of subchains, comprising of one-half the molecular weight of the PEG spacer between hydrophobes. The subchains obey random-walk statistics and occupy an area at the water/lipid interface given by a sphere of diameter [37]

$$\xi_{m_b} = 0.76 m_b^{1/2} \text{ [Å]}, \tag{3}$$

where m_b is the molecular weight of the subchain. The terminal PEG chains are treated differently because they are not bound at both ends; therefore, they occupy an area corresponding to a sphere of diameter twice the molecular weight of the internal subchains. The total number of spheres, N_b, for a polymer is equal to two times the degree of polymerization minus the two ends ($N_b = 2D_p - 2$). The number-averaged diameter for a polymer chain is given as follows [26]:

$$\bar{\xi} = \frac{2}{N_b}(\xi_{2m_b}) + \frac{N_b - 2}{N_b}(\xi_{m_b}). \tag{4}$$

From the number-averaged diameter, the area of the polymer occupied at the lipid interface is determined by $N_b \pi (\bar{\xi}/2)^2$, the number of blobs times the average area of each sphere. For example, HMPEG6k-DP3 has 4 subchains, a number-averaged diameter of 50Å, and occupies an area of 7900Å^2 per polymer. The amount of HMPEG6k-DP3 to cover one square meter lipid, Γ^*, is 21 nmoles or 0.88 mg.

Our view of coverage is based on the mass of polymer needed to cover the exterior liposome surface. From the random walk approximation described above, which we have verified from neutron scattering of the HMPEG polymers [26], we can approximate the area occupied by the HMPEG polymers and the DSPE-PEG5k polymer. Because the polymers differ greatly in molecular weight and number of subchains, we find this the best means for comparison.

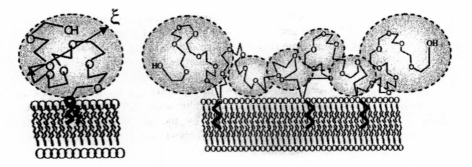

Figure 1. Schematic of (a) DSPE-PEG5k and (b) a hydrophobically-modified PEG with three sites of attachment adsorbed to a 2-dimensional lipid bilayer membrane.

Partitioning model

A partitioning model can be used to describe the association of our hydrophobically-modified polymers to a phospholipid bilayer. If we consider the bilayer and aqueous solution as two phases, we can describe a partition coefficient that relates the mass of bound polymer per external lipid area, $[PL]/[L]$ (in mg/m^2), with the concentration of free polymer, $[P]$ (in mg/ml) such that

$$\frac{[PL]}{[L]} = \frac{\Gamma}{\gamma}[P],$$

(5)

where Γ is the partition coefficient resolved by the change in free energy of the polymer between the two phases and γ is the activity coefficient, representing the deviation from ideality [38]. The plot of $[PL]/[L]$ vs. $[P]$ yields an association isotherm that is linear at low polymer concentrations and has a decreasing slope due to non-ideal interactions ($\gamma > 1$) which depends on the polymer type. A strict thermodynamic analysis will define γ as the ratio of the polymer activity coefficients in he aqueous and bilayer phases such that $\gamma = \gamma_P{}^L/\gamma_P{}^A$. Porcar shows that this partitioning model relates to a basic binding model where $[P]+[L] \leftrightarrow [PL]$. The slope, Γ/γ, is equivalent to the association constant divided by the number of available sites, K/N [38].

Cooperativity

Our definition of cooperativity is based on the higher probability of the polymer chain to remain adsorbed due to increasing the number of attachment sites. We hypothesize that having multiple hydrophobic anchors on a PEG polymer increases the cooperativity of the polymer, resulting in a strong association with the liposome membrane and providing a protective PEG layer. Cooperativity of this polymer series is assessed by comparing the association constants, which evaluates the polymer's ability to desorb from the liposome surface. Each polymer's cooperativity is dependent on its molecular architecture. For DSPE-PEG, with one hydrophobic anchor, an association constant that relates the amount of bound polymer to free polymer is given as follows [39]:

$$K = \frac{[PL]}{[P][L]}. \tag{6}$$

The probability of a polymer desorbing is written in terms of the association constant such that

$$P_{desorption,1} = \frac{1}{K+1}. \tag{7}$$

If K is large, having a strong affinity for binding, then $P_{desorption,1}$ goes to zero. Conversely, if K is small, having a low affinity for adsorption, $P_{desorption,1}$ goes to one. Assuming no correlations exist between the hydrophobes, the probability of a multi-loop polymer desorbing is the probability of all of the anchors desorbing simultaneously, which for a chain with x anchors yields:

$$P_{desorption} = (P_{desorption,1})^x. \tag{8}$$

This model predicts that the effect of cooperativity of the multi-loop PEGs, with 2 to 11 anchors, should be pronounced.

Results and Discussion

Association of HMPEGs with fusogenic liposomes

The series of hydrophobically-modified PEG (HMPEG) polymers was investigated to understand the affect polymer hydrophilicity (molecular weight of PEG) and hydrophobicity (number of anchors) have on the polymers' association with liposomes. We measured the adsorption of HMPEGs to bare liposomes and desorption of PEG from liposomes containing DSPE-PEG5k. Attempts were made at preparing fusogenic liposomes with HMPEGs similar to that of liposomes incorporating DSPE-PEG5k, i.e., adding the polymer to the mixture of lipids prior to solvent removal; however, the lipid/polymer mixture became a gel. HMPEGs were post-added to liposomes at the pertinent concentrations and equilibrated overnight with gentle shaking. For DSPE-PEG5k incorporating liposomes, four formulations were prepared adjusting the molar ratio of DSPE-PEG5k and PC accordingly. The DSPE-PEG5k containing liposomes were equilibrated overnight with gentle shaking at the same lipid concentration. Polymer-coated liposomes were separated from the supernatant which was assayed for polymer and lipid concentration. The error is derived from the precision of the Baleux assay (see Table I) and the phosphate assay (R^2 = 0.992).

We demonstrate that strength of binding is affected by the molecular architecture. The polymer with the strongest binding was HMPEG6k-DP13, with an association constant of 78.1 ± 12.2 (mg/m^2)/(mg/ml). In contrast, HMPEG6k-DP3, with an equivalent size spacer and ratio of PEG to hydrophobe, has an association constant 1/13th as strong. Increasing the cooperativity by approximately four times increases the binding 13 fold. We see an analogous increase in binding with number of anchor for HMPEGs with a 12 kDa spacer, HMPEG12k-DP2.5 and HMPEG12k-DP5.

At constant number of loops, the effect of increasing the polymer's hydrophilicity is determined by changing the molecular weight of the PEG spacer between hydrophobes. By increasing the polymer's hydrophilicity, we change its free energy ($\Delta G=-RT\ln K$), which is related to the equilibrium between the polymer being bound or free in solution. We anticipate that increasing the molecular weight of the PEG spacer will reduce the binding affinity. We observe a decrease in the association constant as we compare the HMPEGs with 6k and 12k spacers, with roughly 3 loops. Yet, the HMPEG35k-DP2.5 polymer has an association strength that lies between HMPEG6k-DP3 and HMPEG12k-DP2.5. This implies that, between 6 and 35 kDa of PEG, there exists a limit where the polymer chains are unable to pack efficiently, resulting in a minimum in the association constant K.

The adsorption isotherms in Figure 2 show consistently increasing slopes, which corresponds to having multiple-layer coverage at high polymer concentrations. Polymer saturation of the surface (or one full unconstrained layer of coverage) was calculated by dividing the total lipid area by the area per chain (from equation 4) to yield the number of polymers required to cover the surface. This number was then multiplied by the molecular weight of the polymer over Avogadro's number to give units of mg/m^2. We calculate full coverage of HMPEG6k-DP13 to be $\Gamma^* = 0.53$ mg/m^2 and we find experimentally the surface coverage Γ to be 0.85 mg/m^2 in equilibrium with $C_p = 0.02$ mg/ml. Multiple layers, or adsorption exceeding full surface coverage, have been observed by Sunamoto et al. [20], who studied the adsorption of hydrophobically-modified polysaccharides to liposomes. They also observed a dependence on binding affinity with increasing number of hydrophobic anchors [20].

In comparison with the strong adsorption of the HMPEGs, we find that DSPE-PEG5k partitions off of the surface. After equilibrating overnight in TES buffer, we observe 21% of the added DSPE-PEG5k is found in the supernatant. PEG covalently bound by a lipid has the weakest association constant {K = 0.41 ± 0.08 (mg/m^2)/(mg/ml)} of the series of polymers tested.

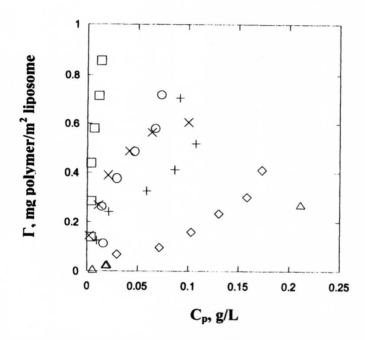

Figure 2. Association of hydrophobically-modified PEG polymers to sucrose encapsulating fusogenic liposomes. Graph depicts surface coverage, Γ, versus the free polymer in solution, C_p, for HMPEG6k-DP3 (○), HMPEG6k-DP13 (□), HMPEG12k-DP2.5 (◊), HMPEG12k-DP5 (×), HMPEG35k-DP2.5 (+), and DSPE-PEG5k (Δ). Each sample contained 1.4mM liposomes and 0.1 to 0.6 mg/ml polymer in 1ml TES buffer. Sucrose encapsulating liposomes incorporating 0.19, 0.37, 1.11, and 3.70 mol% DSPE-PEG5k were equilibrated in TES buffer. Samples were equilibrated for 24 hours. Uncertainty comes from the precision of the Baleux assay (see Table 1) and the precision of the phosphate assay ($R^2 = 0.992$).

In vitro complement depletion assay

The complement assay measures the ability of complement proteins to bind liposomes. These proteins are called complement proteins because they complement or amplify the action of the antibody, which is the principal means of defense against most infections. Complement consists of a series of about 20 proteins that are activated by antibody-antigen complexes or microorganisms that undergo a cascade of proteolytic reactions that result in the assembly of membrane-attack complexes. These complexes form holes in the micro-organisms, resulting in their destruction. Meanwhile, proteolytic fragments promote the defensive mechanism by dilating blood vessels and attracting phagocytic cells to bind, ingest, and destroy the infecting microorganisms [40].

The complement proteins are mainly produced by the liver and circulate in the bloodstream and in extracellular fluid. Most are inactive until triggered by a microorganism or indirectly by an immune response. Complement activation results in the assembly of complement components that assemble into the membrane-attack complex that facilitates microbial cell lysis [40]. We activate sheep erythrocytes with an antibody that triggers complement proteins to assemble membrane-attack complexes resulting in cell lysis. Lysis of sheep erythrocytes, or red blood cells, results in the solution turning red. The extent of complement activation (lysis) can be measured using spectrometry in the visible region.

It has been shown that complement protein adsorption correlates with *in vivo* immune recognition [28]. We assess how well polymer layers made from adsorption of HMPEG and incorporation of DSPE-PEG5k inhibit *in vitro* complement binding. The polymer layer "protects" the liposomes from complement opsonization. The results from the complement assay are graphed as percent hemolysis versus log of the inverse serum dilution, i.e. $\log(1/SD)$. A typical complement assay hemolysis curve (shown in Figure 3) depicts a sigmoidal curve with an abrupt transition from low hemolysis to complete hemolysis as the serum dilution decreases and log of the inverse serum dilution increases. In effect, the hemolysis curve depicts the extent of hemolysis that occurs at serial dilutions of complement serum, whereby at low dilution (-1) the highest concentration of complement is present and thus results in a higher percent hemolysis and extensive dilutions (-3) where low concentrations of serum complement results in no measurable hemolysis.

The CH50 value corresponds to the serum dilution that achieves 50% hemolysis and is directly proportional to the available complement in solution following the initial serum incubation. Shifts in the CH50 indicate changes in the level of protection from complement protein interactions for different polymer-coated liposomes. Bare *N*-C12-DOPE:DOPE (7:3, mol:mol) liposomes without a PEG protective layer have a high degree of complement binding which result in low CH50 values typically under 100. We will arbitrarily define the CH50 for bare liposomes as 0% protection. In comparison, 100% protection

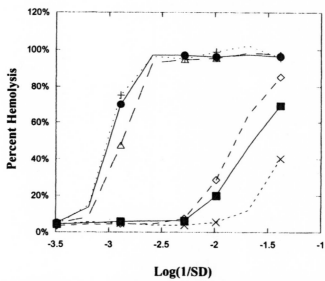

Figure 3. Complement assay for N-C12-DOPE/DOPC 7/3 (mol/mol) liposomes modified with HMPEG6k-DP3. Graph depicts buffer (•), liposomes corresponding to 0.34m² lipid area (■), liposomes with 0.05 mg polymer (◊), 0.15 mg polymer (×), 0.50 mg polymer (□), or 1.50 mg polymer (△). Γ* is 0.88 mg/m². Addition of 0.15 mg HMPEG6k-DP3 shifts the curve to the right instead of towards the left. This is a result of being portrayed on a logarithmic scale, where the error at this dilution is large relative to the difference in hemolysis. To obtain the surface coverage (mg/m²), divide the amount of polymer added by the lipid area, i.e. 1.50 mg HMPEG6k-DP3 / 0.34 m² lipid is 5 mg/m². The precision of each measurement is evaluated based on one standard deviation from the mean for the buffer and liposome controls, 3.1% and 9.1% respectively.

110

will be described as the CH50 value of the buffer control that is generally within the 500 to 1000 range. The HMPEG and DSPE-PEG5k protected liposomes we tested had CH50 values between these two limits. The precision of each measurement is based on one standard deviation from the mean for the buffer and liposome controls, 3.1% and 8.9% respectively. Complement assays (with 30 minute incubation) of HMPEG6k-DP3, HMPEG6k-DP13, HMPEG12k-DP2.5, and DSPE-PEG5k protected liposomes were completed at various levels of polymer coverage and exhibited very good agreement with the data presented.

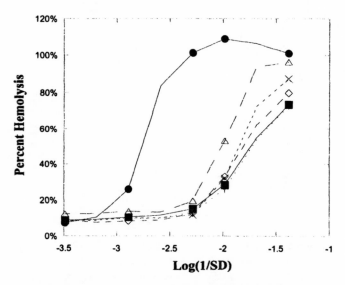

Figure 4. Complement assay for N-C12-DOPE/DOPC 7/3 (mol/mol) liposomes modified with HMPEG6k-DP13. Graph depicts buffer (●), liposomes corresponding to 0.34m² lipid area (■), liposomes with 0.03 mg polymer (◊), 0.09 mg polymer (×), 0.30 mg polymer (□), or 0.90 mg polymer (△). Γ* is 0.53 mg/m².

We outline the results of the complement assay in the following manner: (1) demonstration of how molecular architecture (number of anchors and molecular weight of the PEG spacer) produces cooperativity (Figures 3 and 4); (2) comparison of DSPE-PEG5k (Figure 5) versus the cooperative polymer chains (Figures 3 and 4); and (3) evaluation of the time dependence of the complement assay on the protection of liposomes by HMPEGs (Figure 6) and DSPE-PEG5k (Figure 7).

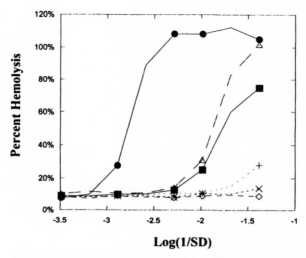

Figure 5. Complement assay for N-C12-DOPE/DOPC 7/3 (mol/mol) liposomes modified with DSPE-PEG5k. Graph depicts buffer (●), liposomes corresponding to 0.34m² lipid area (■), liposomes with 0.02 mg polymer (◊), 0.04 mg polymer (×), 0.10 mg polymer (□), or 0.34 mg polymer (△). Γ* is 0.42 mg/m².

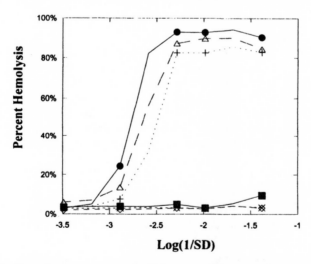

Figure 6. Complement assay with 12 hour incubation with rat sera for N-C12-DOPE/DOPC 7/3 (mol/mol) liposomes modified with HMPEG6k-DP3. Graph depicts buffer (●), liposomes corresponding to 0.34m² lipid area (■), liposomes with 0.05 mg polymer (◊), 0.15 mg polymer (×), 0.50 mg polymer (□), or 1.50 mg polymer (△). Γ* is 0.88 mg/m².

The results shown in Figure 6 are qualitatively similar to the results with a 30-minute incubation shown in Figure 3, which indicate the polymer layer protects against complement binding over the time interval 30 minutes to 12 hours.

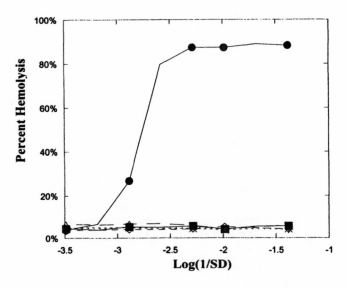

Figure 7. Complement assay with 12 hour incubation with rat sera for N-C12-DOPE/DOPC 7/3 (mol/mol) liposomes modified with DSPE-PEG5k. Graph depicts buffer (●), liposomes corresponding to 0.34m² lipid area (■), liposomes with 0.02 mg polymer (◊), 0.04 mg polymer (×), 0.10 mg polymer (□), or 0.34 mg polymer (Δ). Γ* is 0.59 mg/m². The weak protection shown in (Figure 5) for a 30 minute incubation is eliminated at the 12 incubation, which shows that DSPE-PEG5k does not protect against complement binding over long periods of time.

Number and molecular weight of the PEG spacers

In Figure 3, we demonstrate 98% protection from complement protein binding by adsorption of HMPEG6k-DP3 at levels exceeding full surface coverage: Γ*. For larger PEG spacers, the HMPEG12k series, a degree of polymerization of five (DP5) shows 95% protection at 1.8 times full surface coverage (data not shown); however, it requires five times the calculated surface coverage for HMPEG12k-DP2.5 to exhibit a similar level of protection (data not shown).

HMPEGs with higher molecular weight PEG spacers require more cooperative binding sites to influence protection. We find that lower molecular weight PEG (6k) with low cooperativity (DP=3) exhibits similar results to the 12k PEG with a higher degree of cooperativity (DP=5). This demonstrates a tradeoff between hydrophilicity (size of the PEG spacer) that pulls the polymer off of the surface that balances the number of cooperative sites that keeps the polymer bound. Nevertheless, increased cooperativity alone is insufficient to protect against complement binding. For example, the highly cooperative polymer HMPEG6k-DP13 does not shield against complement binding even at levels five times Γ^*, as depicted in Figure 4.

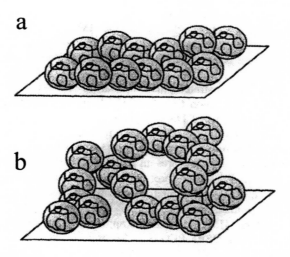

Figure 8. Schematic of polymer blobs associating with a surface (a) at low levels of cooperativity, where the polymer can distribute over the surface, and (b) high levels of cooperativity, which immobilizes the polymer on the surface and also results in inter- and intra-polymer associations that affect polymer coverage.

Our previous studies on this series of highly cooperative polymers [35,36] showed aggregation of the polymers in solution. This suggests that too high a level of cooperativity leads to inter-chain and polymer self-assembly as shown schematically in Figure 8. The hydrophobic interactions are strong and hinder rearrangement of the polymer onto the liposome surface to form an impenetrable layer. Although the mass of polymer to lipid area is high, the structure of the layer is not uniform and allows diffusion of complement protein to the liposome surface. After 24 hours, the HMPEG6k-DP13 coated liposomes aggregate as a result of bridging. However, no aggregation occurred in HMPEG12k-DP2.5

and HMPEG35k-DP2.5 after one week. This is a time-dependent effect that may be controlled using excipients. Further study of polymer-bridging is needed. The HMPEG35k-DP2.5 (data not shown) shows lower than 50% protection at levels five times Γ^*, which is consistent with the high solubility of the PEG chain and low degree of cooperativity. This indicates that strength of binding alone does not dictate complement shielding. The structure of the adsorbed polymer layer plays an important role in protection.

DSPE-PEG5k versus the multiply-attached HMPEG polymers

Figure 5 illustrates that at 2.4 times Γ^* DSPE-PEG5k protects only 15% – compared to 95% at 1.8 times Γ^* for HMPEG12k-DP5. Our previous studies have shown that the 12k PEG loop bound on each end behaves approximately like two terminally grafted 6k PEG chains [26]. Therefore, the disparity in performance between the DSPE-PEG5k and the HMPEG12k-DP5 is not due to the size of the PEG spacer, but rather due to the cooperativity of attachment. The DSPE-PEG5k can be considered the monomer (DP=1) in the sequence of HMPEG12k-DP(n).

It has been demonstrated that liposomes containing DSPE-PEG5k prolongs the *in vivo* circulation time [21,23]. From the PEG coil dimensions, 1.55 mol% (or 0.42 mg/m^2) of DSPE-PEG5k is needed for full coverage of a liposome. Blume and Cevc [23] have shown that upwards of 2.5 mol% DSPE-PEG5k added to a liposome formulation shows a similar degree of protection. The highest surface coverage we present for DSPE-PEG5k is 1.0 mg/m^2 which is equivalent to 3.7 mol%. Therefore, while we have not yet conducted the *in vivo* tests, we can make direct comparisons with the DSPE-PEG5k data in the literature. Our preliminary results form the basis of two hypotheses about *in vivo* experiments: First, HMPEGs with the appropriate architecture will provide greater protection than DSPE-PEG5k. Second, *in vivo* tests will provide evidence of the ability of the *in vitro* complement binding assay to predict PEG-liposome protection *in vivo*. The complement assay has been validated *in vivo* for liposomes with different lipid formulations and modifications [28], but not for PEG-protected liposomes.

Dissociation of the DSPE-PEG5k over time and stability of the HMPEG binding

To determine the ability of complement to penetrate the polymer protective layer, we modified the complement assay by extending the incubation with rat serum from 30 minutes to 12 hours. As shown in Figure 7, the DSPE-PEG5k shows no protection from complement binding after a 12 hour incubation. There are two plausible explanations. The first is that the DSPE-PEG5k layer is dynamic and though the PEG chains are overlapped ($\Gamma^*>1$) fluctuations can expose regions of the liposome surface to which proteins adsorb. The second explanation, which is more likely, is that the weaker association constant of the

DSPE-PEG5k allows dissociation of the polymer from the liposome surface in the presence of serum. Electrostatic repulsion may assist in the desorption of DSPE-PEG5k from the negatively-charged liposome because it is itself negatively-charged. The HMPEG adsorbed layers have slow dynamics as we have shown from Spin Echo Neutron Scattering [26] and therefore would be less prone to protein penetration by the first mechanism. The multiple attachments of HMPEGs slows the dynamics of detachment in the same way that they increase the energetics of attachment. In the literature, there does not appear to have been a thorough study of the dynamics of partitioning of PEG from the liposome surface during *in vivo* tests. Experiments of this kind would aid in the understanding and application of long-circulating liposomes.

In contrast to the DSPE-PEG5k, the HMPEG6K-DP3 shows the same high level of protein shielding after 30 minutes (Figure 3) and 12 hours (Figure 6) of incubation. The stability of the HMPEG polymer to resist either displacement from the liposome surface or penetration by complement protein is remarkable when compared to the DSPE-PEG5k.

Conclusions

We have demonstrated the ability to shield from complement binding by the association of multi-looped, hydrophobically-modified PEG (HMPEG) polymers with liposomes. A series of polymers has been prepared with mono-disperse PEG chains in a strictly alternating copolymer with stearylamide hydrophobes. The polymers are designated: HMPEG"X"-DP"Y", where X is the molecular weight of the PEG spacer in Daltons and Y is the average number of loops. We have introduced the concept of cooperativity, where multiple attachment sites result in a stronger association of the polymer with liposomes which increases with number of anchors. Increasing the molecular weight of the PEG spacer affects cooperativity by controlling the packing efficiency. Some polymers, such as HMPEG6k-DP3 and HMPEG35k-DP2.5, exceed the mass of polymer required to occupy the lipid surface area, producing multiple layers.

HMPEGs provide substantial protection against complement protein interactions as measured by the hemolysis assay. However, the strength of association alone does not correlate with protection. The most tenaciously bound HMPEG, HMPEG6k-DP13, exhibited very low levels of protection. This suggests that the structure of the adsorbed PEG layer plays a key role in protection. The highly cooperative HMPEG6k-DP13 may have strong inter- and intra-polymer associations that hinder the chains from redistributing uni-formly over the liposome surface [41]. The theories of Rubinstein [42] on the association of comb-graft polymers indicate strong intermolecular interactions leading to phase separation for this class of polymers. Further experiments using surface plasmon techniques could identify the importance of intra versus inter chain associations on PEG adsorption and complement protection.

The customary technique to prepare long-circulating liposomes has been to incorporate DSPE-PEG5k lipids into the liposome formulation during the initial bilayer formation process. An advantage of the HMPEG polymers is that they can be post-added to preformed liposomes. This uncouples liposome formulation and encapsulation from protection. Although HMPEG polymers have single acyl chain units (as opposed to two tails of the lipid in DSPE-PEG5k), it is the cooperativity of the acyl chains that associate with the liposome that results in larger association constants for HMPEGs. Each individual hydrophobe can re-equilibrate in solution and distribute on the liposome surface. However, if cooperativity is too high the polymer can become frustrated as predicted by Rubinstein [42]. In addition, more HMPEG polymers can associate with the liposome surface than is possible with DSPE-PEG5k lipids.

Protection from complement binding provided by an adsorbed layer of HMPEG6k-DP3 is constant over 12 hours. In contrast, protection by DSPE-PEG5k decreases from a modest level after the normal 30 minute incubation period to no protection after 12 hours of incubation. The temporal instability of the DSPE-PEG5k may arise from partitioning of the lipid off of the liposome surface or fluctuations in polymer coverage that exposes bare regions on the liposome surface. In either case, HMPEG polymers have neither of these defects.

This study suggests the following simple rules for the structures of HMPEG polymers required for liposome protection. Equivalent protection was observed with HMPEG chains with low degrees of polymerization and low molecular weight PEG (6K) and HMPEG chains with large degrees of polymerization and large PEG (12K). This indicates a tradeoff between solubility of the PEG and cooperativity of anchoring. Polymers with too high a level of cooperativity (DP13) did not perform well, and polymers with too high a molecular weight PEG (35K) did not function well either.

Currently, there is uncertainty as to the role PEG plays on the surface of liposomes that results in decreased liposomal clearance. Although it is assumed that PEG forms a steric barrier that reduces protein binding [16,29,43], Xu and Marchant [44] and Price et al. [43] found similar total protein adsorption profiles on bare liposomes and liposomes incorporating DSPE-PEG5k. Price speculates that the steric barrier may affect the interaction between liposomes and macrophages [43]. Our experiments address complement binding; they do not address why PEG reduces liposomal clearance. This indicates that a direction of future research should be in vivo studies. It may result that the 12k PEG has a longer circulation time because it provides a thicker barrier than the 6k PEG.

Acknowledgements

The authors would like to thank The Liposome Company for funding this research and Elan Pharmaceuticals Inc., who acquired The Liposome Company in 2000, for continuation of this work.

References

[1]   A. Bangham, M.M. Standish, J.C. Watkins, Diffusion of univalent ions across the lamellae of swollen phospholipids, *J. Mol. Biol.* **1965**, *13*, 238-252.

[2]   D.D. Lasic, D. Needham, The "stealth" liposome: A prototypical biomaterial, *Chem. Rev.* **1995**, *95*, 2601-2628.

[3]   T.M. Allen, E.H. Moase, Therapeutic opportunities for targeted liposomal drug delivery, *Adv. Drug Deliver. Rev.* **1996**, *21*, 117-133.

[4]   S.H. Jang, M.G. Wientjes, D. Lu, J.L.-S. Au, Drug delivery and transport to solid tumors, *Pharmaceut. Res.* **2003**, *20*, 1337-1350.

[5]   G. Storm, D.J.A. Crommelin, Liposomes: quo vadis?, *Pharm. Sci. Technol. Today* **1998**, *1*, 19-31.

[6]   R.R.C. New, Introduction, Preparation of Liposomes, Characterization of Liposomes, in: R.R.C. New (Ed.), Liposomes a Practical Approach IRL Press, New York, 1990, 301.

[7]   T. Shangguan, C.C. Pak, S. Ali, A.S. Janoff, P. Meers, Cation-dependent fusogenicity of an N-acyl phosphatidylethanolamine, *Biochim. Biophys. Acta* **1997**, *1368*, 171-183.

[8]   S.C. Semple, A. Chonn, P.R. Cullis, Influence of cholesterol on the association of plasma proteins with liposomes, *Biochemistry* **1996**, *35*, 2521-2525.

[9]   T.M. Allen, A. Chonn, Large unilamellar liposomes with low uptake into the reticuloendothelial system, *FEBS Lett.* **1987**, *223*, 42-46.

[10] A. Chonn, P.R. Cullis, Recent advances in liposome technologies and their applications for systemic gene delivery, *Adv. Drug Deliver. Rev.* **1998**, *30*, 73-83.

[11] R.M. Fielding, Liposomal drug delivery: advantages and limitations from a pharmacokinetic and therapeutic perspective, *Clin. Pharmacokinet.* **1991**, *21*, 155-164.

[12] A. Gabizon, D. Papahadjopoulos, Liposome formulations with prolonged circulation time in blood and enhanced uptake by tumors, *Proc. Natl. Acad. Sci. USA* **1988**, *85*, 6949-6953.

[13] Y.J. Kao, R.L. Juliano, Interactions of liposomes with the reticulo-endothelial system - effects of reticuloendothelial blockade on the

clearance of large unilamellar vesicles, *Biochim. Biophys. Acta* **1981**, *677*, 453-461.

[14] D. Liu, F. Liu, Y.K. Song, Recognition and clearance of liposomes containing phosphatidylserine are mediated by serum opsonin, *Biochim. Biophys. Acta* **1995**, *1235*, 140-146.

[15] J. Marjan, Z. Xie, D. Devine, Liposome-induced activation of the classical complement pathway does not require immunoglobulin, *Biochim. Biophys. Acta* **1994**, *1192*, 35-44.

[16] J. Senior, C. Delgado, D. Fisher, C. Tilcock, G. Gregoriadis, Influence of surface hydrophilicity of liposomes on their interaction with plasma protein and clearance from the circulation: studies with polyethylene glycol coated vesicles, *Biochim. Biophys. Acta* **1991**, *1062*, 77-82.

[17] G. Gregoriadis, Liposomes as drug carriers. Recent trends and progress, J. Wiley and Sons, New York, 1988.

[18] N.Z. Wu, D. Da, T.L. Rudoll, D. Needham, A.R. Whorton, M.W. Dewhirst, Increased microvascular permeability contributes to preferential accumulation of stealth liposomes in tumor tissue, *Cancer Res.* **1993**, *53*, 3765-3770.

[19] M.C. Woodle, D.D. Lasic, Sterically stabilized liposomes, *Biochim. Biophys. Acta* **1992**, *1113*, 171-199.

[20] J. Sunamoto, T. Sato, T. Taguchi, H. Hamazaki, Naturally occurring polysaccharide derivatives which behave as an artificial cell wall on an artificial cell liposome, *Macromolecules* **1992**, *25*, 5665-5670.

[21] A.L. Klibanov, K. Maruyama, V.P. Torchilin, L. Huang, Amphipathic polyethyleneglycols effectively prolong the circulation time of liposomes, *FEBS Lett.* **1990**, *268*, 235-237.

[22] F.K. Bedu-Addo, P. Tang, Y. Xu, L. Huang, Effects of polyethyleneglycol chain length and phospholipid acyl chain composition on the interaction of polyethyleneglycol-phospholipid conjugates with phospholipid: Implications in liposomal drug delivery, *Pharmaceut. Res.* **1996**, *13*, 710-717.

[23] G. Blume, G. Cevc, Liposomes for the sustained drug release in vivo, *Biochim. Biophys. Acta* **1990**, *1029*, 91-97.

[24] A. Chonn, S.C. Semple, P.R. Cullis, Association of blood proteins with large unilamellar liposomes in vivo, *J. Biol. Chem.* **1992**, *267*, 18759-18765.

[25] P.S. Chen, T.Y. Toribara, H. Warner, Anal. Chem. 28 (1956) 1756-1758.

[26] B.S. Yang, J. Lal, J. Kohn, J.S. Huang, W.B. Russel, R.K. Prud'homme, Interaction of surfactant lamellar phase and a strictly alternating comb-graft amphiphilic polymer based on PEG, *Langmuir* **2001**, *17*, 6692-6698.

[27] B. Baleux, Colorimetric determination of nonionic polyethylene oxide surfactants using an Iodine-Iodide solution, *C. R. Acad. Sci. Ser. C* **1972**, *279*, 1617-1620.

[28] P.L. Ahl, S.K. Bhatia, P. Meers, P. Roberts, R. Stevens, R. Dause, W.R. Perkins, A.S. Janoff, Enhancement of the in vivo circulation lifetime of L-a-distearoylphosphatidylcholine liposomes: importance of liposomal aggregation versus complement opsonization, *Biochim. Biophys. Acta* **1997**, *1329*, 370-382.

[29] M. Zhu, Y. Zeng, L. Jiang, P. Huang, Z. Wu, Effects on the amount of total hemolytic complement levels(CH50) and immunoglobulin in serum induced by the implantation of biomaterials into rats, *J. Biomedical Engineering (Chinese)* **1999**, *16*, 275-278.

[30] Z. Xu, G. Smejkal, R.E. Marchant, Interaction of plasma proteins with liposomes of different lipid composition, *Trans. Soc. Biomater.* **1998**, *12*, 276.

[31] M. Koide, S. Shirahama, Y. Tokura, M. Takigawa, M. Hayakawa, F. Furukawa, Lupus erythematosus associated with C1 inhibitor deficiency, *J. Dermatol.* **2002**, *29*, 503-507.

[32] M. Zabaleta-Lanz, R.E. Vargas-Arenas, F. Tapanes, I. Daboin, J. Atahualpa Pinto, N.E. Bianco, Silent nephritis in systemic lupus erythematosus, *Lupus* **2003**, *12*, 26-30.

[33] D.P. Stites, R.P.C. Rogers, Basic and Clinical Immunology, Appleton and Lang, Norwalk, 1991 217-262.

[34] D. Marsh, CRC Handbook of Lipid Bilayers, CRC Press, Boca Raton, 1990 387.

[35] C. Heitz, S. Pendharkar, R.K. Prud'homme, J. Kohn, A new strictly alternating comblike amphiphilic polymer based on PEG 1. Synthesis and associative behavior of a low molecular weight sample, *Macromolecules* **1999**, *32*, 6652-6657.

[36] C. Heitz, R.K. Prud'homme, J. Kohn, A new strictly alternating comblike amphiphilic polymer based on PEG 2. Associative behavior of a high molecular weight sample and interaction with SDS, *Macromolecules* **1999**, *32*, 6658-6667.

[37] J. Brandrup, E.H. Immergut, Polymer Handbook, Wiley and Sons, New York, 1989.

[38] I. Porcar, R. Garcia, V. Soria, A. Campos, Macromolecules in ordered media: 4. Poly(2-vinyl pyridine)-liposome association induced by electrostatic interactions, *Polymer* **1997**, *38*, 3545-3552.

[39] P.G. DeGennes, Scaling theory of polymer adsorption, *J. Phys. Paris* **1976**, *37*, 1445-1452.

[40] B. Alberts, D. Bray, J. Lewis, M. Raff, K. Roberts, J.D. Watson, Molecular Biology of the Cell, Garland, New York, 1989 1219.

[41] Q.T. Pham, W.B. Russel, W. Lau, The effects of adsorbed layers and solution polymer on the viscosity of dispersions containing associative polymers, *The Society of Rheology* **1998**, *41*, 159-176.

[42] M. Rubinstein, A.N. Semenov, Dynamics of entangled solutions of associating polymers, *Macromolecules* **2001**, *34*, 1058-1068.

[43] M.E. Price, R.M. Cornelius, J.L. Brash, Protein adsorption to polyethylene glycol modified liposomes from fibrinogen solution and from plasma, *Biochim. Biophys. Acta* **2001**, *1512*, 191-205.

[44] Z. Xu, R.E. Marchant, Adsorption of plasma proteins on polyethylene oxide-modified lipid bilayers studied by internal reflection fluorescence, *Biomaterials* **2000**, *21*, 1075-1083.

[45] Microsoft Corporation, Microsoft Excel, Copyright 1985-1999 9.0.382 SR-1 (2000).

Chapter 9

Dendrimers Based on Melamine: Vehicles for Drug Delivery?

Eric E. Simanek

Department of Chemistry, Texas A&M University, College Station, TX 77843 (email: simanek@mail.chem.tamu.edu)

Since 2000, we have been investigating dendrimers based on melamine as vehicles for drug delivery. Early efforts dedicated solely to the synthesis of these architectures are now being complemented with inquiries directed toward their intended purpose. In this contribution, we review the synthetic routes pursued to prepare these vehicles and highlight the results of biological investigations.

Introduction

The first generation of polymer-bound therapeutics are in and emerging from clinical trials *(1)*. Often the polymers are linear homopolymers – frequently poly(ethylene) glycol (PEG) – although block copolymers that assemble into micellar structures are also being explored. These polymers convey three principle advantages to the drug. They increase solubility. They increase vascular circulation time. They attenuate toxicity of the drug through selective targeting as a result of ligand/receptor interactions or the enhanced permeability and retention that tumors show for large molecules. The success of these systems offers compelling motivation and proof of concept for the examination of the next-generation of polymer therapeutics.

Dendrimers represent a promising class of molecules for use as vehicles for drug delivery *(2-6)*. The principle advantages that dendrimers offer include *i)* a well-defined (unique and monodisperse) composition, *ii)* multiple sites for manipulation, and *iii)* a globular shape offering a protected hydrophobic interior that can be employed to solubilize guests. Dendrimer-based delivery strategies are not a competing technology. It is unlikely that a single *universal platform* will emerge as the solution for all drug delivery challenges. In truth, each platform – be it dendritic or linear polymer – has advantages and disadvantages. Efforts to understand the appropriate roles for each strategy will require investigations of all systems at all levels.

Our efforts focus on dendrimers based on melamine. Significant energies have been invested in developing efficient syntheses of these targets. Such efforts continue. After discussing synthesis, the results of our preliminary investigations into the use of these vehicles for drug delivery will be presented. We conclude with our plans for the future investigations.

Results and Discussion

Synthesis

The synthesis of dendrimers based on melamine relies on the step-wise and selective displacement of chlorine atoms of cyanuric chloride. As the reaction progresses, the triazine ring becomes increasingly electron rich and accordingly, deactivated towards subsequent substitutions. As a result, tri-substituted triazines can be prepared in one-pot in almost quantitative yields. The temperatures shown in equation 1 correspond to those required for primary amines. These reactions proceed in 5 minutes, 3 h, and 8 h, respectively. Diisopropylethylamine is our base of choice for these reactions. While we have examined both convergent and divergent routes to these dendrimers, we favor the convergent approach in the solution phase. This approach ultimately allows us to prepare "trees with different colored leaves."

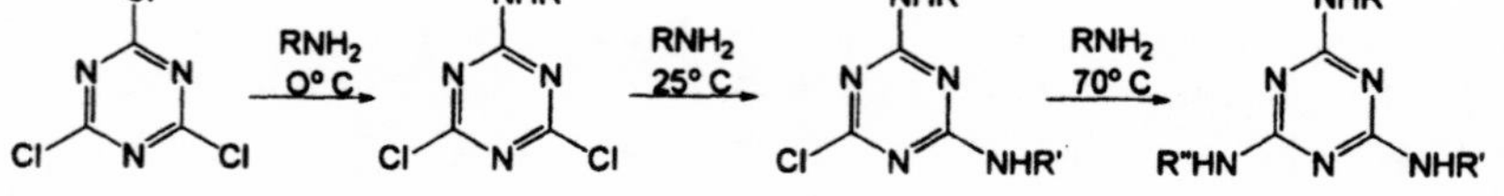

Figure 1. Synthesis scheme of tri-substituted triazines.

Our synthetic endeavors commenced with a proof of concept structure, a third-generation dendrimer of AB$_2$ building blocks wherein the triazines are interconnected with the diamine p-aminobenzylamine (Figure 2) *(7)*. Not only is the convergent synthesis of this architecture tractable, but it does not rely on functional group interconversions or protecting group manipulations. Only the benzylic amine of p-aminobenzylamine reacts with monochlorotriazines. The unreacted aniline NH$_2$ serves as a handle for elaboration upon reaction with the more electrophilic trichlorotriazine in a subsequent step.

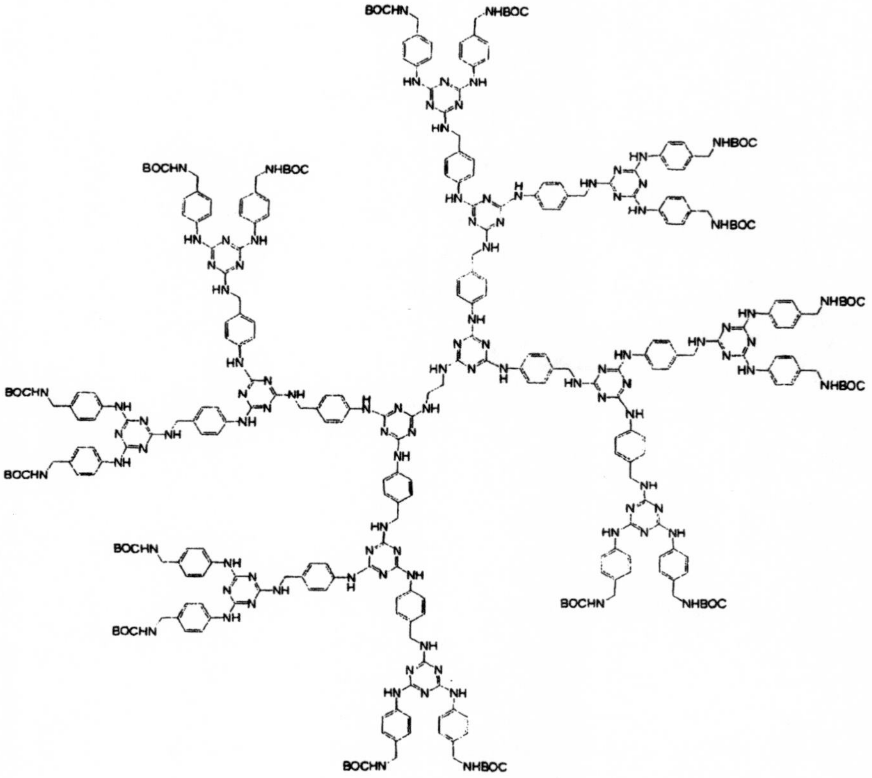

Figure 2. A third-generation dendrimer of AB$_2$ building blocks wherein the triazines are interconnected with the diamine p-aminobenzylamine.

To probe the limits of this chemistry, higher generations of dendrimers were prepared. This exercise revealed the iterative appearance and disappearance of lines in the ^{1}H and ^{13}C NMR spectra (Figure 3) *(8)*. Mass spectrometry has proven valuable as well. We find that third-generation dendrimers can be routinely prepared with no detectable impurities. We can summarize significant energies invested in developing synthetic methods with a list of lessons learned.

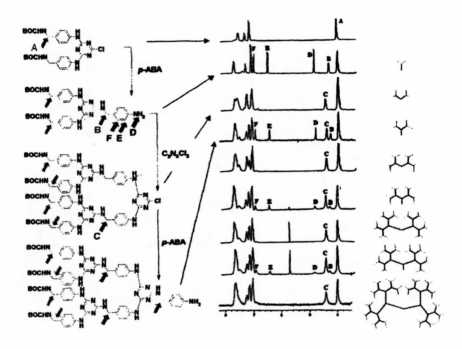

Figure 3. The synthesis of dendrimers followed with the appearance and disappearance of lines in the ^{1}H NMR spectra. (See page 2 of color inserts.)

Lesson 1. Mono- and difunctional dendrimers can be synthesized (9)

Using the convergent synthetic approach, two dendrimers displaying either one or two reactive sites on the periphery were prepared. These targets represent *classical* targets for dendrimer synthesis (Figure 4). In our opinion, the triazine chemistry affords these targets in the most convenient manner and on a scale at which these molecules can be manipulated to form dimers and condensation polymers.

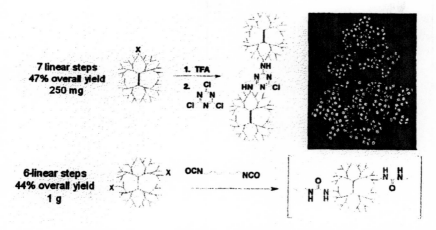

7 linear steps
47% overall yield
250 mg

1. TFA

2.

6-linear steps
44% overall yield
1 g

OCN

NCO

Figure 4. Mono- and di-functionalized targets are readily prepared and can be manipulated. The inset shows the mono-functionalized species and the dimer available by reaction with cyanuric chloride. The piperazine cores are maroon, the peripheral BOC or central triazine are green. (See page 2 of color inserts.)

Lesson 2. Structure-property relationships can be performed (10)

In acidic, aprotic organic solvents, dendrimers comprising triazines linked by *p*-aminobenzylamine with butyl amine peripheral groups form gels. We attribute gelation to the formation of networks intramolecular hydrogen bonds. No gels form when the peripheral butyl amine groups are replaced with piperidine groups or when the interior *p*-aminobenzylamine groups are replaced with piperazines. When some of the interior *p*-aminobenzylamine groups are replaced with piperazine, gelation requires higher concentrations of material. This proof-of-concept exercise offers promise that synthesis can be used to tailor specific properties, and ideally, overcome specific challenges, especially given the wealth of available peripheral and linking groups.

Lesson 3. A series of chemoselective linking groups are available (11)

Using competition reactions with a model monochlorotriazine and mixtures of amine nucleophiles, the relative reactivity of amines has been determined (Figure 5). From this data, novel linkers have been designed by incorporating two groups with significantly different reactivities into the same molecule. We have moved from *p*-aminobenzylamine to aminomethylpiperidine, a group with

126

similar reactivity differences, but better shelf-life. Compounds containing an aniline discolor over time.

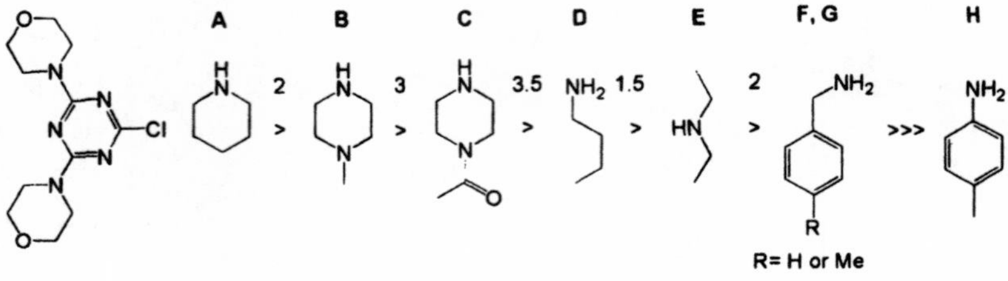

Figure 5. Competition reactions reveal relative rates for substitution.

Lesson 4. Compositionally diverse dendrimers accessible at moderate scale

We have completed the synthesis of dendrimers that display five or six orthogonally reactive sites by applying Wong's orthogonally protected carbohydrates (Figure 6) (12,13).

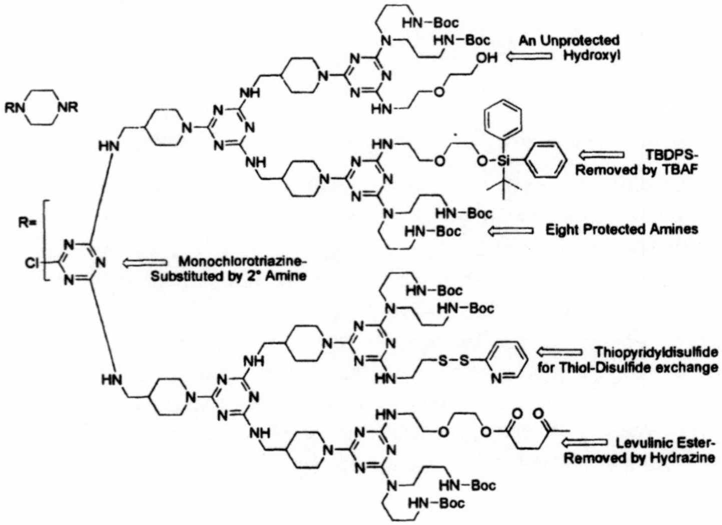

Figure 6. Dendrimers with selectively accessible reactive sites.

A free hydroxy group is available for immediate acylation. Hydroxy groups masked as *tert*-butyldiphenylsilyl ethers or levulic acid esters can be unmasked with tetrabutylammonium fluoride or hydrazine, respectively. The monochlorotriazine is available for nucleophilic aromatic substitution. The pyridyldisulfide is available for thiol-disulfide exchange. The BOC-protected amines offer a final site for manipulation. These targets are available on a 5 g scale.

Lesson 5. Convergent approaches reducing the number of synthetic steps (14)

As we move to scalable synthetic routes, we have explored classically-convergent strategies wherein a multivalent core is prepared divergently and coupled to multiple convergently-assembled peripheral groups *(15)*. The strategy has been used for the synthesis of a dendrimer with 48 reactive surface groups. This target is available in 5 linear steps and can be produced on 5 g scale with only four chromatographic purifications. While this does not yet fully satisfy our desires for kg-scale, column-free syntheses, it represents our state-of-the-art.

Opportunities for Drug Delivery

Using these vehicles for drug delivery presents a significant number of challenges in the context of efficiacy, absorbtion, distribution, metabolism, excretion and toxicity. Our understanding of triazine chemistry has risen to the state such that if the appropriate composition of a third-generation dendrimer were known – for instance, 12 2-kDa PEG chains, a 6:4 ratio of two different drugs, 4 peptides for targeting and a small molecule reporter – such a target likely could be reasonably assembled in one month. The challenge to us is that the molecular basis for efficacy is unknown. Accordingly, we are addressing the multiple facets of the drug delivery problem simultaneously with the hope that the lessons learned can be combined to arrive at a molecular solution. Our broad target is cancer therapy. The lessons learned are summarized.

Lesson I. Hydrophobic guests (drugs) can be sequestered (16)

A third-generation dendrimer comprising piperazine linking groups and cationic amines on the periphery can sequester hydrophobic guests in water. These guests include pyrene (0.2 molecules/dendrimer), 10-hydroxy-camptothecin (4 molecules/dendrimer), and a bisindolemethane (5 molecules/dendrimer). We attribute sequestration to a "dendritic phase" that resembles organic solvent into which these drugs preferentially partition over water. This partitioning is reversible. In cell culture, the bisindolemethane delivered in DMSO *(aq)* has similar activity (upregulation of luciferase in a PPARγ gene expression construct) as the molecule delivered in the dendrimer.

128

Lesson II: The position of a biolabile disulfide determines its exchange rate (17)

Disulfide reduction represents one opportunity for drug delivery because the intracellular concentration of thiols including cysteine, homocysteine, and glutathione are 100-1000x higher than that of the vasculature. We have found that dendrimers with mulitple disulfide-linked fluorophores undergo thiol-disulfide exchange with an exogenous thiol (dithiothreitol) at different rates (Figure 7).

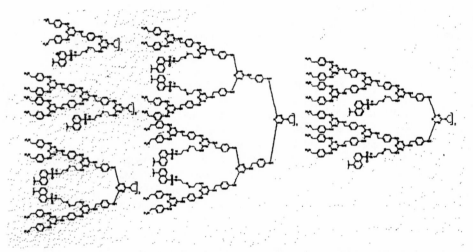

Figure 7. Exchange rates with dithiothreitol decrease across this series.

The half-life of exchange for groups buried within the interior of the dendrimer (as predicted from a 2-D representation of the molecule) is 2.2-fold slower than groups on smaller dendrimers. Groups that appear accessible on larger dendrimers have half-lives 1.8-fold slower than those groups on smaller dendrimers. Interestingly, as the number of disulfides exchanged for a given dendrimer increases, the rate for exchange of remaining disulfides increases according to mathematical models derived from relative population data collected by mass spectometry.

Lesson III. Peptide-dendrimer and drug-dendrimer conjugates can be prepared by thiol-disulfide exchange (18)

Motivation for peptide-dendrimer constructs can be found in peptides that (i) selectively target tumors, (ii) facilitate cellular transport and nuclear locali-zation, and (iii) stimulate the immune system. By incorporating either four or eight pyridyldisulfide groups into a dendrimer, we find that complete exchange occurs with the small thiol-containing molecule captopril. Exchange of four groups with a cysteine-containing nonapeptide proceeds to completion, but exchange of eight groups does not, presumably due to steric hindrance. Instead, the dendrimer displays six nonapeptides (Figure 8).

Lesson IV. DNA-dendrimer conjugates can be prepared (19)

Following similar inspiration, we envisioned that exchange with thiol-terminated DNA should afford similar products (Figure 9). This exercise has been frought with technical challenges of scale, execution and analysis. We are able to obtain the desired conjugate with great effort. Mass spectrometry corroborates disulfide bond formation. Gel electrophoresis shows that the product of the reaction mixture contains both DNA and dendrimer as the band observed stains with both ethidium bromide and coomassie, respectively. The conjugate can be purified by HPLC to provide amounts of material suitable for hydridization to surfaces displaying complementary oligonucleotides.

In vivo Analysis

Our preliminary forays into animal models have been designed to address fundamental hurdles of toxicity and antigenicity that this class of molecules might suffer from.

Lesson A. A peptide-dendrimer conjugate can induce an immune response.

The sera from rabbits immunized with this construct were compared with sera from rabbits immunized with a BSA-peptide conjugate (Figure 10). Both BSA conjugates elicit an immune response, but only one of the two rabbits challenged with peptide carried on the dendrimer was effective as determined by direct and competitive ELISA assays using a panel of antigens. Unlike BSA, which is recognized as an antigen by ELISA, the dendrimer is immunologically silent.

Toxicological assays show that the dendrimer-peptide conjugate had no effect on renal and hepatic functions based on blood chemistry analysis and is not hemolytic at concentrations used in the study.

Lesson B. Cationic dendrimers are tolerated... poorly (20)

Our first generation vehicle for drug delivery was a cationic dendrimer displaying primary amines on the periphery proved to be hemolytic at low concentrations, but tolerable when delivered by intraperitoneal (i.p.) injection up

130

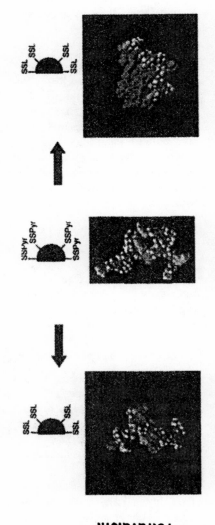

Tetravalent

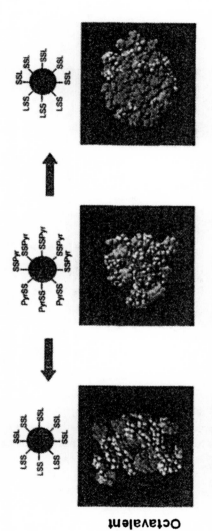

Figure 8. Thiol-disulfide exchange of thiopyridyl groups (center, yellow) with captopril (left, green) or nonapeptides (right, blue) occurs with tetravalent (top row) and octavalent (bottom row) dendrimers. (See page 3 of color inserts.)

132

Figure 9. Oligonucleotide incorporation through thiol-disulfide exchange.

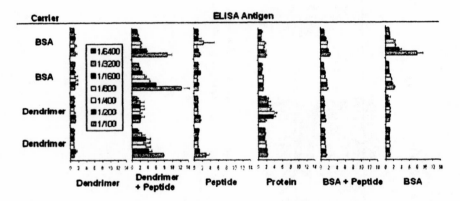

Figure 10. ELISA assays with antigens of serum from four rabbits challenged with an antigenic peptide presented either on BSA or on dendrimer. The ratio of absorbances of serum collected before and after antigenic challenge.

to 40 mg/kg in mice before the onset of mortality at 160 mg/kg. The liver toxicity of methotrexate and 6-mercaptopurine were attenuated when coadministered with this dendrimer compared to free drug. Subchronic dosing *i.p.* showed onset of significant liver toxicity.

Lesson C: Synthesis can overcome initial toxicity challenges (14)

A family of second generation dendrimers was prepared to address toxicity and hemolysis observed with the first dendrimer candidate (Figure 11). This target (1) contains AB_2 groups on the interior and AB_4 groups on the periphery to afford a generation three dendrimer with 48 surface groups that can be functionalized.

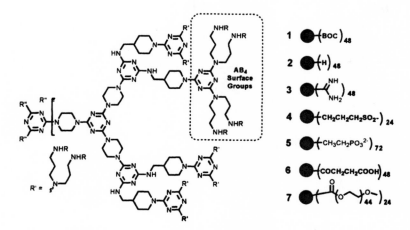

Figure 11. A family of second generation dendrimers, prepared to address toxicity and hemolysis.

Cytotoxicity and hemolysis were probed with two cationic dendrimers, three anionic dendrimers and a PEGylated species. Cell viability using normal cells (Clone 9) was determined using the MTT assay and found to be significant in the cationic species, but markedly reduced in the anionic and neutral dendrimers (Figure 12). Hemolysis was also determined and similar trends were observed (Figure 13).

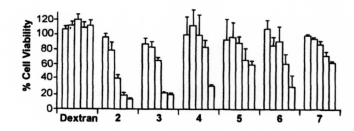

Figure 12. Viability measured at concentrations that increase from left to right: 0.001 mg/mL; 0.01 mg/mL; 0.1 mg/mL; 1 mg/mL; 10 mg/mL.

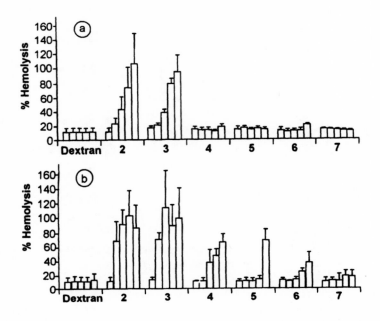

Figure 13. The impact of surface modification on the hemotoxicity of 2-7 at 1 h (a) and 24 h (b) at doses that increase from left to right: 0.001 mg/mL; 0.01 mg/mL; 0.1 mg/mL; 1 mg/mL; 10 mg/mL.

The PEGylated target was evaluated *in vivo*. In single bolus injections to mice, we observe no anomolous blood chemistry using blood urea nitrogen (kidney damage indicator) or alanine transaminase levels (liver damage) at doses up to 2.6 g/kg *i.p.* or 1.3 g/kg delivered *i.v.* to the tail vein.

Future Directions

Moving directly to tumor growth inhibition studies or regression studies is wholly unattractive to us given our currently level of understanding. While a "therapeutic regime" might be uncovered, it would largely represent serendipity. Instead, we are pushing towards a system where substantive feedback is provided. Such a system best encompasses real-time imaging with suitably labeled agents. Our current goal is to monitor distribution of dendrimers *in vivo* in real time (or close to it) to determine the molecular basis for tumor targeting. Once targeted, we hope that *in vitro* analysis of release rates of drugs can be extrapolated to the *in vivo* cancer model. Monitoring simultaneously the delivery of vehicle and drug and correlating this data to tumor volume changes represents our next substantive challenge.

Acknowledgements

The work described was performed by a talented group of coworkers including Dr. Hui-ting Chen, Megan E. McLean, Michael F. Neerman, Mackay B. Steffensen, Alona P. Umali, and Dr. Wen Zhang. Dr. Shane E. Tichy is an intergral collaborator from the mass spectrometry facility at Texas A&M University. Professor Alan R. Parrish is our collaborator from the Department of Medical Pharmacology at the Texas A&M University Health Science Center. The DNA-dendrimer conjugates were inspired by Professors Rob Corn, Bob Hamers, and Lloyd Smith at the University of Wisconsin and pursued with Professor Dick Crooks of Texas A&M. Funding for these projects came from N.I.H., D.A.R.P.A., and the Center for Microencapsulation and Drug Delivery at Texas A&M University.

References

1. Duncan, R. The Dawning Era of Polymer Therapeutics. *Nature Drug Discovery* **2003**, *2*, 347-360.
2. Stiriba, S.-E.; Frey, H.; Haag, R. Dendritic Polymers in Biomedical Applications: From Potential to Clinical Use in Diagnostics and Therapy. *Angew. Chem. Int. Ed.* **2002**, *41*, 1329.
3. Aulenta, F.; Hayes, W.; Rannard, S. Dendrimers: A new class of nanoscopic containers and delivery devices. *Eur. Poly. J.* **2003**, *39*, 1741.
4. Cloninger, M.J. Biological applications of dendrimers. *Curr. Opin. Chem. Biol.* **2002**, *6*, 742.
5. Esfand R.; Tomalia D. A. Poly(amidoamine) (PAMAM) dendrimers: from biomimicry to drug delivery and biomedical applications. *Drug Discov. Today* **2001** 6, 427-436.
6. Patri, A.K.; Majoros, I.J.; Baker, J.R. Jr. Dendritic polymer macromolecular carriers for drug delivery. *Curr. Opin. Chem. Biol.*, **2002**, *6*, 466.
7. Zhang, W.; Simanek, E.E. Dendrimers Based on Melamine. Divergent and Orthogonal, Convergent Syntheses of a G3 Dendrimer. *Org. Lett.* **2000**, *2*, 843-845.
8. Zhang, W.; Simanek, E.E. Synthesis and Characterization of Higher Generation Dendrons Based on p-Aminobenzylamine. Evidence for Molecular Recognition of Cu(II). *Tetrahedron Lett.* **2001**, *42*, 5355-5357.

9. Zhang, W.; Nowlan, D.T.III; Thomson, L.M.; Lackowski, W.M.; Simanek, E.E. Orthogonal, Convergent Syntheses of Dendrimers Based on Melamine with One or Two Surface Sites for Manipulation. *J. Am. Chem. Soc.* **2001**, *123*, 8914-8922.

10. Zhang, W.; Gonzalez, S.O.; Simanek, E.E. Structure-Activity Relationships in Dendrimers Based on Triazines: Gelation Depends on Choice of Linking and Surface Groups. *Macromolecules* **2002**, *35*, 9015-9021.

11. Steffensen, M.B.; Simanek, E.E. Chemoselective building blocks for dendrimers from relative reactivity data. *Org. Lett.* **2003**, *5*, 2359-2361.

12. Steffensen, M.B.; Simanek, E.E. Synthesis and Manipulation of Orthogonally Protected Dendrimers: Building Blocks for Library Synthesis. *Angew. Chem. Int. Ed.* **2004**, *43*, 5178-5180.

13. Wong, C.-H.; Ye, X.-S.; Zhang, Z. Assembly of Oligosaccharide Libraries with a Designed Building Block and an Efficient Orthogonal Protection-Deprotection Strategy. *J. Am. Chem. Soc.* **1998**, *120*, 7137-7138.

14. Chen, H.-T.; Neerman, M.F.; Parrish, A.R.; Simanek, E.E. Cytotoxicty, Hemolysis and Acute In Vivo Toxicity of Dendrimers Based on Melamine, Candidate Vehicles for Drug Delivery. *J. Am. Chem. Soc.* **2004**, *126*, 10044-10048.

15. Grayson, S. M.; Fréchet, J. M. J. Convergent Dendrons and Dendrimers: from Synthesis to Applications. *Chem. Rev.* **2001**, 101, 3819-68.

16. Zhang, W.; Jiang, J.; Qin, C.; Thomson, L.M.; Parrish, A.R.; Safe,S.H.; Simanek, E.E. Triazine Dendrimers for Drug Delivery: Evaluation of Solubilization Properties, Activity in Cell Culture, and In Vivo Toxicity of a Candidate Vehicle. *Supramol. Chem.* **2003**, *15*, 607-615.

17. Zhang, W.; Tichy, S.E.; Pérez, L.M.; Maria, G.; Lindahl, P.A.; Simanek, E.E. Evaluation of Multivalent Dendrimers Based on Melamine: Kinetics of Thiol-Disulfide Exchange Depends on the Structure of the Dendrimer. *J. Am. Chem. Soc.* **2003**, *125*, 5086-5094.

18. Umali, A.P.; Simanek, E.E. Preparation of Multivalent Dendrimers Through Thiol-disulfide Exchange. *Org. Lett.* **2003**, 5, 1245-1247.

19. Bell, S.A.; McLean, M.E.; Oh, S.-K.; Tichy, S.E.; Corn, R.M.; Crooks, R.M.; Simanek, E.E. Synthesis and Characterization of DNA-Dendrimer Conjugates. *Bioconj. Chem.* **2003**, *14*, 488-493.

20. Neerman, M.R.; Zhang, W.; Parrish, A.R.; Simanek, E.E. In vivo evaluation of a triazine dendrimer: a potential vehicle for drug delivery. *Int. J. Pharm.* **2004**, *1*, 390-393.

Chapter 10

Targeted and Non-targeted Polymer Drug Delivery Systems

K. Bruce Thurmond II[1], John McEwan[2], Dan G. Moro[1],
John R. Rice[1], Gregory Russell-Jones[2], John V. St. John[1],
Paul Sood[1], Donald R. Stewart[1] and David P. Nowotnik[1,*]

[1]Access Pharmaceuticals, Inc., 2600 Stemmons Freeway, Suite 176,
Dallas, TX 75207
[2]Access Pharmaceuticals Australia Pty Ltd, 15–17 Gibbes Street,
Chatswood, NSW 2067, Australia
*Corresponding author: dpn@accesspharma.com

Water soluble polymers used for the delivery of platinum
drugs to tumor tissue, vitamin mediated polymer-drug
targeting, and hydrogel nanoparticle aggregates as depot
controlled-release devices are discussed. Platinum (II)
complexes have been bound to poly(hydroxypropyl-
methacrylamide) (HPMA) based materials, and superior tumor
growth inhibition in mice was observed when compared to the
small molecule analogs. Attachment of vitamin B_{12} or folic
acid targeting agents to poly(HPMA-*co*-HPMA-GFLG-
methotrexate) conjugates gave superior tumor growth
inhibition relative to the untargeted polymer conjugates.
Narrow polydispersed hydrogel nanoparticles were synthesized with actives (small molecules or macromolecules)
trapped within or between the individual particles of the
aggregate network. Data are presented which show that
erosion and degradation rates of the aggregate networks can
be tailored to control the release of the entrapped molecules.

Introduction

The use of polymeric materials for the administration of drugs represents an important application in the delivery of pharmaceutical products, allowing site specific targeting, localized drug release, and reduction in the side effect profile (*1-4*). These potential benefits have lead to the study of polymer therapeutics for use in a variety of disease states by numerous research groups around the world (*5-6*). The reasons for using a polymer scaffold for drug delivery varies somewhat from application to application, but there are several features of the polymer approach which apply across all therapeutic areas. Firstly, attaching the small molecule drug to a polymer scaffold increases the *in vivo* circulation time of the drug. Secondly, the hydrophilic polymer also serves to solubilize often insoluble small molecules. Additionally, conjugation to the polymer may provide *in vivo* stability since the drug can be shielded from attack, and selective release may be achieved given that the small molecule is not completely available for uptake or binding in the bloodstream. Furthermore, the polymer may serve as a targeting agent, especially for tumor directed drug delivery. Lastly, the hydrophilic polymer can be cleared through the kidneys thereby preventing any long-term concerns about the polymer remaining in the body (*7-8*).

Conjugation of the drug to the polymer is only one method by which a polymer can serve as a delivery vehicle for drug molecules. Alternative methods include encapsulation or embedding of the drug molecule within a polymeric matrix to deliver small or macromolecular drugs (*9-10*). For these polymeric systems the rate of delivery is controlled by the rate of degradation of the polymer or the diffusion rate of the drug through the polymeric matrix. While obviously effective, such methods are not applicable to all types of drugs and indications; thus alternative approaches must be considered.

Access Pharmaceuticals is pursuing several polymeric approaches to deliver drug molecules, both small and large. Specifically, polymer-drug conjugates and hydrogel nanoparticle aggregates with encapsulated or embedded drug molecules have been synthesized. This chapter provides an overview of some of the technologies currently being developed by Access Pharmaceuticals.

Materials and Methods

Polymer Platinum Agents

AP5280

The random co-polymer poly(HPMA-*co*-MA-GFLG-ONp) (20-25 kDa) was produced in 75 % yield according to existing methods (*11*). In the designation of the polymer, HPMA corresponds to ± 2-hydroxypropyl-methacrylamide, MA is methacryl, GFLG is the tetrapeptide glycyl-phenyl-alanyl-leucyl-glycine, and ONp is the *p*-nitrophenylester attached through the carboxyl group of the tetrapeptide. The chelating group, diethyl amino-malonate, was coupled to the polymer *via* replacement of the ONp groups, and the polymer conjugate was isolated in 90 % yield (*12*). Hydrolysis of the diethyl ester groups followed by complexation with diaqua *cis*-Platinum(II) diammine cation $((NH_3)_2Pt(OH_2)_2^{2+})$ gave the O,O'-amidomalonate chelate, which was converted to the N,O-amidomalonate chelate in phosphate-buffered saline. Purification *via* tangential flow filtration (TFF) afforded pure poly(HPMA-*co*-MA-GFLG-Ama=Pt(NH$_3$)$_2$ N,O-chelate) as confirmed by ^{195}Pt NMR spectroscopy and other analyses (*13*).

AP5346

The random co-polymer poly(HPMA-*co*-MA-GG-QNp) (20-25 kDa) was produced in 80 % yield similar to existing methods (*11*). The GG corresponds to the dipeptide glycyl-glycine. BOC-protected glycine was coupled to diethyl aminomalonate and deprotected in 73 % overall yield. This chelating group was coupled to the polymer *via* replacement of the ONp groups, and the polymer conjugate was isolated in 85 % yield. Hydrolysis of the diethyl ester groups followed by complexation with 1*R*,2*R*-DACH-Pt(OH$_2$)$^{2+}$ cation (DACH corresponds to diaminocyclohexane) resulted in the formation of the O,O'-amidomalonate chelate, which was converted to the N,O-amidomalonate chelate in phosphate-buffered saline. Purification *via* TFF afforded pure poly(HPMA-*co*-MA-GGG-Ama=Pt-1*R*,2*R*-DACH N,O-chelate) as confirmed by ^{195}Pt NMR spectroscopy and other analyses.

Vitamin Targeted Polymers

Poly(HPMA) was fractionated to give a polymer with a M_w >100 kDa. After coupling N-protected GFLG, and methotrexate (Mtx), the drug-peptide conjugate was bound to the polymer *via* a carbamate with the secondary alcohol of the HPMA groups (~4 %w/w) to give poly(HPMA-*co*-HPMA-GFLG-Mtx). Following purification *via* dialysis, the targeting agents (TA), aminohexyl-vitamin B$_{12}$ (cyanocobalamin) or aminoethyl-folate (*14*), were attached *via* a carbamate with the secondary alcohol of the HPMA groups *(~5 %w/w)* to afford poly(HPMA-*co*-HPMA-GFLG-Mtx-*co*-HPMA-TA). The final conjugates were purified *via* dialysis and lyophilized.

Nanoparticle Aggregates

Poly(HEMA) nanoparticles were synthesized from a HEMA ethylene-glycol dimethacrylate (EGDMA) ratio of 99 : 1 at 4 % monomers by weight in water. Sodium dodecyl sulfate (SDS) (0.03 wt%) was used as a surfactant, and potassium persulfate (0.01 wt%) was used as the radical initiator. The solution was heated at 50 °C for 16 h before purification *via* TFF to remove SDS and residual monomers. Particles were characterized by laser light scattering.

Synthetic variations can include changing the ratio of monomers SDS resulting in larger or smaller nanoparticles, the addition of methacrylic acid (MAA) as a comonomer from 1-50 mole%, or the addition of a fluorescent comonomer to establish the aggregates *in vivo* fate *via* fluorescence microscopy.

Aggregates are formed by ultracentrifugation or precipitation of nano-particles from solution. Proteins are loaded into aggregates by trapping during aggregate formation. Small molecules are loaded in a similar fashion. Aggregates containing particles with MAA as a comonomer erode at physiological pH. The rate of erosion is controlled by the amount of MAA comonomer in the particles and the ratios of particle types.

Results and Discussion

A. Polymer Platinum Conjugates

Following the synthesis and testing of a large number of polymer-platinum conjugates, two drugs, AP5280 and AP5346, were advanced to clinical development. Each polymeric system is composed of a Pt(II) drug chelated to a 20–25 kDa polymer through an amidomalonate group. The polymer-drug conjugates are expected to capitalize on the enhanced permeation and retention (EPR) effect for tumor targeting. The EPR effect is known to provide increased accumulation of the polymer, and therefore drug, to the tumor site. This occurs as the polymer is transferred from the bloodstream through the leaky endothelium of a tumor directly into the tumor mass where it accumulates due to poor lymphatic drainage. Endocytosis of the polymer occurs, followed by drug release and activation, which results in tumor cell death. This specific, increased tumor accumulation also limits the uptake of the cytotoxic compound by normal cells (*15-16*).

The polymer precursors for AP5280 and AP5346 are synthesized in similar fashion but differ in the peptide linker (GFLG vs. GGG) and the small molecule platinum cation used $((NH_3)_2Pt(OH_2)^{2+}$ vs. $1R,2R$-DACH-Pt(OH$_2$)$_2^{2+}$ (Figure 1). The GFLG tetrapeptide was selected for AP5280 as it is a cleavable linker (*7*). A simpler triglycine linker proved to be very effective in AP5346.

Both of these compounds exhibit tumor growth inhibition (TGI), which in some cases, is superior to that shown by their small molecule analogs. For example, in a TGI study directly comparing AP5280 and carboplatin, with both being dosed at their maximum tolerated dose (MTD), tumor growth inhibition was greater for AP5280 (Figure 2).

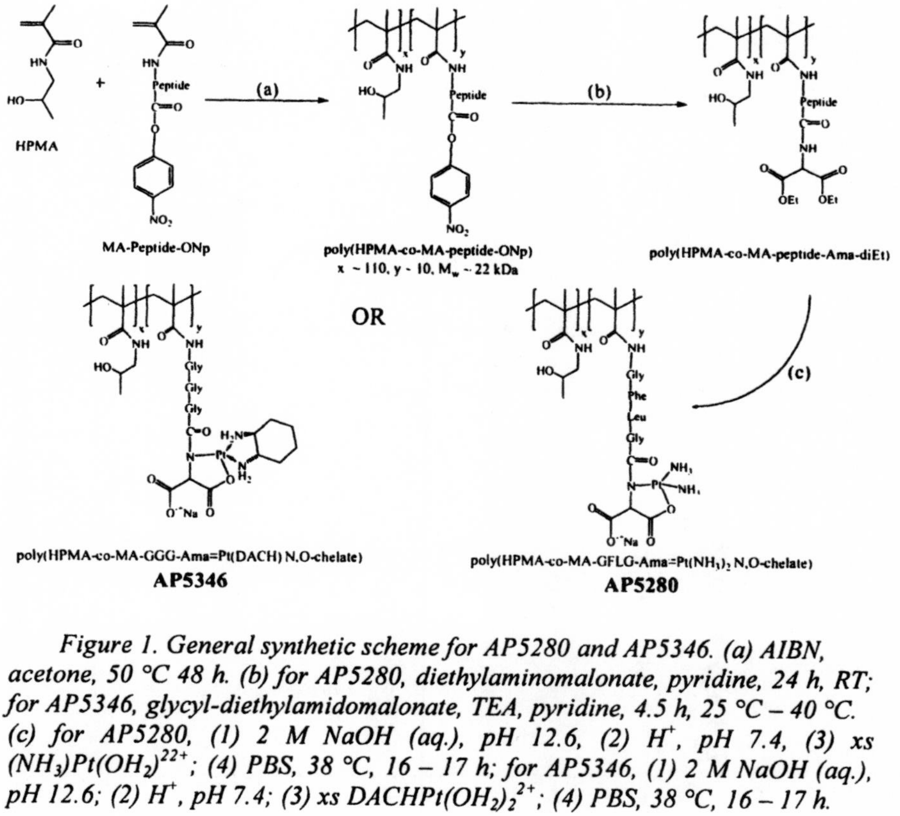

Figure 1. General synthetic scheme for AP5280 and AP5346. (a) AIBN, acetone, 50 °C 48 h. (b) for AP5280, diethylaminomalonate, pyridine, 24 h, RT; for AP5346, glycyl-diethylamidomalonate, TEA, pyridine, 4.5 h, 25 °C – 40 °C. (c) for AP5280, (1) 2 M NaOH (aq.), pH 12.6, (2) H^+, pH 7.4, (3) xs $(NH_3)Pt(OH_2)^{22+}$; (4) PBS, 38 °C, 16 – 17 h; for AP5346, (1) 2 M NaOH (aq.), pH 12.6; (2) H^+, pH 7.4; (3) xs $DACHPt(OH_2)_2^{2+}$; (4) PBS, 38 °C, 16 – 17 h.

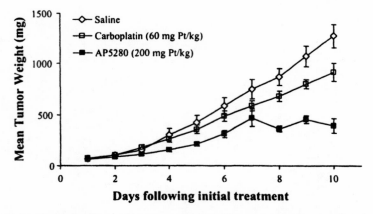

Figure 2. TGI study in B16 tumor bearing mice comparing AP5280 and carboplatin administered with 5 daily IV doses at their respective MTD.

The enhanced tumor inhibition was likely due in large part to increased Pt-DNA adduct formation (*17*). In a study in mice bearing B16 melanoma tumors, an 8-fold increase in Pt-DNA adducts was observed for AP5280 versus carboplatin in tumor tissue (Figure 3).

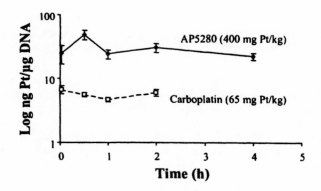

Figure 3. Comparison of platinum-DNA adducts from AP5280 and carboplatin following a single IV injection in B16 tumor bearing mice.

This increase is significant because it indicates the increased amount of platinum being delivered, released, and maintained with the polymeric delivery approach. It also shows that drug delivery with this polymer allows more platinum to be safely administered and results in more Pt-drug uptake by the tumor. These data clearly indicate the benefit provided with the polymeric approach to delivering small molecule cytotoxic compounds.

The success of AP5280 was extended to an alternative polymer-Pt species, AP5346. A DACH-Pt complex was developed because DACH-platinum agents are known to have a different spectrum of activity when compared to cisplatin or carboplatin. Oxaliplatin, a small molecule 1R,2R-DACH-Pt, is known to be active in colon cancer; thus it was reasoned that a polymer DACH-platinum conjugate might result in improved efficacy in this indication with perhaps lower toxicity or an altered toxicity profile. Initial testing indicated the small molecule DACH-Pt released from AP5346 remained active; thus additional studies were pursued.

As with AP5280, a TGI study comparing the polymeric drug conjugate (AP5346) to its small molecule analog (oxaliplatin) at the MTD of each was carried out (Figure 4). Again, substantial tumor growth inhibition was observed for the polymer platinum (AP5346) compared to the small molecule analog, oxaliplatin. Animals treated with AP5346 exhibited essentially no tumor growth for 30 days while tumor growth was nearly unchecked in the vehicle and oxaliplatin groups.

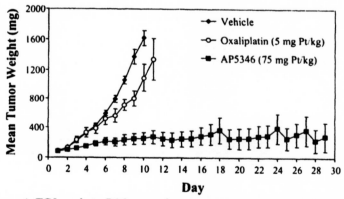

Figure 4. TGI study in B16 tumor bearing mice comparing AP5346 and oxaliplatin administered once weekly for 3 weeks at their respective MTD.

In numerous other models (Colo-26, Lewis lung, 2008 human ovarian xenograft, HT-29 xenograft, and HCT-116 xenograft) there was equivalent or superior activity for AP5346 versus oxaliplatin and carboplatin. For both of these drug candidates (AP5280 and AP5346), early stage clinical trials are on going to examine their safety and efficacy.

B. Vitamin Mediated Targeted Delivery

While the EPR effect does provide effective tumor targeting, it may be enhanced further by employing a receptor mediated approach. It is well known that the demand for certain vitamins increases in many disease states; therefore, the attachment to the polymer of vitamin B_{12} (VB_{12}), biotin, and/or folic acid (FA), which appear to be essential for tumor growth, should provide enhanced tumor delivery (18-19). Binding to these receptors would allow the polymer to be taken up in a receptor mediated endocytotic process which takes the disease modifying drug to a specific location within the cell and lowers the risk of uptake by normal cells. By combining a targeting agent with the polymer's ability to localize in tumor vasculature, a two pronged attack would be carried out to insure drug uptake in tumor cells.

In order to determine which cell lines overexpress vitamin receptors, in vitro studies were carried out to see which tumor lines underwent vitamin specific receptor mediated endocytosis. Each cell line was treated with the control polymer, poly(HPMA-co-HPMA-Rhodamine) (5 % Rhodamine by weight), and targeting agent (VB_{12}, FA, biotin) conjugates of the control polymer. In separate experiments, polymers with and without targeting agent were injected into tumor bearing mice, tumor samples were taken after 5 h, and frozen sections were prepared and imaged to determine the uptake using a Zeiss Axioplan Fluorescent microscope. Three cell lines demonstrated upregulation of the folic acid receptor, and three demonstrated upregulation of the TCII receptor (for VB_{12}). However, what was most interesting from these data was that biotin uptake was enhanced whenever either folate or VB_{12} uptake was enhanced (Table I).

Table I. Vitamin receptor mediated uptake of various tumor cell lines.

Tumor	Polymer	Folate	VB_{12}	Biotin
Colo-26	-	+/-	++	+++
P815	-	+/-	++	+++
L1210FR	-	++	+	+++
O157	+/-	+/-	+/-	+/-
BW5147	+/-	+/-	+/-	+/-
M109	-	+	+++	+++
B16	-	-	-	-
LL-2	-	-	-	-
Ov 2008	-	+++	-	++
HCT-116	+	-	-	-
ID8	-	+++	-	++

The reason why a biotin receptor is upregulated whenever either folate or VB_{12} is upregulated is unclear. This may be occurring because the biotin receptor is believed to be intimately involved in lipid synthesis; therefore, as new cells are forming and new membranes are being formed, more biotin will be upregulated (*20*).

A comparative study of active vitamin targeting *in vivo* was carried out using a cell line which demonstrates folate upregulation but not TCII (VB_{12}) upregulation (Figure 5). The effect of untargeted poly(HPMA)-based material to deliver methotrexate is obvious (Figure 5, B vs. C), but the effect of the targeting agent (TA) is more subtle. The VB_{12} targeted material provides no enhancement over the methotrexate polymer (Figure 5, C vs. D), and this was not surprising as VB_{12} is not upregulated for this cell line as indicated in Figure 5. The use of folate does result in a statistically significant improvement in tumor growth inhibition when compared to the methotrexate polymer (Figure 5, C vs. F) which was expected since the cell line was positive for upregulation of the folate receptor. These data not only indicate the benefit of a polymeric approach over its small molecule counterpart, but also show that vitamin targeting can provide further enhancement of delivery of polymer therapeutics to tumors.

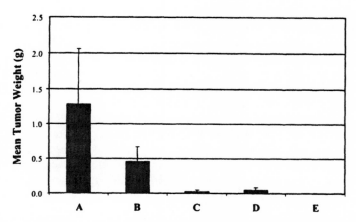

Figure 5. Comparison of the effect on mean tumor weight of vitamin targeted polymers in L1210 tumor cells after 3X i.v. injection treatment (days 1, 2, and 3,) with 10 mg/kg Mtx. Tumors were measured at day 13. (A) Control, (B) Mtx, (C)poly(HPMA-co-HPMA-GFLG-Mtx), (D) poly(HPMA-co-HPMA-GFLG-Mtx-co-HPMA- VB12), (E) poly(HPMA-co-HPMA-GFLG-Mtx-co-HPMA-FA).

C. Hydrogel Nanoparticle Aggregates

The last area of research to be discussed involves the use of hydrogel nano-particle aggregates for drug delivery. Hydrogel materials are quite attractive for potential *in vivo* use as a drug delivery or tissue engineering system because they are generally considered inert materials and have a long history of biocompatibility and safety (*21-22*). The hydrogel nanoparticle aggregates being developed at Access have several unique properties and characteristics: (a) they consist of narrow polydispersed nano- or micro-particles; (b) they are shape retentive materials, extrusion capable, porous, elastic, and tough; (c) actives can be trapped within or between particles in an innocuous environment (water); (d) these materials are uniquely suited for incorporating, stabilizing, and releasing macromolecules; and (e) most importantly, erosion and degradation rates can be tailored and controlled to provide the appropriate therapeutic dose for a specific active.

The synthesis of hydrogel particle aggregates with hydroxyethylmetha-crylate (HEMA) monomer and ethyleneglycoldimethacrylate (EGDMA) affords a non-degradable, non-erodible aggregate. Variations, which allow for degra-dation or erodibility, are possible through the incorporation of a degradable cross-linker and/or ionizable co-monomer. It is also possible to incorporate a fluorescent co-monomer, which allows the degradation of the aggregates to be followed using fluorescence microscopy. Currently, actives have been incorpo-rated and categorized into three sets of nanoparticle aggregates; non-erodible and non-degradable (Figure 6A), erodible and degradable (Figure 6B), and erodible and non-degradable (Figure 6C).

The non-erodible, non-degradable nanoparticles (Figure 6A) were loaded with horseradish peroxidase (HRP), placed into a PBS solution (pH 7.4) at 37 °C, and the activity of the released enzyme was measured. Activity was maintained at 90 % or greater over a 48 h period (Figure 7A). In a separate experiment, HRP loaded nanoparticles were placed into a PBS solution (pH 7.4) at 37 °C, and after 24 h of incubation, trypsin, which is known to cleave and reduce the activity of HRP, was added for 3 h. A significant decrease in the activity of the released HRP was observed (Figure 7B). This was expected since the HRP was in solution with trypsin around the aggregate. Following the removal of the trypsin solution and washing of the aggregate, the activity of subsequently released HRP was restored because the aggregate had provided protection against cleavage for the embedded HRP (Figure 7B).

Additional data have been obtained that indicate the release of macro-molecules from the pores of a non-erodible aggregate can be controlled by changing the pore size. This is accomplished by changing the size of the poly(HEMA) nanoparticles within the aggregate. Using different sized nano-particles, 25 nm and 120 nm, two aggregates containing 256 kDa dextran were synthesized. A zero order release of 70 % of the total dextran was observed over

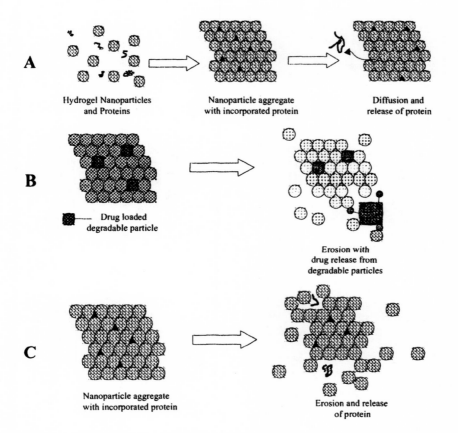

Figure 6. Three different methods from which an active can be released from a hydrogel nanoparticle aggregate. (A) Diffusion from non-degradable, non-erodible aggregate; (B) Erosion of the aggregate with drug release from within the degradable particles; (C) Erosion of the aggregate releases macromolecules trapped within the pores of the aggregate.

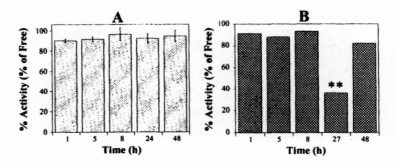

Figure 7. 60-nm poly(HEMA) nanoparticle aggregates loaded with HRP and incubated in a PBS solution (pH 7.4) at 37 °C. (A) Activity of released HRP; (B) Activity of released HRP before and after trypsin addition (**).

6 days for the 25 nm particle aggregate versus 60 % release of the total dextran within the first day for the 120 nm particle aggregate. These data indicate that controlling the size of the nanoparticles allows for the tailored release rates of macromolecules from a non-erodible, non-degradable nanoparticle aggregate.

Using an erodible nanoparticle aggregate with (Figure 6B) and without (Figure 6C) degradable crosslinkers, it is possible to release potential drug actives. The UV-chromaphore, Bromocresol Green, was encapsulated or embedded in degradable and non-degradable erodible nanoparticles and the release was measured spectrophotometrically (Figure 8). Simply changing from a non-degradable crosslinker (EGDMA) to a degradable one (Succinate-(MA-PEG-Glycotate)$_2$, a proprietary Access crosslinker) effects the release of a small molecule from the aggregate. The degradable aggregate releases ~30 % more of the small molecule over a 24 h period than the non-degradable aggregate. Adjusting the percentage of cross-linker as well as controlling the size of the nanoparticle allows the release rate to be tailored depending upon the drug and the application.

The erosion rate of a nanoparticle aggregate can be controlled further by altering the concentration of an ionizable co-monomer (Figure 6C), such as methacrylic acid (MAA). Increasing the concentration of MAA comonomer in a poly(HEMA) nanoparticle results in a rapid release of model protein, BSA (Figure 9). This is simply due to the increased erodibility of the aggregates as the concentration of MAA is increased, which can be easily observed by SEM (Figure 10).

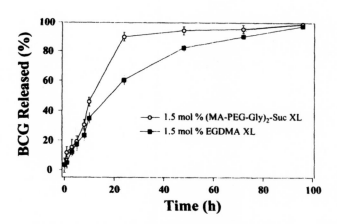

Figure 8. Bromocresol Green (BCG) release from 90 : 10 poly(HEMA): poly(MAA) nanoparticle aggregates.

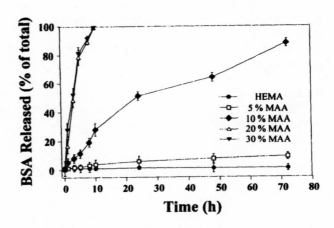

Figure 9. Release of bovine serum albumin (BSA) from erodible nanoparticle (100 nm) aggregates suspended in PBS at 37 °C.

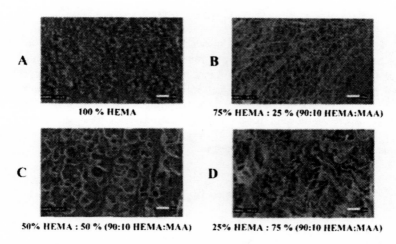

Figure 10. SEM images of nanoparticle aggregates of similar size after 24 h at 37 °C in PBS. Scale bar in lower right corner is 10 μm.

152

For a 100 % poly(HEMA) nanoparticle aggregate (Figure 10A), the surface shows no signs of erosion or degradation, but as the percentage of carboxylate groups incorporated in the nanoparticle is increased (Figures 10B-D), the SEM images show there is an increased rate of erosion and degradation as more and more individual particles break free from the aggregate over time.

While porous enough to allow for cellular infiltration and growth, these materials still maintain their structural integrity which may allow for their use as tissue engineering scaffolds (23). This is one example showing how easily the release can be controlled and tailored to specific applications based on alteration of a single component of the aggregate.

Conclusion

Throughout this chapter, data of successful polymeric drug delivery agents have been given. Specifically, the polymer platinum program has resulted in two drugs (AP5280 and AP5346) in clinical trials; both of which out-perform their small molecule analogs in preclinical studies. Vitamin targeted delivery shows the increased benefit of targeting a polymer-drug conjugate to a specific disease site by further inhibiting tumor cell growth. Lastly, hydrogel nano-particle aggregates can be formed in such a fashion as to protect, control, and tailor the release of a wide range of molecules, both small and macromolecular.

Acknowledgements

The authors would like to thank the Access Pharmaceuticals' staff for all of their hard work in obtaining these results.

References

1. Greenwald, R.B.; Zhao, H.; Xia, J.; Wu, D.; Nervi, S.; Stinson, S.F.; Majerova, E.; Bramhall, C.; Zaharevitz, D.W. Poly(ethylene glycol) prodrugs of the CDK inhibitor, alsterpaullone (NSC 705707): synthesis and pharmacokinetic studies. *Bioconjugate Chem.* **2004**, *15*, 1076-1083.
2. Duncan, R.; Gac-Breton, S.; Keane, R.; Musila, R.; Sat, Y.N.; Satchi, R.; Searle, F. Polymer-drug conjugates, PDEPT and PELT: basic principles for design and transfer from the lab to the clinic. *J. Control. Rel.* **2001**, *74*, 135-146.
3. Kopecek, J.; Kopeckova, P.; Minko, T.; Lu, Z.R.; Peterson, C.M. Water soluble polymers in tumor targeted delivery. *J. Control. Rel.* **2001**, *74*, 147-158.
4. Langer, C.J. CT-2103: emerging utility and therapy for solid tumors. *Expert Opin. Investig. Drugs* **2004**, *13*, 1501-1508.

5. Wang, D.; Li, W.; Pechar, M.; Kopeckova, P.; Bromine, D.; Kopecek, J. Cathepsin K inhibitor-polymer conjugates: potential drugs for the treatment of osteoporosis and rheumatoid arthritis. *Int. J. of Pharma.* **2004,** *277,* 73-79.

6. Rodriguez, G.; Gallardo, A.; Fernandez, M.; Rebuelta, M.; Bujan, J.; Bellon, J.M.; Honduvilla, N.G.; Escudero, C.; San Roman, J. Hydrophilic polymer drug from a derivative of salicylic acid: synthesis, controlled release studies and biological behavior. *Macromol. Biosci.* **2004,** *4,* 579-586.

7. Vasey, P.A.; Kaye, S.B.; Morrison, R.; Twelves, C.; Wilson, P.; Duncan, R.; Thomson, A.H.; Murray, L.S.; Hilditch, T.E.; Murray, T.; Burtles, S.; Fraier, D.; Frigerio, E.; Cassidy, J. Phase I Clinical and Pharmacokinetic Study of PK1 [N-(2-Hydroxypropyl)methacrylamide Copolymer Doxorubicin]: First Member of a New Class of Chemotherapeutic Agents-Drug-Polymer Conjugates. *Clin. Cancer Res.* **1999,** *5,* 83-94.

8. Seymour, L.W.; Duncan, R.; Strohalm, J.; Kopecek, J. Effect of molecular weight (M_w) of N-(2-hydroxypropyl)methacrylamide copolymers on body distribution and rate of excretion after subcutaneous, intraperitoneal, and intravenous administration to rats. *J. Biomed. Mater. Res.* **1987,** *21,* 1341-1358.

9. Avgoustakis, K.; Beletsi, A.; Panagi, Z.; Klepetsanis, P.; Karydas, A.G.; Ithakissios, D.S. PLGA-mPEG nanoparticles of cisplatin: *in vitro* nanoparticle degradation, *in vitro* drug release and *in vivo* drug residence in blood properties. *J. Control. Rel.* **2002,** *79,* 123-135.

10. Huang, G.; Gao, J.; Hu, Z.; St. John, J.V.; Ponder, B.C.; Moro, D. Controlled drug release from hydrogel nanoparticle networks. *J. Control. Rel.* **2004,** *94,* 303-311.

11. Kopecek, J.; Rejmanova, P.; Strohalm, J.; Ulbrich, K.; Rihova, B.; Chytry, V.; Lloyd, J.B.; Duncan, R. Synthetic polymeric drugs. U.S. Pat. 5,037,883, 1991.

12. Gianasi, E.; Buckley, R.G.; Latigo, J.; Wasil, M.; Duncan, R. HPMA Copolymer Platinates Containing Dicarboxylato Ligands. Preparation, Characterization, and *In Vitro* and *In Vivo* Evaluation. *J. Drug Targeting* **2002,** *10,* 549-556.

13. Stewart, D.R.; Rice, J.R. St. John, J.V. N,O-amidomalonate platinum complexes. U.S. Pat. 6,692,734, 2004.

14. McEwan, J.F.; Veitch, H.S.; Russell-Jones, G.J. Synthesis and Biological Activity of Ribose-5'-Carbamate Derivatives of Vitamin B_{12}. *Bioconjugate Chem.* **1999,** *10,* 1131-1136.

15. Maeda, H.; Wu, J.; Sawa, T.; Matsumura, Y.; Hori, K. Tumor vascular permeability and the EPR effect in macromolecular therapeutics: a review. *J. Control. Rel.* **2000,** *65,* 271-284.

16. Kissel, M.; Peschke, P.; Subr, V.; Ulbrich, K.; Schuhmacher, J.; Debus, J.; Friedrich, E. Synthetic Macromolecular Drug Carriers: Biodistribution of Poly[(N-2-hydroxypropyl)methacrylamide] Copolymers and Their Accumulation in Solid Rat Tumors. *PDA J. of Pharma. Sci. & Tech.* **2001**, *55*, 191-201.
17. Mishima, M.; Samimi, G.; Kondo, A.; Lin, X.; Howell, S.B. The cellular pharmacology of oxaliplatin resistance. *Eur. J. Cancer* **2002**, *38*, 1405-1412.
18. Russell-Jones, G.J. The potential use of receptor-mediated endocytosis for oral drug delivery. *Adv. Drug Del. Rev.* **2001**, *47*, 21-37.
19. Wang, S.; Low, P.S. Folate-mediated targeting of antineoplastic drugs, imaging agents, and nucleic acids to cancer cells. *J. Control. Rel.* **1998**, *53*, 39-48.
20. Ramaswamy, K. Intestinal absorption of water soluble vitamins. Focus on molecular mechanism of the intestinal biotin transport process. *Am. J. Physiol.* **1999**, *277*, C603-604.
21. Hoffman, A.S. Hydrogels for biomedical applications. *Adv. Drug Del. Rev.* **2002**, *43*, 3-12.
22. Liu, Z.L.; Hu, H.; Zhuo, R.X. Konjac Glucomannan-*graft*-Acrylic Acid Hydrogels Containing Azo Crosslinker for Colon-Specific Delivery. *J. Polym. Sci. Part A: Polym. Chem.* **2004**, *42*, 4370-4378.
23. Orban, J.M.; Marra, K.G.; Hollinger, J.O. Composition Options for Tissue Engineered Bone. *Tissue Eng.* **2002**, *8*, 529-539.

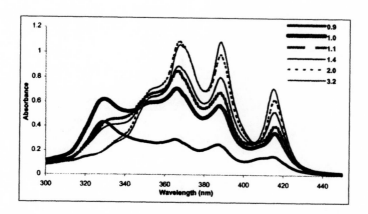

Plate 2.3. Absorption spectra of AmB solubilized by PEG-DSPE micelle at varied mole ratios of PEG-DSPE to AmB (14 µg/mL).

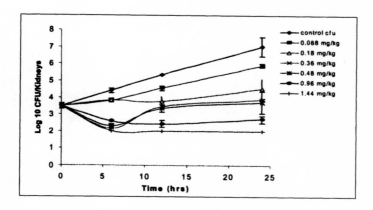

Plate 2.5. In vivo time-kill plot for AmB in PEG-DSPE micelles (1.34 mol/mol) as a function of dose in a neutropenic murine model of candidiasis.

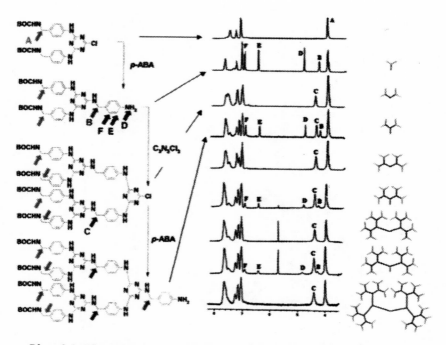

Plate 9.3. The synthesis of dendrimers followed with the appearance and disappearance of lines in the ^1H NMR spectra.

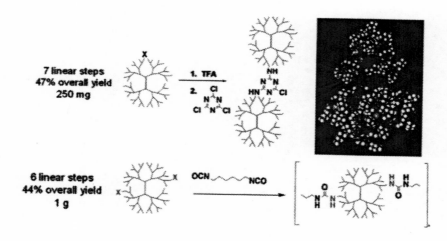

Plate 9.4. Mono- and di-functionalized targets are readily prepared and can be manipulated. The inset shows the mono-functionalized species and the dimer available by reaction with cyanuric chloride. The piperazine cores are maroon, the peripheral BOC or central triazine are green.

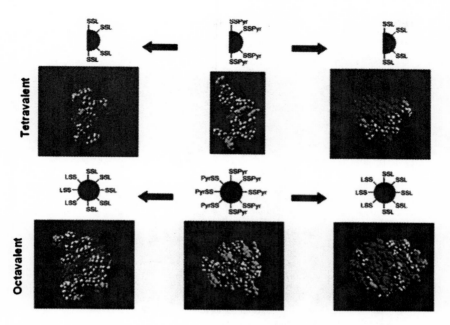

Plate 9.8. Thiol-disulfide exchange of thiopyridyl groups (center, yellow) with captopril (left, green) or nonapeptides (right, blue) occurs with tetravalent (top row) and octavalent (bottom row) dendrimers.

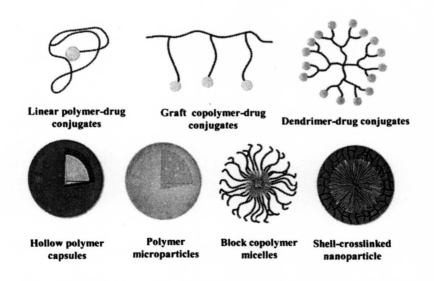

Plate 18.2. Polymer architectures and encapsulation techniques for drug delivery.

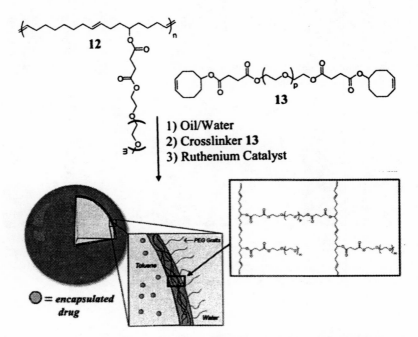

12

13

1) Oil/Water
2) Crosslinker **13**
3) Ruthenium Catalyst

= encapsulated drug

Plate 18.5. Schematic illustration of interfacial assembly of polycyclooctene-g-PEG and crosslinking by ring-opening cross-metathesis with a bis-cyclooctene crosslinker 13.

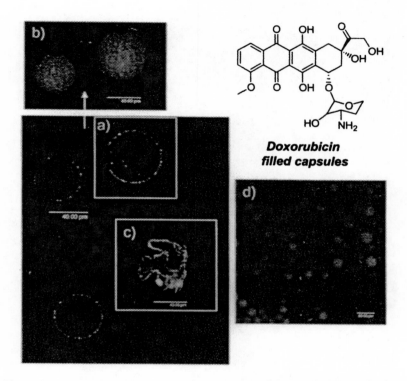

Doxorubicin
filled capsules

Plate 18.6. Laser scanning confocal microscopy of fluorescent-labeled graft
copolymer capsules. (a) cross-sectional slice; (b) 3-D projection image of
polymer covered capsules; (c) collapsed capsule after introduction of ethanol;
(d) projection image of DOX-filled capsules.

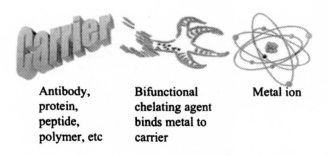

| Antibody, protein, peptide, polymer, etc | Bifunctional chelating agent binds metal to carrier | Metal ion |

Plate 20. 1. Roll of a bifunctional chelant in a conjugate.

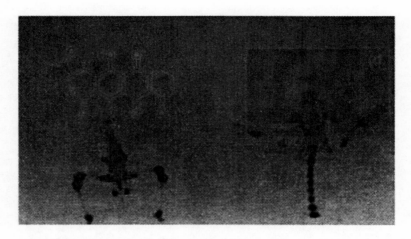

Plate 20.3. ^{153}Sm-EDTMP scintillation scan 3 hours post injection in a rabbit.

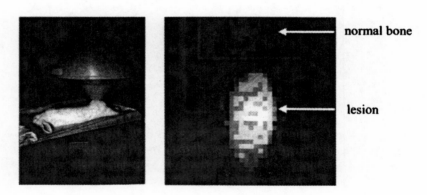

normal bone

lesion

Plate 20.4. Rabbit drill-hole model. Left - scintillation camera. Right - image shows selective uptake of ^{153}Sm-EDTMP in drill-hole lesion.

Sites of metastatic disease

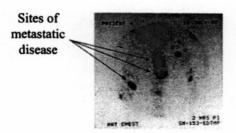

Plate 20.5. Image from human trial of ^{153}Sm-EDTMP showing sites of metastatic bone tumors.

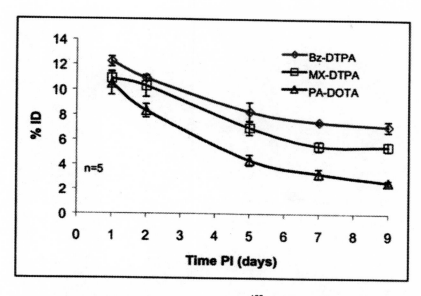

Plate 20.9. Biodistribution study of three ^{177}Lu-labeled bioconjugates.

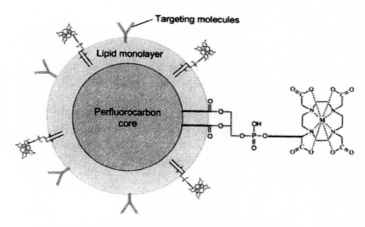

Plate 20.13. Targeting lipid nanoparticle.

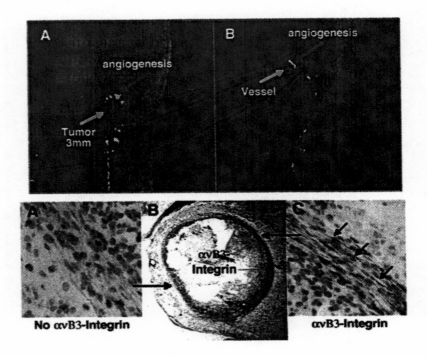

Plate 21.7. Top: Neovasculature of nascent Vx-2 tumor imaged with $\alpha_v\beta_3$-targeted paramagnetic nanoparticles (left), and the venous advential source of tumor neovasculature noted a few millimeters away. Bottom: Immunohistology of the Vx-2 tumor (imaged above) revealing asymmetric location of neovasculature adjacent to muscle plane (B) and high magnification insets showing the presence and sparse distribution of $\alpha_v\beta_3$-integrin expression within an imaging voxel.

Chapter 11

Peptide Carriers for Protein Transduction: How to Generate a Drug from Your Favorite Protein

M. C. Morris, J. Depollier, S. Deshayes, F. Heitz, and G. Divita[*]

Centre de Recherches de Biochimie Macromoléculaire, Department of Molecular Biophysics and Therapeutic, FRE–2593 CNRS, IFR–122, 1919 Route de Mende, 34293 Montpellier, France
[*]Corresponding author: gilles.divita@crbm.cnrs-mop.fr

The development of peptide-drugs and therapeutic proteins is limited by the poor permeability and the selectivity of the cell membrane. We have developed a new peptide-based strategy (Pep-1) for protein transduction into cells. Pep-1 is an amphipathic peptide consisting of a hydrophobic domain and a hydrophilic lysine-rich domain. Pep-1 efficiently delivers a variety of fully biologically active peptides and proteins into cells, without the need for prior chemical cross-linking or denaturation steps. The mechanism through which Pep-1 delivers macromolecules does not involve the endosomal pathway and the dissociation of the Pep-1/macromolecule particle occurs immediately after it crosses the cell membrane. Pep-1 has recently been applied to the screening of therapeutic peptides *in vivo* and presents several advantages: stability in physiological buffer, lack of toxicity and of sensitivity to serum. Pep-1 technology could contribute significantly to the development of fundamental and therapeutic applications.

Introduction

In order to circumvent the technological problems of gene delivery, an increasing interest is being taken in designing novel strategies that allow the delivery of peptides and full-length proteins into a large number of cells *(1-4)*. Recently, substantial progress has been made in the development of cell penetrating peptide-based drug delivery systems which are able overcome both extracellular and intracellular limitations *(2,5-7)*. The Cell Penetrating Peptide (CPP) family includes several peptide sequences: synthetic and natural cell-permeable peptides, protein transduction domain (PTD) and membrane-translocating sequences, which all have the potency to translocate the cell membrane independently of transporters or specific receptors *(2)*. A series of small protein domains, termed Protein Transduction Domains (PTDs), have been shown to cross biological membranes efficiently, and to promote the delivery of peptides and proteins into cells *(1-5)*. The use of PTD-mediated transfection has proven that "protein therapy" can have a major impact on the future of therapies in a variety of viral diseases and cancers. Peptides derived from the transactivating regulatory protein (TAT) of human immunodeficiency virus (HIV) *(8-11)*, the third alpha-helix of Antennapedia homeodomain protein *(4,12)*, VP22 protein from Herpes Simplex Virus (13), poly-arginine peptide sequence *(14,15)* as well as Transportan and derivates *(16)* have been success-fully used to improve the delivery of covalently-linked peptides or proteins into cells and shown to be of considerable interest for protein therapeutics *(1-3)*.

Recently, the uptake mechanism of PTDs and cell penetrating peptides (CPPs) has become very controversial, mainly due to the fact that very little is known concerning the active conformation of CCPs associated with membrane penetration and translocation *(17,18)*. The cellular uptake mechanism of PTD has been completely revised and shown to be associated to endosomal pathway. This was clearly demonstrated for TAT-peptide by monitoring the cell entry of a biologically active protein, in order to avoid any artifact associated to fixation or to the use of fluorescence probes or cellular markers. Cellular uptake occurs through lipid-raft-dependent macropinositosis *(19)*. However, it seems clear that there are several routes for cellular uptake as TAT-peptide attached to liposome enter the cells independently of the endosomal pathway *(20)*.

Protein Transduction Domains display a certain number of limitations in that they all require cross-linking to the target peptide or protein. Moreover, protein transduction using PTD-TAT-fusion protein system may require unfolding of the protein prior to delivery, introducing an additional delay between the time of delivery and intracellular activation of the protein. In order to offer an alternative to the covalent PTDs technology, we have designed a new strategy for protein delivery based on a short amphipathic peptide carrier, Pep-1. This peptide carrier is able to efficiently deliver a variety of peptides and proteins into several cell lines in a fully biologically active form, without the need for prior chemical covalent coupling or denaturation steps *(21)*. In the present manuscript, we demonstrate the potency of the non-covalent Pep-1 technology for therapeutic applications, by delivering antibodies and the cell cycle inhibitor protein p27^{kip1}. The internalization mechanism of Pep-1 is discussed on the basis of the structural requirement for its cellular uptake.

Materials and Methods

Peptide Synthesis and Analysis

All peptides were synthesized by solid phase peptide synthesis using AEDI-Expansin resin with a 9050 Pepsynthetizer (Millipore UK) according to the Fmoc/tBoc method, and purified as already described *(22)*. Pep-1 is acetylated at its N-terminus and contains a cysteamine group at its C-terminus so as to enable coupling of Fluorescein-maleimide *(21)*.

Proteins

The cell cycle inhibitor protein p27^{kip1} encoded by pET21b vector (Novagen) was expressed in *E coli*. His-tag protein was purified using Ni-NTA chromatography followed by ions exchange then size exclusion chromatography. FITC-labeled anti-tubulin antibody was obtained from Santa Cruz .

Cell Culture and Pep-1 Mediated Delivery

Adherent fibroblastic HS-68 and Hela cell lines were cultured in Dulbecco's Modified Eagle's Medium (DMEM) supplemented with 2mM glutamine, 1% antibiotics (streptomycin 10'000 µg/ml, penicillin, 10'000 IU/ ml) and 10% (w/v) fetal calf serum (FCS), at 37°C in a humidified atmosphere containing 5% CO_2 as described previously *(22)*. Pep-1/protein complexes with a molecular ratio of 5/1 or 20/1, were formed in phosphate buffer or DMEM (500 µl of medium containing 0.5 µg of protein) and incubated for 30 min at 37 °C. Cells grown to 60% confluency were overlaid with these preformed com-plexes. After 30 min incubation at 37°C, 1 ml of fresh DMEM supplemented with 10% FBS was added to the cells, without removing the overlay of Pep-1/protein. For cell cycle studies, HS-68 fibroblasts were synchronized by serum privation for 40 hrs, then restimulated to enter the cycle by addition of fresh DMEM supplemented with 20 % FCS for 4 hrs. Cells were then analyzed by fluorescence microscopy to monitor antibody delivery or sorted by flow cyto-metry to determine the effect of p27^{kip1}. Experiments were performed on a FacsCalibur (Becton Dickinson) using CellQuest software. For each individual sample 10,000 cells were counted.

158

Results and Discussion

Pep-1 Design and Properties

We designed a 21 residue peptide carrier, Pep-1 (KETWWETWWTEW-SQP-KKKRKV) consisting of three domains: a hydrophobic tryptophan-rich motif (KETWWETWWTEW), required for efficient targeting to the cell membrane and for forming hydrophobic interactions with proteins, a hydrophilic lysine-rich domain derived from the Nuclear Localization Sequence (NLS) of SV40 large T antigen (KKKRKV), required to improve intracellular delivery and solubility of the peptide vector, and a spacer domain (SQP), separating the two domains mentioned above, containing a proline residue, which improves the flexibility and the integrity of both the hydrophobic and the hydrophilic domains *(21)*.

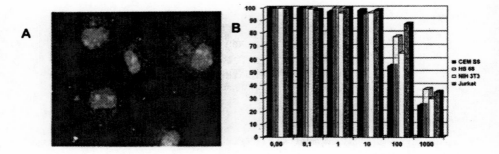

Figure 1. Cellular localization and cytotoxicity of Pep-1. (A) Cellular localization of Pep-1. FITC-labeled Pep-1 (0.1 μM) was applied onto HS-68 cells, for 10 min at 37 °C. Cellular localization of Pep-1 was monitored by FITC fluorescence microscopy on live cells. (B) Cytotoxicity of Pep-1. The toxicity of Pep-1 was monitored in different cell line, using MTT staining assay.

Moreover, Pep-1 contains a cysteamine group at their C-terminus, required for its protein transduction properties. As reported in Figure 1A, fluorescently labeled Pep-1 is able to penetrate cells rapidly in less than 10 min and to localize to the nucleus. An important criterion for a potent carrier is low toxicity. We showed that Pep-1 does not exhibit any toxicity up to a concentration of 500 μM (reference *(21)* and Figure 1B).

Mechanism of Internalization of Pep-1

Recent work has clearly demonstrated that the efficiency of CPPs is directly correlated to their cellular uptake mechanism. The major concern in the development of new PTD strategy is to avoid any endosomal pathway and/or to facilitate the escape of the cargoes from the early endosomes in order to limit its degradation *(17-19)*. Therefore, an important criterion to be considered in the process of investigating the mechanism of PTDs or CCPs is the structural requirement for its cellular uptake. We had combined a variety of physical and spectroscopic approaches to gain insight into the structure(s) involved and in the interactions of Pep-1/cargo complexes with lipids, and thus to characterize its mechanism of cellular internalization. Pep-1 is able to rapidly associate with peptides or proteins in solution through non-covalent hydrophobic interactions and forms stable complexes independently of a specific peptidyl sequence *(21)*. We have demonstrated that Pep-1 does not undergo conformational changes upon formation of a particle with a cargo peptide or protein, in contrast inter-action with the membrane components such as phospholipids induced helical folding of the carrier, which is essential for the uptake mechanism. We suggested that the outer part of the Pep-1 "shell" surrounding the cargo is involved in interactions with membrane and form transient trans-membrane helical pore-like structure, which facilitates insertion in the membrane and initiates the translocation process. We have proposed a four step mechanism (Figure 2): *(a)* formation of the Pep-1/cargo complexes involving hydrophobic and electrostatic interactions depending on the nature of the cargo, *(b)* inter-action of the complex with the external side of the cell *(c)* insertion of the complex into the membrane associated with partial conformational changes and pore formation *(d)* release of the complex into the cytoplasm with partial "de-caging" of the cargo *(23)*.

Pep-1 Mediated Transduction of Antibody

We evaluated the ability of Pep-1 to deliver therapeutic peptides and proteins into mammalian cell lines. We previously demonstrated that Pep-1 is able to rapidly form stable complexes with peptide and protein through non-covalent hydrophobic interactions *(21)*. Getting fully active antibody into cells constitutes of a main interest for cell therapy. To this aim, we investigated the ability of Pep-1 to deliver antibodies into a human fibroblastic cell line. FITC-conjugated anti-tubulin antibodies were used at a concentration of 0.1 µM and incubated with a 20-fold molar excess of Pep-1 (2 µM). Cells were incubated with Pep-1/antibody complexes for 1 hr in serum-free medium and were then extensively washed prior to observation by confocal microscopy. As fixation procedures have sometimes been reported to cause artifactual uptake, experiments were performed on live cells. As reported in Figure 3, characteristic labeling of tubulin were observed: decoration of cells in telophase or in meta-

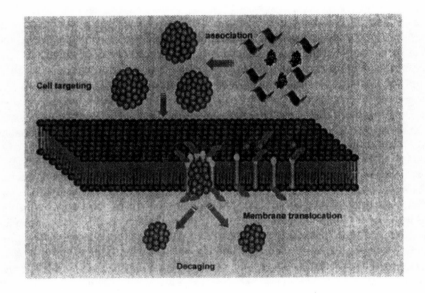

Figure 2. Proposed schematic model for translocation of the Pep-1/cargo complex through phospholipid bilayers. (Reproduced form reference (23), Copyright 2004 American Chemical Society)

phase. These results confirmed that Pep-1 is able to deliver antibodies into cells, whilst preserving their ability to recognize antigens within cells. Counting fluorescent cells reveals a transduction efficiency of about 80%. Finally, efficient transfection of antibodies was equally observed when transfection was performed at low temperature, supporting the idea that Pep-1-mediated transfection is independent of the endosomal pathway. That Pep-1 is able to promote delivery and proper localization of antibodies to their target antigens confirms that Pep-1 does not affect the appropriate sub-cellular localization of the proteins it delivers and supports the fact that the "de-caging" process namely the dissociation between Pep-1 and cargo occurs rapidly in the cytoplasm once the complexes enter the cells.

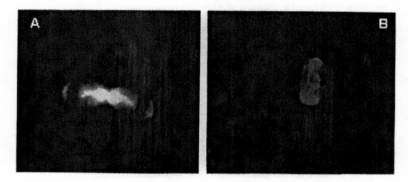

Figure 3. Pep-1 mediated anti-tubulin antibody delivery into mammalian cells. Cells were incubated in the presence of Pep-1/FITC-labeled-antibody complexes for 1 hr, then extensively washed and directly observed by fluorescence microscopy. Panel A: tubulin decoration of a cell in telophase, panel B: tubulin decoration of a cell in metaphase.

Pep-1 Mediated Transduction of the Cell Cycle Inhibitor p27^{kip1}

The cell cycle inhibitor p27^{Kip1} binds to and inhibits Cdk/cyclin complexes involved in the G1/S transition such as Cdk2/cyclin E and Cdk2/cyclin A *(24)*. Due to its important role in the control of the cell cycle progression, p27^{kip1} has been proposed as a good candidate for arresting cancer cell proliferation. Covalent strategy based on the TAT-peptide has been successfully used for the delivery of p27^{kip1} into cultured cells and *ex-vivo (25)*. Here we have used the non-covalent Pep-1 approach to deliver p27^{kip1} protein into cells and have investigated its biochemical mechanism *in cellulo*. Human fibroblasts grown to a confluence of 50% were synchronized by serum deprivation for 40 hrs, then restimulated to enter the cycle by addition of fresh DMEM supplemented with

20 % FCS for 4 hrs. Cells were then overlaid with Pep-1/p27^{kip1} complexes (molar ratio 20:1) and analyzed by flow cytometry 10, 15 and 20 hrs after transduction. Pep-1 or p27^{kip1} alone have no effect on cell cycle progression, as most of the cells incubated in the presence of Pep-1 (5µM) or P27^{kip1} (1 µM) enter mitosis after 20 hrs (Figure 4A). In contrast, upon p27^{kip1} delivery by Pep-1 (Figure 4B), analysis of the distribution of cells show that more than 80% of the cells are immediately blocked in G1, after release from serum privation. These results demonstrated that Pep-1 efficiently delivers p27^{kip1} protein in a biologically active form without the need of cross-linking. The cell cycle arrest induced by Pep-1 mediated delivery of p27^{kip1} was confirmed on several cell lines including cancer cells, suggesting that Pep-1 can be a potent technology for the delivery of therapeutic protein *in vivo*.

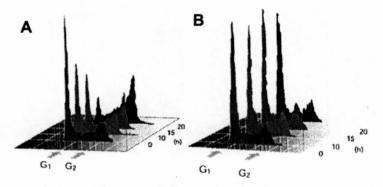

Figure 4. Pep-1 mediated p27^{kip1} delivery into mammalian cells. Cells released from synchrony were treated with Pep-1 or p27^{kip1} (A) or with Pep-1/ p27^{kip1} (B) and analyzed by flow cytometry 10, 15 and 20 hrs later.

Conclusions

Over the past ten years, substantial progress has been made in the design of new technology to improve cellular uptake of therapeutic compounds. This evolution was directly correlated with the dramatic acceleration in the production of new therapeutic molecules, whilst the cell delivery systems described until then were restricted by very specific issues. CPPs constitute one of the most promising generation of tools for delivering biologically active molecules into cells and thereby can have a major impact on future of treatments *(2,6)*. They have been shown to efficiently introduce drugs, antisense DNAs, PNAs, oligonucleotides and small proteins into cells both *in vivo* and *in vitro*. Peptide carriers present several advantages in that they are modulable, lack immunogenicity, are easy to produce and can incorporate a number of specific attributes required for efficient cargo delivery. Moreover, in addition to their usefulness in the laboratory, they are promising reagents for therapeutic screens and may potentially enable direct targeting of specific events throughout the cell cycle.

Most of the PTDs or CPPs described so far need a cross-linking step for delivery of the protein, which may be a limitation to their applications. We propose a new potent strategy for the delivery of full-length proteins and peptides into mammalian cells, based on a short peptide carrier, Pep-1, which allows the delivery of several distinct proteins and peptides into different cell lines without the need for cross-linking or denaturation steps *(21)*. Here, we demonstrate that Pep-1 facilitates the delivery of antibodies and therapeutic proteins in a fully biologically form. Moreover, this vector presents several advantages including the rapid deliver of proteins into cells with very high efficiency, stability in physiological buffer, lack of toxicity and lack of sensitivity to serum *(21)*. In term of cellular uptake mechanism, Pep-1 behaves significantly differently from other similarly designed cell-penetrating peptides. The Pep-1 cell translocation mechanism is independent of the endosomal pathway and involves the formation of transient transmembrane helical pore-like structure.

Several studies have demonstrated the potency of Pep-1 technology for the delivery of peptides and proteins both *in vitro* and *in vivo (21,26-28)*. Recently, this technology has been successfully applied for the delivery of antisense PNA into mammalian cells *(29)*. Taken together, these results reveal that Pep-1 technology is a powerful tool for basic research, for studying the role of proteins and for targeting specific protein/protein interactions *in vitro* and *in vivo*, as well as in a therapeutic context for the screening of potential therapeutic proteins and peptides. Therefore, Pep-1 technology constitutes a great alternative to covalent strategy and can have a major impact on the use of protein and peptide for future therapy.

Acknowledgements

This work was supported in part by the CNRS, Active-Motif (Carlsbad, CA) and by grants from the Agence Nationale de Recherche sur le SIDA (ANRS), the EU (Grant QLK2-CT-2001-01451) and the Association pour la Recherche sur le Cancer to MCM (ARC-4326) and to GD (ARC-5271). We thank K. Hondorp, J. Archdeacon for constructive discussions on Chariot project, and P. Travo for technical advice on microscopy.

164

References

1. Ford, K.G.; Souberbielle, B.E.; Darling, D; Farzaneh, F. Protein transduction: an alternative to genetic intervention? *Gene Therapy* **2001**, *8*, 1-4.
2. Langel, U. *Cell Penetrating peptides: Processes and application*, **2002**, CRC press, Pharmacology & Toxicology series.
3. Wadia, J.S.; Dowdy, S.F. Protein transduction technology. *Curr Opin Biotechnol* **2002**, *13*, 52-56
4. Prochiantz, A. Messenger proteins : homeoproteins, TAT and others. *Curr. Opin. Cell Biol.* **2000**, *12*, 400-406.
5. Schwarze, S.R.; Dowdy, S.F. *In vivo* protein transduction: intracellular delivery of biologically active proteins, compounds and DNA. *Trend Pharmacol. Sci.* **2000**, *21*, 45-48
6. Morris M.C.; Chaloin, L.; Heitz, F.; Divita, G. Translocating peptides and proteins and their use for gene delivery. *Curr. Opin. Biotechnol.* **2000**, *11*, 461-466.
7. Gariepy, J.; Kawamura, K. Vectorial delivery of macromolecules into cells using peptide-based vehicles. *Trends Biotechnol.*, **2000**, *19*, 21-26
8. Frankel, A.D.; Pabo, C.O. Cellular uptake of the tat protein from human immunodeficiency virus. *Cell* **1998**, *55*, 1189-1193.
9. Fawell, S. *et al.* Tat-mediated delivery of heterologous proteins into cells. *Proc. Natl. Acad. Sci. USA* **1994**, *91*, 664-668.
10. Vives, E.; Brodin, P.; Lebleu, B. A truncated HIV-1 Tat protein basic domain rapidly translocates through the plasma membrane and accumulates in the cell nucleus. *J Biol Chem.* **1997**, *272*, 16010-16017.
11. Schwarze, S.R.; Ho, A.; Vocero-Akbani, A.; Dowdy, S.F. *In vivo* protein transduction : delivery of a biologically active protein into the mouse. *Science* **1999**, *285*, 1569-1572.
12. Derossi, D.; Joliot, A.H.; Chassaings, G.; Prochiantz, A. The third helix of the antennapedia homeodomain translocates through biological membranes. *J. Biol. Chem.* **1994**, *269*, 10444-10450.
13. Elliott, G.; O'Hare, P. Intercellular trafficking and protein delivery by a Herpesvirus structural protein. *Cell* **1997**, *88*, 223-233.
14. Futaki, S.; Suzuki, T.; Ohashi, W.; Yagami, T.; Tanaka, S.; Ueda, K.; Sugiura, Y. Arginine-rich peptides. An abundant source of membrane-permeable peptides having potential as carriers for intracellular protein delivery. *J Biol Chem.* **2001**, *276*, 5836-5840
15. Wender P.A. The design, synthesis, and evaluation of molecules that enable or enhance uptake :peptoid molecular transporters. *Proc. Natl. Acad. Sci. U.S.A.* **2000**, *97*, 13003-13008.

16. Pooga, M.; Kut, C.; Kihlmark, M.; Hallbrink, M.; Fernaeus, S.; Raid, R.; Land, T.; Hallberg, E.; Bartfai, T.; Langel, U. Cellular translocation of proteins by transportan. *FASEB J* **2001**, *8*, 1451-1474

17. Richard J.P.; Melikov, K.; Vives, E.; Ramos, C.; Verbeure, B.; Gait, M.J.; Chernomordik, L.V.; Lebleu, B. Cell-penetrating peptides. A reevaluation of the mechanism of cellular uptake. *J Biol Chem.* **2003**, *278*, 585-590.

18. Lundberg, M.; Wikstrom, S.; Johansson, M. Cell surface adherence and endocytosis of protein transduction domain. *Mol. Ther.* **2003**, *8*, 143-150

19. Wadia, J.; Stan, R.V.; Dowdy, S. Transducible TAT-HA fusogenic peptide enhances escape of TAT-fusion proteins after lipid raft macropinocytosis *Nat. Med.* **2004**, *10*, 310-315

20. Torchilin, V.P.; Levchenko, T.S.; Rammohan, R.; Volodina, N.; Papahad-jopoulos-Sternberg, B.; D'Souza, G.G. Cell transfection in vitro and in vivo with nontoxic TAT peptide-liposome-DNA complexes. *Proc Natl Acad Sci U S A* **2003**, *100*, 1972-1977

21. Morris, M.C.; Depollier, J.; Mery, J.; Heitz, F.; Divita, G. A peptide carrier for the delivery of biologically active proteins into mammalian cells. *Nat Biotechnol.*, **2001**, *19*, 1173.

22. Morris, M.C.; Chaloin, L.; Mery, J.; Heitz, F.; Divita G. A novel potent strategy for gene delivery using a single peptide vector as a carrier. *Nucleic Acids Res.* **1999**, *27*, 3510-3517

23. Deshayes, S.; Heitz, A.; Morris, M.C.; Charnet, P.; Divita, G.; Heitz, F. Insight into the Mechanism of Internalization of the Cell-Penetrating Carrier Peptide Pep-1 through Conformational Analysis. *Biochemistry*, **2004**, *43*, 1449-1457.

24. Morgan D.O. Cyclin-dependent kinases: engines, clocks, and micropro-cessors. *Annu Rev Cell Dev Biol.* **1997** *13*, 261-291.

25. Snyder, EL; Meade, B.R.; Dowdy, S.F. Anti-cancer protein transduction strategies: restitution of p27 tumor suppresor function. *J. Control. Release* **2003**, *91*, 45-51

26. Gallo, G.; Yee, H.F.; Letourneau P.C. Actin turnover is required to prevent axon retraction driven by endogenous actomyosin contractility. *J. Cell Biol.* **2002**, *158*, 1219-1228.

27. Pratt R.L.; Kinch M.S. Activation of the EphA2 tyrosine kinase stimulates the MAP/ERK kinase signaling cascade. *Oncogene* **2002**, *21*, 7690-7699.

28. Aoshiba, K.; Yokohori, N.; Nagai, A. Alveolar wall apoptosis causes lung destruction and amphysematous changes. *Am. J. Respir. Cell. Mol. Biol.* **2003**, *28*, 555-562

29. Morris' M.C; Chaloin L.; Choob M.; Archdeacon J.; Heitz F.; Divita G. The combination of a new generation of PNAs with a peptide-based carrier enables efficient targeting of cell cycle progression. *Gene Ther.* **2004**, *11*, 757-764.

Chapter 12

Molecular Understanding of Cellular Uptake by Arginine-Rich Cell Penetrating Peptides

Jonathan B. Rothbard[*], Theodore C. Jessop, and Paul A. Wender

Department of Chemistry, Stanford University, Stanford, CA 94305
[*]Corresponding author: rothbardj@yahoo.com

A mechanistic hypothesis is presented for how water-soluble guanidinium-rich transporters attached to small cargos (MW ca. <3000) can migrate across the non-polar lipid membrane of a cell and enter the cytosol. Positively-charged and water soluble arginine oligomers can associate with negatively-charged, bidentate hydrogen bond acceptor groups of endogenous membrane constituents, leading to the formation of membrane soluble ion pair complexes. The resultant less polar, ion pair complexes partition into the lipid bilayer and migrate in a direction, and with a rate, influenced by the membrane potential. The complex dissociates on the inner leaf of the membrane and the transporter conjugate enters the cytosol. This mechanism could also be involved in the translocation of guanidinium-rich molecules that are endocytosed due to their size or the conditions of the assay, across the endosomal membrane.

Introduction

Biological membranes have evolved in part to prevent xenobiotics from passively entering cells *(1)*. Numerous organisms have developed proteins, many of which are transcription factors, that breach these biological barriers through a variety of mechanisms *(2)*. The protein HIV tat, for example, when used *in vitro* rapidly enters the cytosol (and nucleus) of a wide spectrum of cells after endocytosis *(3)*. However, the nine amino acid peptide required for the uptake of HIV tat, residues 49-57 (RKKRRQRRR), appears itself to utilize an additional mechanism as evident from its uptake even at 4 °C, by a route differentiated from the intact protein *(4)*. We have found that guanidinium-rich oligomers enter suspension cells more effectively than the tat nonamer *(5)*, often without the production of observable endocytotic vesicles *(6,7)*. We describe herein studies on the cellular uptake mechanism of guanidinium-rich transporters conjugated to small molecules (MW ca. <3000).

Pertinent to the formulation of a mechanism, our previous studies demonstrated that the guanidinium head groups of tat 49-57 are critical for its uptake into cells. Replacement of any of the arginine residues with alanine diminished uptake. Conversely, replacement of all non-arginine residues in the tat nonamer with arginines provided transporters that exhibit superior rates of uptake. Charge itself is necessary, but not sufficient, as is evident from the comparatively poor uptake of lysine nonamers *(5,6)*. The number of arginines is also important, with optimal uptake for oligomers of 7-15 residues *(6)*. Backbone chirality is not critical for uptake. Even the position of attachment and length of the side chains can be altered as shown with guanidinium-rich peptoids that exhibit highly efficient uptake. Changes in the backbone composition and in the side chain spacing also can increase uptake *(6,8-11)*. Even highly branched guanidinium-rich oligosaccharides and dendrimers are efficient transporters *(7,12-14)*. In contrast to receptor-mediated uptake, an increase in conformational flexibility generally favors uptake.

Materials and Methods

Fluoresceinated peptides were synthesized on an Applied Biosystems (Foster City, CA) model 433 automated peptide synthesizer using solid-phase Fmoc chemistry and HATU as the peptide coupling reagent. The peptides were cleaved from the resin using 95:5 TFA/triisopropylsilane for eight hours. The peptides were precipitated with diethyl ether, purified by RP-HPLC with molecular weight confirmed by electrospray or laser desorption mass spectrometry. The purity of the peptides was >95% as determined by analytical reverse-phase HPLC.

The synthesis of methylated arginine oligomers was accomplished by per-guanidinylation of oligoornithines with $1H$-pyrazole-N-methyl-1-carboxamidine or $1H$-pyrazole-N,N-dimethyl-1-carboxamidine).[15]

Octanol / water partitioning used the hydrochloride salts of Fl-aca-Orn$_8$-CONH$_2$ and Fl-aca-Arg$_8$-CONH$_2$, prepared by ion exchange through a column of Dowex 1 x 8 200 chloride resin. An aqueous solution of the peptide (44 μM, 0.5 mL) and octanol (0.5 mL) were combined in a 2 mL glass vial. For experiments involving sodium laurate, 2.3 eq of lauric acid in octanol and an equivalent molar amount of sodium hydroxide in water was added. The mixture was shaken for 30 seconds, centrifuged (~2000 rpm, 1 min), and photographed. Quantitative analysis of partitioning was accomplished by removal of an aliquot of each layer and UV analysis.

Cellular uptake of the fluorescent peptides was measured by flow cytometry. Varying concentrations of fluorescein aminocaproic acid-D-Arg$_8$-CONH$_2$ and fluorescein-aminocaproic acid-RKKRRQRRR-CONH$_2$ (tat 49-57), between 50 and 0.4 micromolar in phosphate buffered saline, were incubated in triplicate, with the human T lymphocyte cell line, Jurkat, or the macrophage cell lines, J774 or RAW 264, for five minutes. At the end of the incubation period the cells were spun, washed twice with PBS, transferred to tubes and analyzed on a Caliber flow cytometer (Becton-Dickinson) using the 480 argon laser. The mean fluorescence of 10,000 viable cells was measured for each concentration of peptide. The data was plotted as an average of the triplicates at each concentration with standard error.

Potassium PBS (K+PBS) was prepared by replacement of the sodium salts in PBS with equivalent potassium salts. Buffers with intermediate potassium concentrations can be prepared by varying the ratio of PBS and K+PBS mixed together. These buffers are referred to by their millimolar content of potassium, i.e., Ko40 is 40 mM in potassium. For experiments with differing amounts of potassium ions in the buffer, the cells were washed twice with the buffer. The cells were then exposed to the peptide in that buffer, washed with the same buffer, and finally resuspended in that buffer for analysis by flow cytometry.

Uptake in the presence of either Gramicidin A or Valinomycin was accomplished as follows. Jurkat cells (3 x 10^6/ml) in PBS were treated with 1 μM gramicidin A (CalbioChem) or 50 μM valinomycin (CalbioChem) for thirty minutes and treated with fluorescein aminocaproic acid- D-Arg$_8$-CONH$_2$ as above.

To demonstrate that endocytosis was not inhibited by K+PBS, the macrophage cell line, J774 (3 x 10^5 cells), was aliquoted into separate chambers of a Lab-tek chamber slide and allowed to adhere and multiply for 14 hours. The slide was cooled to 3°C and 100 μl of 0.2 mg/ml of Texas red transferrin (Molecular Probes) was added to each chamber and allowed to incubate for 20 minutes on ice. The chambers were rinsed with either PBS or K+PBS, cooled to

3°C, and 200 µl of ice cold buffer was added. The slide was transferred to a 37°C incubator and incubated for 15 minutes. The slide was washed with 0.1M glycine pH 3.0, containing 0.15M NaCl. The chambers were removed and a coverslip was added. The cells were visualized using a Zeiss Axiovert 2 fluorescent microscope equipped with 530 long pass filter.

Results and Discussion

Several mechanisms could accommodate the observed structure-function relationships for guanidinium-rich transporters (see introduction) and some could operate concurrently. A receptor-mediated process is inconsistent with the broad range of structural modifications that promote uptake, and especially the observation that more flexible systems work better. Conventional passive diffusion across the non-polar interior of the plasma membrane is difficult to reconcile with the polarity of the arginine oligomers and the dependency of uptake on the number of charges. In contrast to passive diffusion, in which a migrating conjugate maintains its polarity, the polar, positively charged guanidinium oligomers could adaptively diffuse into the non-polar membrane by recruiting negatively charged cell surface constituents to transiently produce a less polar, ion pair complex. Indeed, the polarity and bioavailability of many proteins and peptides can be changed by the intentional pre-formation of a non-covalent complex with anionic agents such as SDS *(16)*. To test whether a highly water-soluble guanidinium-rich oligomer could be rendered lipid-soluble through ion pair formation, a fluoresceinated arginine octamer (Fl-aca-D-Arg$_8$-CONH$_2$) was added to a bilayer of octanol and water. Not surprisingly, the highly polar charged system partitioned almost exclusively (>95%) into the water layer (Figure 1). When, however, a surrogate for a membrane-bound fatty acid salt, namely sodium laurate, was added to this mixture, the transporter partitioned completely (99%) into the octanol layer *(17)*. The relative partitioning was quantified by separation of the layers and analysis of the dissolved agents.

While other polycations, like short ornithine oligomers, might partition through a similar ion pair mechanism, they are observed to be significantly less effective in cellular uptake than the arginine oligomers. This difference could arise, in part, from the more effective bidentate hydrogen bonding possible for guanidinium groups versus monodentate hydrogen bonding for the ammonium groups. Consistent with this analysis, when ornithine oligomers were submitted to the above two phase partitioning experiments, they preferentially (>95%) stayed in the aqueous layer even with added sodium laurate (A and C).

Critical to the adaptive diffusion mechanism is the special ability of guanidinium groups to transiently form bidentate hydrogen bonds with cell surface hydrogen bond acceptors. While incorporating the high basicity and dispersed cationic charge of a guanidinium group, alkylated guanidiniums would

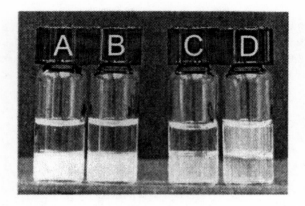

Figure 1. Octanol / water (upper and lower phase, respectively) partitioning of Fl-aca-Orn$_8$-CONH$_2$ and Fl-aca-D-Arg$_8$-CONH$_2$ alone (A, B) and after addition of sodium laurate (C, D). In vials A, B and C, more than 98% of the the peptide remains in the aqueous (lower) phase. Vial D has more than 95% of Fl-aca-D-Arg$_8$-CONH$_2$ in the octanol (upper) phase.

have an attenuated ability to form hydrogen bonds with phosphates, carboxylates or sulfates and thus would be expected to be less effective transporters. This proved to be the case. When octamers of mono- and dimethylated arginine (Fl-aca-Arg$^m_8$-CONH$_2$, Fl-aca-Arg$^{mm}_8$-CONH$_2$), synthesized from the corresponding ornithine octamer, were assayed for cell entry, uptake of the former was reduced by 80% and the latter by greater than 95% when compared with an unalkylated arginine octamer (Figure 2).

If the guanidiniums of a transporter form bidendate ion pairs with anions on the cell surface, then the counterions associated with the guanidinium groups must be exchanged on the cell surface, and not transported into the cell. To visualize the proposed ion exchange, fluorescein was used as a counterion by adding a molar equivalent of fluorescein to the free base of an octamer of arginine, formed by eluting from a hydroxide anion exchange column, to afford Ac-D-Arg$_8$-CONH$_2$ • 1x fluorescein salt. When human T cells were treated with this salt for five minutes, washed, and analyzed by flow cytometry, fluorescent microscopy, and fluorescent confocal microscopy, a pattern was seen distinctly different than that observed when the cells were treated with the transporter covalently attached to the fluorescein. In contrast to the bimodal distribution of fluorescence, ranging over three orders of magnitude, exhibited by cells treated with Fl-aca-D-Arg$_8$-CONH$_2$ when analyzed by flow cytometry (Figure 3A), cells treated with Ac-D-Arg$_8$-CONH$_2$ • 1x fluorescein salt exhibited a relatively uniform distribution of fluorescence (Figure 3C).

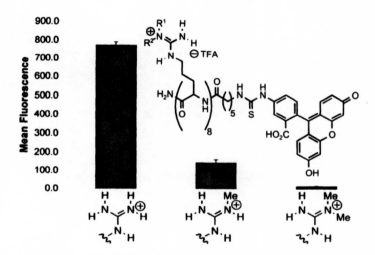

Figure 2. Reduced uptake of arginine oligomers with mono- and dimethylated guanidiniums. Mean fluorescence of Jurkat cells after treatment (5 min, 50 µM) with Fl-aca-D-Arg$_8$-CONH$_2$ (left), Fl-aca-Arg$^m{}_8$-CONH$_2$ (center), and Fl-aca-Arg$^{mm}{}_8$-CONH$_2$ (right); Fl = Fluorescein-HNC(S)-, aca = aminocaproic acid, Arg^m = N^G-methylarginine, Arg^{mm} = N^G,N^G-dimethylarginine.

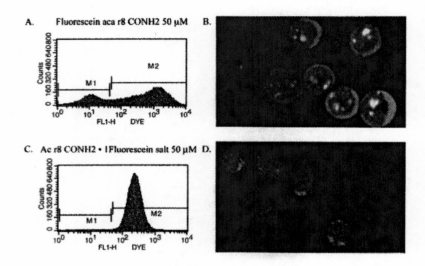

Figure 3. The cellular staining of lymphocytes differs when treated with either fluorescein covalently conjugated, or complexed as a salt, to guanidinium-rich transporters. Histogram of cellular fluorescence after treatment of the human T lymphocyte cell line, Jurkat, after being incubated with Fl-aca-D-Arg₈-CONH₂ (panel A) or Ac-HN-D-Arg₈-CONH₂ • 1x fluorescein salt (panel C) (50 µM, for 5 minutes). Representative fluorescent micrographs of the treated lymphocytes isolated using gate M2 in panels A and C shown in B and D, respectively.

When the highly stained population resulting from treatment with the covalent conjugate was isolated by fluorescence activated cell sorting, and visualized by fluorescent microscopy and fluorescent confocal microscopy, the peptide was localized in the nuclei, the cytosol and most prominently, the nucleoli (Figure 3B). In contrast, microscopic analysis of cells treated with the salt revealed solely cell surface-staining with a small degree of patching and capping (Figure 3D). Moreover, unlike cells treated with the covalent conjugate Fl-aca-D-Arg$_8$-CONH$_2$, the fluorescence of cells treated with Ac-D-Arg$_8$-CONH$_2$ • 1x Fluorescein salt was not diminished by pretreatment with either 1% sodium azide or K$^+$PBS. These data support the hypothesis that the polyguanidinyl transporter binds to anions on the cell surface with a concomitant ion exchange. If the counterion of the guanidinium is polar and water soluble, it will diffuse into the medium. If the counterion is hydrophobic, such as fluorescein, it will partition into the membrane.

While charge complementation with endogenous membrane constituents allows for adaptive entry into the membrane, it does not explain the driving force for passage through the membrane and the energy dependency of uptake observed in some studies. Given that phospholipid membranes in viable cells exhibit a membrane potential, the maintenance of which requires ATP, we reasoned that uptake of cationic-rich transporters might be driven by the voltage potential across most cell membranes (18). To test this hypothesis, the membrane potential was reduced to close to zero by incubating the cells with an isotonic buffer with potassium concentrations equivalent to that found intracellularly. The intracellular concentration of K^+ in lymphocytes is ~140 mM, the extracellular concentration is ~5 mM, and it is primarily this concentration gradient that maintains the transmembrane potential (18). Replacement of the sodium salts in PBS with equimolar amounts of the equivalent potassium salts afforded what was called K^+PBS.

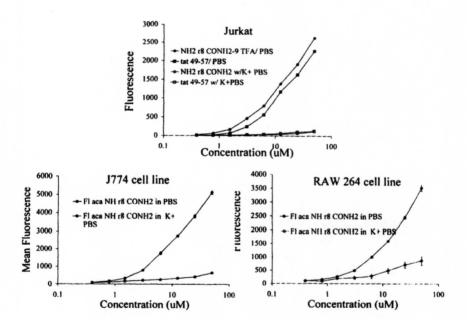

Figure 4. Cellular uptake of Fl-aca-D-Arg$_8$-CONH$_2$ (circles) and Fl-aca-tat$_{49-57}$-CONH$_2$ (squares) in PBS (solid markers) and K^+PBS (outlined markers) in a lymphocyte (panel a) and two macrophage cell lines (panels b and c). The macrophage cell lines were trypsinized for 5 minutes, lymphocytes and macrophages were washed three times with either K^+PBS or PBS and incubated with varying concentrations of the fluorescent peptides for five minutes. The cells were washed and analyzed by flow cytometry.

174

To test whether the membrane potential in lymphocytes was a factor in transport of guanidinium-rich transporters, fluorescently labeled tat 49-57 (Fl-aca-tat$_{49-57}$-CONH$_2$) and an octamer of D-arginine (Fl-aca-D-Arg$_8$-CONH$_2$) were incubated individually with Jurkat cells for 5 minutes in either PBS or K$^+$PBS. The cells were washed and analyzed by flow cytometry. Uptake was reduced by greater than 90% at all concentrations when the assay was done in the presence of a buffer with a high concentration of potassium (Figure 4A). The observed inhibition of uptake was equivalent to that seen when the cells were pretreated with sodium azide. The effects of high concentrations of potassium are not limited to lymphocytes as the uptake of Fl-aca-D-Arg$_8$-CONH$_2$ into two macrophage cell lines, J774 and RAW 264 (Figure 4B and C), was also inhibited.

To determine whether the reduction of membrane potential could affect endocytosis, the macrophage cell line, J774, was treated with Texas red transferrin, under conditions that permitted endocytosis, and in the presence, or absence, of K$^+$PBS. Visualization of the cellular fluorescence under these conditions revealed no significant difference in the amount of transferrin endocytosed (Figure 5). The ability to significantly limit cellular penetration of guanidinium-rich transporters by preincubation with K$^+$PBS is in sharp contrast with its failure to block the uptake of transferrin, providing further evidence that an alternative mechanism to endocytosis is involved in the uptake of guanidinium-rich transporters.

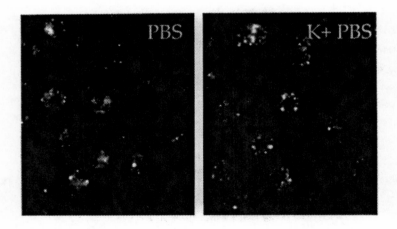

Figure 5. Demonstration that incubation with high potassium buffers does not inhibit endocytosis in a macrophage cell line. Fluorescent micrographs of J774 cells incubated with Texas red transferrin at 37°C for 30 minutes in either PBS or K$^+$PBS.

To determine whether peptide translocation required a minimum voltage or varied continuously with the membrane potential, the uptake experiment of the labeled arginine oligomer was repeated using a series of buffers whose potassium ion concentration varied between 140 mM and zero. The results demonstrated that uptake decreased with an increase in the external concentration of potassium (Figure 6A). The uptake as measured by cellular fluorescence appeared to vary linearly with the potassium Nernst potential[19] calculated across the range of extracellular potassium ion concentrations (Figure 6B).

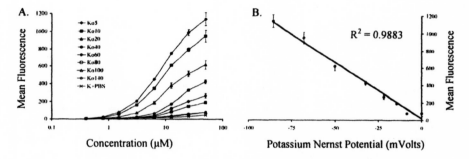

Figure 6. Differential inhibition of cellular uptake of Fl-aca-D-Arg₈CONH₂ in the human lymphocyte cell line, Jurkat, by incubation with isotonic buffers containing increasing amounts of potassium ions. The concentration (mM) of the potassium ions in each of the buffers correspond to the K_o values (A). Cells were washed three times with the listed buffers, incubated with varying concentrations of Fl-aca-D-Arg₈CONH₂ for five minutes in the same buffer. The cells were washed with and kept in the appropriate buffer and analyzed by flow cytometry. The uptake as measured by cellular fluorescence appeared to vary linearly with the K^+ Nernst potential calculated across the range of extracellular K^+ concentrations (B). Calculation of the K^+ Nernst potential (E_K) in cells incubated with buffers with varying K^+ concentrations was done using the Nernst equation, $E_K = (RT/F) \ln (K_o/K_i)$ where R = the gas constant, T = temperature, F = Faraday's constant, K_o = extracellular potassium ion concentration and K_i = intracellular potassium ion concentration.

To explore whether high potassium buffers inhibited uptake by modulating the membrane potential, or by an alternative effect, lymphocytes were pretreated with gramicidin A, a pore-forming peptide known to reduce membrane potential *(20)*, prior to the addition of Fl-aca-D-Arg₈-CONH₂. This procedure reduced cellular uptake by more than 90% (Figure 7). The reciprocal experiment, hyperpolarizing the cell to increase uptake, was accomplished with valinomycin, a peptide antibiotic that selectively shuttles potassium ions across the membrane *(21)*. When Jurkat cells were preincubated with 50 µM valinomycin, the uptake of Fl-aca-D-Arg$_8$-CONH$_2$ was significantly increased (Figure 7).

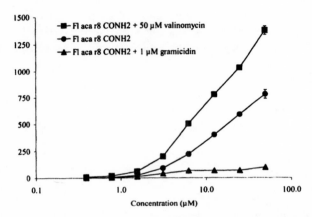

Figure 7. Cellular uptake of Fl-aca-D-Arg$_8$-CONH$_2$ into the human T lympho-cyte cell line, Jurkat, alone or with preincubation with valinomycin or gramicidin A.

Conclusions

The development of strategies to breach biological barriers is a major challenge as science moves to exploit the opportunities arising from genomic studies, proteomics, systems biology, combinatorial chemistry, and rational drug design. Most therapeutic agents and probes with intracellular targets enter cells through passive diffusion and as such they generally must conform to a rather restricted log P range, allowing for solubility in both the polar extra-cellular milieu and non-polar membrane. Notwithstanding the attention given to structural diversity in drug discovery, the relative lack of physical property diversity limits the range of agents that can be used as therapeutics and probes. The studies discussed in this review provide a mechanistic rationale for how a freely water soluble, polar transporter can change its effective polarity, enter non-polar membranes through adaptive non-covalent association with membrane constituents, and move into the cytosol under the influence of a membrane potential. In this process, guanidinium-rich transporters only weakly associated with their counterions in the polar extracellular milieu *(22)* readily exchange their counterions to form bidendate hydrogen bonds with phosphates, sulfates and/or carboxylates anchored on the cell surface. This hydrogen-bonding converts otherwise polar cationic and anionic functionalities into lipophilic ion pairs, whose hydrogen bond associations strengthen as they enter the non-polar lipid bilayer *(22)*. The directionality and rate of passage of these ion pairs through the bilayer is dictated by the membrane potential. The ion pair complexes dissociate on the inner leaf of the membrane and the transporters enter the cytosol in the microscopic reverse of the cell surface association. Too few guanidinium groups diminish cell surface adherence and therefore entry; too many could inhibit escape from the inner leaf of the membrane. A range of approximately 7-20 guanidinium groups works well for most cargos with a preference for 8 dictated by the cost of goods and a novel method of preparation of octamers *(23)*.

This mechanism is consistent with the diverse spectrum of guanidinium-rich structures that rapidly enter cells, emphasizing that the key feature for entry into the cell is the formation of the bidentate hydrogen bonds between the guanidinium head groups and anionic structures on the cell surface and not the structural and conformational details of the transporter. Consequently, the stereochemistry, the composition of the backbone, and the spacing of the guanidiniums extending either from, or along, the backbone all can be modified with only modest changes in transporter activity. The essential structural requirement appears to be a minimum number of guanidiniums, which explains why even nonlinear dendrimers and oligosaccharides decorated with sufficient number of guanidiniums are transported effectively.

The importance of the membrane potential suggests that the transporter complex has an overall positive charge, which could arise either from incomplete ion pairing with anionic cellular components or by ion pair formation with zwitterionic species. Incomplete ion pair formation is entropically favored as the number of guanidiniums increases and would be consistent with the requirement of a minimum of guanidinium head groups for effective transport. In the proposed mechanism, the large number of potential binding sites on the cell surface also explains why inhibition of uptake at sensible molar excesses of a competitive inhibitor is not observed, and also why neither prokaryotic nor eukaryotic transport negative mutants have been generated (unpublished data). Inhibition of uptake by sodium azide is better explained by the dissipation of the membrane potential or inhibition of ion export rather than inhibition of an energy dependent receptor mediated pathway of uptake.

An attractive aspect of the mechanism proposed above is that each of the central steps has precedent. Modification of the polarity of highly basic peptides by formation of ion pair complexes with hydrophobic anions is fundamental to ion pair chromatography and has been used to increase the stability of biologically active peptides *(16,24-26)*. Furthermore, the importance of ion pair formation in the cellular uptake of cationic peptides has been considered by others *(17)*. Movement of cations, as well as arginines, across biological mem-

178

branes in the direction of the membrane potential also has significant precedent. Wilson and colleagues established that the movement of a cationic cyanine dye into or across the membrane was proportional to membrane potential and is sufficiently quantitative that the dyes are used in flow cytometry to measure the membrane potential of cells (27,28).

It has been argued that guanidinium-rich transporters enter cells exclusively by endocytosis (29,30). Clearly, however, multiple mechanisms of entry are possible given the variety of cargo sizes, cell types, and experimental conditions. For example, for larger cargoes an endocytotic pathway would be favored (4,31) as diffusion of an ion pair complex would be expected to decrease with size. However, even here, endocytotic entry results in an endosomal bound conjugate that has not traversed a barrier but must do so to become freely available in the cytosol. In contrast, the mechanism discussed above allows for adaptive diffusion across the membrane to produce a freely available, cytosolic conjugate. Interestingly, endosomes retain a membrane potential, therefore the mechanism discussed above could also be operative in the translocation of guanidinium-rich molecules through endosomes into the cytosol. This premise is supported by two recent papers demonstrating that the endosomal exit of FGF-1 and FGF-2, which both contain arginine-rich domains, required the endosomal membrane potential (32,33). Consequently, the voltage-dependent, lipid-mediated uptake mechanism can be invoked in crossing of both the endosome and the plasma membrane lipid bilayers.

In summary, the data in this paper summarizes evidence supporting a voltage-dependent, adaptive translocation mechanism for the cellular uptake of guanidinium-rich molecular transporters attached to small molecular weight cargos. The proposed mechanism emphasizes the importance of ion pair formation based on bidentate hydrogen bonding and electrostatics that allow for adaptive translocation of a polar transporter into a non-polar membrane and transmembrane migration under the influence of the membrane voltage potential. This mechanism provides an improved molecular understanding of the translocation of guanidinium-rich transporters across phospholipid membranes. *In vivo* studies have established that arginine-rich peptides are widely distributed in tissue, which indicates that these transporters are capable of crossing a variety of phospholipid and non-phospholipid barriers (34).

References

1. Alberts, B.; Lewis, J.; Raff, M.; Roberts, K.; Watson, J. *Molecular Biology of the Cell*; Garland: New York, N.Y., 1994.
2. Joliot, A.; Prochiantz, A. "Transduction Peptides: From Technology to Physiology" *Nat. Cell Biol.* **2004**, *6*, 189-196.

3. Mann, D.A.; Frankel, A.D. Endocytosis and Targeting of Exogenous HIV-1 Tat Protein. *Embo J.* **1991**, *10*, 1733-1739.
4. Silhol, M.; Tyagi, M.; Giacca, M.; Lebleu, B.; Vives, E. Different Mechanisms for Cellular Internalization of the HIV-1 Tat-Derived Cell Penetrating Peptide and Recombinant Proteins Fused to Tat. *Eur. J. Biochem.* **2002**, *269*, 494-501.
5. Wender, P.A.; Mitchell, D.J.; Pattabiraman, K.; Pelkey, E.T.; Steinman, L.; Rothbard, J.B. The Design ; Synthesis ; and Evaluation of Molecules That Enable or Enhance Cellular Uptake: Peptoid Molecular Transporters. *Proc. Nat. Acad. Sci.* **2000**, *97*, 13003-13008.
6. Mitchell, D.J.; Kim, D.T.; Steinman, L.; Fathman, C.G.; Rothbard, J.B. Polyarginine Enters Cells More Efficiently Than Other Polycationic Homopolymers. *J. Pep. Res.* **2000**, *56*, 318-325.
7. Luedtke, N.W.; Carmichael, P.; Tor, Y. Cellular Uptake of Aminoglycosides, Guanidinoglycosides, and Poly-Arginine. *J. Am. Chem. Soc.* **2003**, *125*, 12374-12375.
8. Wender, P.A.; Rothbard, J.B.; Jessop, T.C.; Kreider, E.L.; Wylie, B.L. Oligocarbamate Molecular Transporters: Design, Synthesis, and Biological Evaluation of a New Class of Transporters for Drug Delivery. *J. Am. Chem. Soc.* **2002**, *124*, 13382-13383.
9. Rothbard, J.B.; Kreider, E.; Vandeusen, C.L.; Wright, L.; Wylie, B.L.; Wender, P.A. Arginine-Rich Molecular Transporters for Drug Delivery: Role of Backbone Spacing in Cellular Uptake. *J. Med. Chem.* **2002**, *45*, 3612-3618.
10. Umezawa, N.; Gelman, M.A.; Haigis, M.C.; Raines, R.T.; Gellman, S.H. Translocation of a Beta-Peptide across Cell Membranes. *J. Am. Chem. Soc.* **2002**, *124*, 368-369.
11. Rueping, M.; Mahajan, Y.; Sauer, M.; Seebach, D. Cellular Uptake Studies with Beta-Peptides. *Chembiochem* **2002**, *3*, 257-259.
12. Vandeusen, C.L. In *Department of Chemistry*; Stanford University: Stanford, 2003, p 335.
13. Futaki, S.; Nakase, I.; Suzuki, T.; Youjun, Z.; Sugiura, Y. Translocation of Branched-Chain Arginine Peptides through Cell Membranes: Flexibility in the Spatial Disposition of Positive Charges in Membrane-Permeable Peptides. *Biochemistry* **2002**, *41*, 7925-7930.
14. Chung, H.H.; Harms, G.; Seong, C.M.; Choi, B.H.; Min, C.; Taulane, J.P.; Goodman, M. Dendritic Oligoguanidines as Intracellular Translocators. *Biopolymers* **2004**, *76*, 83-96.
15. Bernatowicz, M.S.; Matsueda, G.R. An Improved Synthesis of N(G)-Allyl-(L)-Arginine. *Synthetic Communications* **1993**, *23*, 657-661.

16. Powers, M.E.; Matsuura, J.; Brassell, J.; Manning, M.C.; Shefter, E. Enhanced Solubility of Proteins and Peptides in Nonpolar-Solvents through Hydrophobic Ion-Pairing. *Biopolymers* **1993**, *33*, 927-932.

17. Sakai, N.; Matile, S. Anion-Mediated Transfer of Polyarginine across Liquid and Bilayer Membranes *J. Am. Chem. Soc.* **2003**, *125*, 14348-14356.

18. Aidley, D.; Stanfield, P. *Ion Channels*; Cambridge University Press: Cambridge, England, 1996.

19. Weiss, T. *Cellular Biophysics*; MIT Press: Boston, MA, 1996; Vol. 1.

20. Urban, B.W.; Hladky, S.B.; Haydon, D.A. Ion Movements in Gramicidin Pores. An Example of Single-File Transport. *Biochim. Biophys. Acta* **1980**, *602*, 331-354.

21. Harada, H.; Morita, M.; Suketa, Y. K+ Ionophores Inhibit Nerve Growth Factor-Induced Neuronal Differentiation in Rat Adrenal Pheochromocytoma Pc12 Cells. *Biochim. Biophys. Acta* **1994**, *1220*, 310-314.

22. Onda, M.; Yoshihara, K.; Koyano, H.; Ariga, K.; Kunitake, T. Molecular Recognition of Nucleotides by the Guanidinium Unit at the Surface of Aqueous Micelles and Bilayers. A Comparison of Microscopic and Macroscopic Interfaces. *J. Am. Chem. Soc.* **1996**, *118*, 8524-8530.

23. Wender, P.A.; Jessop, T.C.; Pattabiraman, K.; Pelkey, E.T.; Vandeusen, C.L. An Efficient, Scalable Synthesis of the Molecular Transporter Octaarginine Via a Segment Doubling Strategy. *Org. Lett.* **2001**, *3*, 3229-3232.

24. Hancock, W.S.; Bishop, C.A.; Prestidge, R.L.; Harding, D.R.; Hearn, M.T. Reversed-Phase, High-Pressure Liquid Chromatography of Peptides and Proteins with Ion-Pairing Reagents. *Science* **1978**, *200*, 1168-1170.

25. Gennaro, M.C."Reversed-Phase Ion-Pair and Ion-Interaction Chromatography. *Adv. Chrom.* **1995**, *35*, 343-381.

26. Schill, G. High-Performance Ion-Pair Chromatography" *Journal of Biochemical and Biophysical Methods* **1989**, *18*, 249-270.

27. Wilson, H.A.; Seligmann, B.E.; Chused, T.M. Voltage-Sensitive Cyanine Dye Fluorescence Signals in Lymphocytes-Plasma-Membrane and Mitochondrial Components. *J. Cell. Phys.* **1985**, *125*, 61-71.

28. Wilson, H.A.; Chused, T.M. Lymphocyte Membrane-Potential and Ca2+-Sensitive Potassium Channels Described by Oxonol Dye Fluorescence Measurements. *J. Cell. Phys.* **1985**, *125*, 72-81.

29. Vives, E. Cellular Uptake of the Tat Peptide: An Endocytosis Mechanism Following Ionic Interactions. *J. Mol. Recog.* **2003**, *16*, 265-271.

30. Fuchs, S.M.; Raines, R.T. Pathway for Polyarginine Entry into Mammalian Cell. *Biochemistry* **2004**, *43*, 2438-2444.

31. Wadia, J.S.; Stan, R.V.; Dowdy, S.F. Transducible Tat-Ha Fusogenic Peptide Enhances Escape of Tat-Fusion Proteins after Lipid Raft Macropinocytosis. *Nat. Med.* **2004**, *10*, 310-315.

32. Malecki, J.; Wiedlocha, A.; Wesche, J.; Olsnes, S. Vesicle Transmembrane Potential Is Required for Translocation to the Cytosol of Externally Added FGF-1. *Embo J.* **2002**, *21*, 4480-4490.
33. Malecki, J.; Wesche, J.; Skjerpen, C.S.; Wiedlocha, A.; Olsnes, S. Translocation of FGF-1 and FGF-2 across Vesicular Membranes Occurs During G(1)-Phase by a Common Mechanism. *Mol. Biol. Cell.* **2004**, *15*, 801-814.
34. Rothbard, J.B.; Garlington, S.; Lin, Q.; Kirschberg, T.; Kreider, E.; Mcgrane, P.L.; Wender, P.A.; Khavari, P.A. Conjugation of Arginine Oligomers to Cyclosporin a Facilitates Topical Delivery and Inhibition of Inflammation. *Nat. Med.* **2000**, *6*, 1253-1257.

Chapter 13

Recent Advances in Poly(ethyleneimine) Gene Carrier Design

Darin Y. Furgeson[1] and Sung Wan Kim[2,*]

[1]Department of Biomedical Engineering, Box 90281, Duke University,
Durham, NC 27708–0281
[2]Department of Pharmaceutics and Pharmaceutical Chemistry, Center
for Controlled Chemical Delivery, University of Utah, Salt Lake
City, UT 84112–5820
*Corresponding author: rburns@pharm.utah.edu

Poly(ethylenimine) (PEI) is the leading nonviral gene carrier
described in the literature today; moreover, it has been modified with a number of groups and modalities. Modifications
include chemical groups for shielding of the cationic charge,
targeting groups for specific cells, and biodegradable linkers
for increased biocompatibility. We will review the major advances in PEI design and modification for both *in vitro* and *in
vivo* applications.

Introduction

Proteins and/or drugs are administered to treat disease; however, gene
therapy has the potential to improve disease treatment and to emerge as a leader
in molecular medicine. Gene therapy essentially delivers a genetic sequence
encoding therapeutic proteins that are expressed within the pathogenic cell.
Furthermore, the potential for the treatment of such a wide range of genetic
abnormalities could lead to the ability to correct a genetic defect *in utero*, and
therefore, cure diseases such as Duchenne muscular dystrophy *(1)*, cystic
fibrosis *(2)*, and Smith-Magenis syndrome *(3)*.

Although there are multiple methods for transgene expression, viral and
nonviral gene carriers are the most common. The number of clinical trials with
viral carriers overshadows that of nonviral carriers *(4)* due to their inherently

high degree of transfection and their increased persistence of gene expression. However, these carriers are prone to host immunogenic responses and are limited in the transgene size. Nonviral gene carriers consist primarily of a cationic region to condense the anionic therapeutic plasmid DNA (pDNA), commonly referred to as a polyelectrolyte, thereby protecting the DNA pro-drug cargo from deleterious nucleases. The condensation of pDNA is primarily through electrostatic interactions between cationic groups on the carrier and the anionic phosphate backbone of the pDNA. The formation of these electrostatically condensed complexes may also be a function of hydrogen bonds, especially with linear poly(ethylenimine), solvent exclusion, and hydrophobic interaction (a strong factor in cationic lipid/pDNA complex formation). Upon cellular internalization and nuclear import, the transgene is transcribed into messenger RNA (mRNA) encoding the target protein. The mRNA is exported into the cytosol and translated into the therapeutic protein. This protein may then exert its therapeutic effect locally, and/or in the systemic treatment of disease.

The cost of cytotoxicity and immunogenicity for nonviral gene carriers is less than viral carriers; however, the transfection efficiency and persistence of gene expression are lower compared to the evolutionary advanced viral gene carriers. Viral gene carriers provide efficient gene transfection, but with a significant risk including mutagenesis and cancer (5). Viral gene carriers contain domains to promote receptor-mediated endocytosis, protein transduction, fusogenicity, and nuclear translocation. These domains have since been attached to nonviral gene carriers to induce comparable degrees of targeting, endosomal escape, and/or karyophily. The cytotoxicity of nonviral gene carriers, as seen with branched poly(ethylenimine) (BPEI) MW 25k, is due to the predominant cationic charge. This high cationic charge state facilitates unfavorable interactions with anionic species in vivo such as erythrocytes and albumin; moreover, the excessive positive charge may contribute to destabilization of the cellular membrane.

The numerous drawbacks of viral carriers have led to the development of a number of nonviral gene carriers and techniques including cationic polymers, cationic lipids, optimization of naked pDNA constructs, chemical precipitation, and mechanical means of efficient transfection. Nonviral gene carriers are fast becoming the vectors of choice for gene delivery applications due to increased ability for transgene delivery and expression, decreased potential for immunogenicity, cost, and ease of production. Poly(ethylenimine) (PEI), a stalwart nonviral gene carrier, consists of a wide-range of varying molecular weights, branched and linear architectures, chemical modifications, and targeting modalities. The impetus towards modifying PEI is to primarily increase gene transfer efficiency without a loss in biocompatibility; consequently, researchers have turned to molecular weight optimization, the inclusion of chemical groups for targeting and surface charge modification, and specific biodegradation sites.

Here we will present the major modifications to both BPEI and linear poly(ethylenimine) (LPEI) found in the literature.

Results and Discussion

Poly(ethyleneimine) PEI

To date no other polymer has been studied as widely in nonviral gene delivery than PEI. Typically, the synthesis of PEI is through acid-catalyzed, ring-opening cationic polymerization of aziridine, resulting in the ($NHCH_2CH_2$) monomer unit with molecular weights varying from 423 Da to 800 kDa in both branched and linear conformations. The monomer unit of PEI ($NHCH_2CH_2$) positions each amine unit in close proximity (4.48 Å) to other neighboring amine units *(6)*. Subsequent protonation of these amines confirms PEI with the highest cationic density of any synthetic polymer for the purpose of DNA condensation, subsequent protection from deleterious nucleases, and escape from the lysosomal compartment. However, this high cationic density is not without cost. A highly cationic PEI gene carrier may also attract anionic components in the blood stream such as erythrocytes; moreover, these unfavorable interactions may result in aggregation and removal from the body through the reticuloendothelial system (RES). Consequently, there is an intricate balance between optimal cationic condensation of pDNA without excess cationic charge. PEI architecture comes in one of two forms, BPEI or LPEI. Both of these geometries have certain advantages and disadvantages, but the majority of PEI research reported in the literature has been conducted with BPEI, possibly due to its increased number of primary amines that are widely available for conjugation to other modalities.

Branched Poly(ethyleneimine) BPEI

Background

Commercially available BPEI is commonly produced by cationic ring-opening polymerization resulting in molecular weights from 1200 Da to 800 kDa. While displaying a high cationic density, BPEI particles are polyelectrolytes that are attracted to anionic surfaces, proteins, and corresponding anionic polyelectrolytes such as DNA. The synthesis of PEI results in the formation of a highly branched PEI with primary, secondary, and tertiary amines. BPEI is also unique in that it is readily soluble in both organic solvents (e.g., methylene chloride and chloroform) and water. The branching points of this polymer lie at the tertiary amines and the relative amounts of amines are 1:2:1 primary: secondary:tertiary amines. The pKa of the primary amines of BPEI (~9.65) allows for efficient pDNA condensation; moreover, the number of protonated amines in BPEI increases from 19 to 46% at pH 7 and pH 5, respectively, indicating pH buffering capability at the secondary and tertiary amine sites *(7)*. However, these same cationic groups also provide areas for unfavorable electrostatic interactions with negatively charged serum proteins such as albumin. In addition, the excess cationic charge may also destabilize the cell membrane, thereby inducing local toxicity and cell death. The secondary and tertiary amines are capable of further protonation, the mechanism of which is exploited for endosomal release. An endocytosed BPEI/pDNA complex with a net cationic charge will first be bombarded by protons because of a decrease in pH within the lysosomal compartment. These protons will be buffered by the "proton sponge effect", in which the secondary and tertiary amines of the polymer will become protonated *(8)*. The ability of PEI to act as a weak base allows for the endosome to osmotically swell and rupture due to the rapid influx of chloride counterions; consequently, the polymer/pDNA complexes and free pDNA are released from the deleterious environment of the lysosome.

Modifications of BPEI

In order to disperse surface charge, poly(ethylene glycol) (PEG) has been conjugated to BPEI. In addition, small MW BPEI has been linked with bio-degradable (hydrolytically labile) bonds in order to diffuse the cationic charge. Furthermore, BPEI does not contain active groups for tissue or cell targeting, and therefore, researchers have attached a number of targeting modalities (both chemical and biological) to move beyond passive targeting. A review of the literature shows BPEI modification with numerous groups including the following:

- acetylation *(9)*
- antiCD3 *(10)*
- antiGD2 *(11)*
- biodegradable *(12,13)*
- cetylation *(14)*
- cholesterol *(15-20)*
- diacrylate cross-linkers *(13)*
- Fab′ antibody fragment *(21)*
- folate *(22)*

- glycosylation *(11,23-33)*
- melittin *(34)*
- palmitoylation *(20)*
- Pluronic *(35)*
- poly(NIPAAM) *(36,37)*
- PEG *(11,20-22,35,38-44)*
- thiolation *(45)*
- transferrin *(44,46-51)*

A more detailed listing of known modifications to BPEI for *in vitro* studies is shown in Table I, showing the derivative and degree of modification in either molar ratios or % amine conjugated.

Table I Chemical and biological modifications to BPEI for *in vitro* studies.

BPEI MW [Da]	Derivative and Modification Degree (PEI:derivative molar ratio, otherwise % modified PEI)	Transfected Cell Line	pDNA	Ref
25k	Acetylation (15-43%)	C2C12 et al.	pLuc	(9)
600-25k	Cetylation (1:3)	COS-1	pGFP	(14)
1800	Cholesterol (1:1)	A7R5	pLuc	(17)
1800	Cholesterol (1:0.5)	CT26 et al.	pLuc	(15)
1800-10k	Cholesterol (1:1) (2° amine)	Jurkat	pEGFP	(19)
25k	Dextran (1:10-75)	-	pEGFP	(33)
800	Diacrylate cross-linkers	-	-	(13)
25k	Folate-PEG-Folate (1:2.3)	SMC, CT26	pLuc pmIFN-γ	(22)
25k	Glycosylation (5%)	ΣCFTE29o-	pLuc,	(28)
25k	Glycosylation (5-25%) antiGD2 (25%)	BNL Cl.2 HeLa, et al.	pLuc	(11)
25k	Melittin (1:10)	K562 et al.	pLuc	(34)
25k	Palmitoylation (14%) PEGylation/palmitoylation (8% PA, 3% PEG) Cholesterol (1:0.5)	A431 A549	pEGFP	(20)
25k	PEGyation (350-1.9k) (1:2-57)	C3	pLuc	(43)
25k	PEGylation (2k) (1:10) Fab′ (1:0.5)	OVCAR-3	pLuc	(21)
25k	PEGylation (2k) (1:1-14.5)	NT2	pLuc	(40)
25k	PEGylation (550) (14-22%)	3T3, HepG2	pβ-gal	(42)
25k	PEGylation (550-20k) (1:1-35)	-	-	(41)
2k	PEGylation (8k) (1:1.7)	-	ODN	(39)
25k	PEGylation 550-20k (1:1-35)	3T3	pLuc	(38)
2k-25k	PEO (1:1-1.7) Pluronic 123 (1:37-61)	COS-7	pLuc	(35)
2k-25k	poly(NIPAAM)	HeLa	pEGFP	(36, 37)
25k	Thiolation (15%)	A549	pLuc	(45)
800k	Transferrin	B16F10 et al.	pLuc, phIL-2	(48)
800k	Transferrin AntiCD3	N2A, B16, et al.	pmIL-2	(10)

The most frequent modification of BPEI found in the literature is the incorporation of PEG. The use of PEG in nonviral gene carriers may be for any of the following reasons: (i) increased safety profile, (ii) decreased protein adsorption, (iii) targeting ligand spacer, and (iv) increased pDNA condensation through volume exclusion *(52)*. Kissel *et al.* have extensively incorporated PEG into PEI-mediated gene transfer studies *(38,53,54)*. In one study, activated linear PEG (MW 550 Da-20 kDa) was grafted to BPEI (MW 25 kDa) in various degrees of substitution; moreover, the degree of PEGylation and MW of the PEG were found to affect pDNA condensation and the bioactivity of the BPEI-*g*-PEG/pDNA complexes *(38)*. Cellular internalization of PEI primarily occurs through two mechanisms: pinocytosis and adsorptive endocytosis; however, with the use of targeting ligands, receptor-mediated endocytosis is also possible. The success of BPEI modification led to a number of *in vivo* studies for biodistribution, tumor regression, and biocompatibility. Table II details a number of *in vivo* studies, including the animal/disease model, pDNA vector of choice, and optimal ratio (e.g. N/P) between the polymer and pDNA.

Linear Poly(ethyleneimine) LPEI

Background

An alternative to BPEI is LPEI, in which the degree of excess cationic charge is somewhat controlled at physiological pH. Whereas BPEI has been modified by a number of groups for the reasons previously outlined *(11,22,55-59)*, the modifications of LPEI are few in number but have included cholesterol conjugation *(60-65)* and attachment to poly(D,L-lactide-*co*-glycolide) *(66)*. The synthesis of LPEI is more complicated in that the linear conformation requires cessation and purification of the polymer at multiple steps in the synthesis in addition to a lower temperature.

In 1972 the first successful attempt of LPEI synthesis was reported, describing the unique solubility characteristics of LPEI in that it was insoluble in cold water but soluble in hot water. The intermolecular hydrogen bonds between LPEI backbones prevented solubility in water at normal conditions; consequently, heat was needed to break the intermolecular hydrogen bonds and replace them with hydrogen bonds to water molecules. Alternatively, protonation at low pH could break these intermolecular bonds. The differences between the methylene and amine hydrogens were seen with ^{1}H-NMR that showed δ 2.70 (4H) (methylene) and δ 1.57 (1H) (secondary amine) *(67)*. LPEI consists entirely of secondary amines that act as both proton donors and acceptors. The solubility behavior of LPEI is much different from BPEI in that LPEI precipitates in water as various insoluble crystallohydrates *(68-70)*.

Table II. *In vivo* model studies for modified BPEI gene carriers.

BPEI MW	Modification and Degree	Animal Model (animal; R.A., study)	pDNA (amount)	Ref
1800	Cholesterol (1:1)	BALB/c mice; S.C., CT-26 tumor		
1800	Cholesterol (1:1)	New Zealand white rabbit; LV injection	pLuc (20 µg)	(17)
25k	Palmitoylation (14%) PEGylation/palmitoylation (8% PA, 3% PEG) Cholesterol (1:0.5)	CD-1 mice; IV injection for biocompatibility and liver expression	pEGFP (50 µg)	(20)
25k	PEGylation (2k) (1:1-14.5)	Wistar rats Lumbar injection	pLuc (4 µg)	(40)
2k – 25k	PEO (1:1-1.7) Pluronic 123 (1:37-61)	C57Bl/6 mice I.V., Biodistribution	pLuc (50 µg)	(35)
800k	Transferrin	A/J mice with S.C. Neuro2a tumors I.V. injection	pLuc (50 µg)	(46, 47)
800k	Transferrin	A/J mice with SC Neuro2a or M3 tumors; intratumoral inj.	pluc (20 µg)	(47)
25k	Transferrin	A/J mice with SC Neuro2a tumors; DBA/2 mice with SC M-3 tumors, BALB/c mice with SC MethA cells	pLuc PTNF-α (50 µg)	(49)
25k-800k	Transferrin	B6/CBA/F1-hybride mice; in utero injection	pLuc (3 µg)	(51)
25k	Transferrin (1:1-1.1)	A/J mice with Neuro2a tumors; I.V. injection, biodistribution	pLuc pβ-gal (50 µg)	(50)
25k	Transferrin with/without PEGylation	A/J mice with SC Neuro2a tumors; I.V. injection, optical imaging	pLuc (50 µg)	(44)

Anhydrous LPEI was the first synthetic polymer found to be a double helix with an NH-group core and a hydrophobic methylene shell *(69)*. Upon hydration and absorption of water, LPEI extends into a fully extended trans- or zigzag-form. In addition to the hydrogen bonding capability of LPEI, it is also limited by its formation of hydrate compounds, in which it was discovered that water molecules place themselves in between the LPEI backbones after heating or dropping the pH *(71)*. Despite these shortcomings, LPEI shows favorable characteristics beyond that of BPEI of the same molecular weight, including:

1. Higher transfection efficiency than BPEI *(72)*
2. Lower cell cycle dependence than BPEI *(72)*
3. LPEI is better tolerated *in vivo* than BPEI *(50)*
4. Transgene expression in pneumocytes beyond the pulmonary epithelium *(73)*
5. LPEI/pDNA complexes range in size from 100 – 275 nm.

Thus, LPEI displays physical and biological characteristics different than BPEI, which has led to further developments with passive and active targeting characteristics, biodegradability, and PEGylation.

Modifications of LPEI

The number of modifications with LPEI in the literature lacks the numbers compared with BPEI; however, some advances have been made in this area. The most striking difference between LPEI and BPEI chemical derivatives is the number of PEGylation studies with LPEI. Due to the decreased toxicity seen with LPEI compared to BPEI carriers *(63)*, we hypothesize that PEGylation is not indicated with LPEI as it is with BPEI. The small number of LPEI derivatives includes only chemical modifications and include the following:

- cholesterol *(60-65)*
- glycosylation *(33)*
- poly(D,L-lactide-*co*-glycolide) (PLGA) *(66)*
- poly(NIPAAM) *(37)*

Further details are provided in Table III for these LPEI modifications, including the degree of chemical coupling for *in vitro* studies. Table IV shows *in vivo* LPEI derivatives. The number of reported modifications of LPEI is only beginning to show, with such species as PEGylated LPEI, biodegradable LPEI, and cholesterol-modified LPEI *(60-64)*. LPEI derivatives containing various monomer amounts of ethylenimine or *N*-propylethylenimine have also been developed *(74)*. The majority of LPEI modifications involve both cholesterol-modified LPEI MW 25 kDa and ExGen 500, a LPEI MW 22 kDa molecule, manufactured by PolyPlus-Transfection, France. LPEI MW 22 kDa has also been mixed with BPEI, conjugated to transferrirn (Tf-BPEI), for increased *in vivo* expression *(50)*. ExGen 500 has since been modified with targeting groups and nuclear localization sequences (NLS) and is supplied by Qbiogene (Carlsbad, CA). PEGylation of PEI gene carriers can decrease or abrogate the immunogenic response of bound anionic proteins and thus increase the circulation time of the polyplexes.

Table III. Modifications to LPEI for *in vitro* studies.

LPEI MW	Derivative and Modification Degree (PEI:derivative molar ratio, otherwise % modified PEI)	Transfected Cell Line	pDNA	Ref
423	Cholesterol (1:1)	CT-26	pmIL-12	(60)
25k	Cholesterol, (1:1-2)	B16-F0 Renca	pmIL-12	(62-64)
25k	Dextran [1:1-10]	-	pEGFP	(33)
423	PLGA [1:0.90-0.97]	HaCat	-	(66)
25k	poly(NIPAAM)	HeLa	pEGFP	(37)

Table IV. Modified LPEI derivatives·used for *in vivo* studies.

LPEI MW	Modification and Degree	Animal Model (animal; R.A., study)	pDNA (amount)	Ref
423	Cholesterol [1:1]	BALB/c; IV injection, biodistribution	pmIL-12 (20 µg)	(61, 62)
25k	Cholesterol [1:1-2]	BALB/c mice bearing Renca SC tumors and induced pulmonary metastases (Renca)	pmIL-12 (20 µg)	(62, 65)

To further explore the biophysical basis for behavioral differences between BPEI and LPEI, we performed a head-to-head comparison between BPEI and LPEI of the same MW, 25 kDa. LPEI was modified with cholesterol (LPC) in three different geometries: linear, T-shaped, and a combined linear/T-shaped. These carriers were complexed with pmIL-12 and transfected B16-F0 (murine melanoma) cells that are known to be easily transfected with PEI *(48)*. From this study we found a significant difference at the ratio of nitrogen residues to phosphate residues (N/P) 20/1 between transgene expression with LPEI, LPC, and BPEI, as shown in Table V. There were no apparent effects with BPEI-cholesterol conjugates. Similar results were obtained with *in vitro* Renca cell transfection *(63)*, the cell line used in subsequent *in vivo* tumor studies: S.C. tumors treated with local (peritumoral) injections and induced pulmonary metastases treated with intravenous injections. Recently, we have shown linear poly(ethylenimine)-cholesterol (LPC) complexed with mIL-12 revealed cessation and regression of neoplasm growth in ectopic experimental tumors in mice (65). Furthermore, these studies showed a substantial and significant ($p < 0.001$) (ANOVA) drop in the number of pulmonary metastases following weekly injections of 20 mg pDNA, complexed with a number of carriers. The distribution of the samples (n=5) from the *in vivo* pulmonary metastases study is shown in Figure 1.

Table V. Protein expression levels of mIL-12 p70 following transfection (N/P 20/1, 20 mg) of B16-F0 and Renca cells. Data reported as mean ± SD, n = 8 ($p < 0.05$ ANOVA), adapted from *(63)*.

	B16-F0	Renca
Sample	mIL-12 p70 (pg/mL)	mIL-12 p70 (pg/mL)
Naked pmIL-12	230 ± 80	93 ± 13
BPEI/pmIL-12	2581 ± 62	149 ± 86
LPEI/pmIL-12	4667 ± 38	321 ± 36
LPC-L	5393 ± 138	1204 ± 368
LPC-T	4899 ± 348	1975 ± 201
LPC-LT	482 ± 135	-

Conclusions

PEI-mediated gene delivery has been successful in a wide variety of cell lines and a number of *in vivo* systems. The majority of research has been focused on modifying BPEI, but with the recent successes of LPEI-mediated gene transfer, modified LPEI is emerging as a strong alternative. Our previous studies with cholesterol conjugation provided substantial changes in LMW BPEI uptake with a number of cell lines *(15,17,18)*; moreover, these effects provided the motivation to evaluate cholesterol conjugation to LPEI. From these PEI modifications we are better able to treat a wide range of diseases including cancer, diabetes, and ischemic myocardium. As clinicians begin to turn more towards molecular medicine, nonviral gene therapy led by modified PEI gene carriers will be at the forefront.

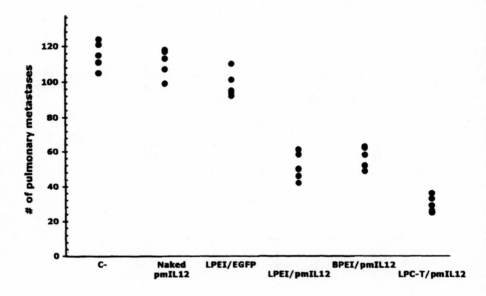

*Figure 1. Distribution of samples in the pulmonary metastases regression study.
BALB/c mice bearing Renca induced pulmonary metastases were injected
weekly with 20 mg of pDNA complexed with the listed carriers. Data reported
are individual points at N/P 20/1 following cessation of the study, n = 5 (p<0.05
ANOVA), adapted from (65).*

193

References

1. Wells, D.J.; Wells, K.E. Gene transfer studies in animals: what do they really tell us about the prospects for gene therapy in DMD? *Neuromuscul. Disord.* **2002**, *12 Suppl 1*, S11-22.
2. Ferrari, S.; Geddes, D.M.; Alton, E.W. Barriers to and new approaches for gene therapy and gene delivery in cystic fibrosis. *Adv. Drug Deliv. Rev.* **2002**, *54*, 1373-1393.
3. Smith, A.C. *et al.* Interstitial deletion of (17)(p11.2p11.2) in nine patients. *Am. J. Med. Genet.* **1986**, *24*, 393-414.
4. http://www.wiley.co.uk/genmed/clinical/ accessed on May 17, 2004.
5. Teichler Zallen, D. U.S. gene therapy in crisis. *Trends Genet.* **2000**, *16*, 272-275.
6. Dean, J.A. Lange's Handbook of Chemistry, McGraw-Hill, Inc., New York, (1992).
7. Suh, J.; Paik, H.-J.; Hwang, B.K. Ionization of poly(ethylenimine) and poly(allylamine) at various pH's. *Bioorg. Chem.* **1994**, *22*, 318-327.
8. Boussif, O. *et al.* A versatile vector for gene and oligonucleotide transfer into cells in culture and *in vivo*: polyethylenimine. *Proc Natl Acad Sci USA* **1995**, *92*, 7297-7301.
9. Forrest, M.L. *et al.* Partial acetylation of polyethylenimine enhances in vitro gene delivery. *Pharm. Res.* **2004**, *21*, 365-371.
10. Kircheis, R., *et al.* Coupling of cell-binding ligands to polyethylenimine for targeted gene delivery. *Gene Ther* **1997**, *4*, 409-418.
11. Erbacher, P. *et al.* Transfection and physical properties of various saccharide, poly(ethylene glycol), and antibody-derivatized polyethylenimines (PEI). *J. Gene Med.* **1999**, *1*, 210-222.
12. Ahn, C.H. *et al.* Biodegradable poly(ethylenimine) for plasmid DNA delivery. *J Control Release* **2002**, *80*, 273-282.
13. Forrest, M.L.; Koerber, J.T.; Pack, D.W. A degradable polyethylenimine derivative with low toxicity for highly efficient gene delivery. *Bioconjug Chem* **2003**, *14*, 934-940.
14. Yamazaki, Y. *et al.* Polycation liposomes, a novel nonviral gene transfer system, constructed from cetylated polyethylenimine. *Gene Ther* **2000**, *7*, 1148-1155.
15. Han, S.-O.; Mahato, R.I.; Kim, S.W. Water-soluble lipopolymer for gene delivery. *Bioconj Chem* **2001**, *12*, 337-345.
16. Yockman, J.W. *et al.* Tumor regression by repeated intratumoral delivery of water soluble lipopolymers/p2CMVmIL-12 complexes. *J Control Release* **2003**, *87*, 177-186.
17. Lee, M. *et al.* Water-soluble lipopolymer as an efficient carrier for gene delivery to myocardium. *Gene Ther* **2003**, *10*, 585-593.

18. Mahato, R.I. *et al.* Intratumoral delivery of p2CMVmIL-12 using water-soluble lipopolymers. *Mol Ther* **2001**, *4*, 130-138.
19. Wang, D.A. *et al.* Novel branched poly(ethylenimine)-cholesterol water-soluble lipopolymers for gene delivery. *Biomacromolecules* **2002**, *3*, 1197-1207.
20. Brownlie, A.; Uchegbu, I.F.; and Schatzlein, A.G. PEI-based vesicle-polymer hybrid gene delivery system with improved biocompatibility. *Int J Pharm* **2004**, *274*, 41-52.
21. Merdan, T. *et al.* Pegylated polyethylenimine-Fab' antibody fragment conjugates for targeted gene delivery to human ovarian carcinoma cells. *Bioconjug Chem* **2003**, *14*, 989-996.
22. Benns, J.M. *et al.* Folate-PEG-folate-graft-polyethylenimine-based gene delivery. *J Drug Target* **2001**, *9*, 123-139.
23. Bettinger, T.; Remy, J.S.; Erbacher, P. Size reduction of galactosylated PEI/DNA complexes improves lectin-mediated gene transfer into hepatocytes. *Bioconjug Chem* **1999**, *10*, 558-561.
24. Diebold, S.S. *et al.* Mannose polyethylenimine conjugates for targeted DNA delivery into dendritic cells. *J Biol Chem* **1999**, *274*, 19087-19094.
25. Diebold, S.S. *et al.* Efficient gene delivery into human dendritic cells by adenovirus polyethylenimine and mannose polyethylenimine transfection. *Hum Gene Ther* **1999**, *10*, 775-786.
26. Fajac, I.; Briand, P.; Monsigny, M. Gene therapy of cystic fibrosis: the glycofection approach. *Glycoconj J* **2001**, *18*, 723-729.
27. Fajac, I. *et al.* Uptake of plasmid/glycosylated polymer complexes and gene transfer efficiency in differentiated airway epithelial cells. *J Gene Med* **2003**, *5*, 38-48.
28. Grosse, S. *et al.* Lactosylated polyethylenimine for gene transfer into airway epithelial cells: role of the sugar moiety in cell delivery and intracellular trafficking of the complexes. *J Gene Med* **2004**, *6*, 345-356.
29. Kunath, K. *et al.* Galactose-PEI-DNA complexes for targeted gene delivery: degree of substitution affects complex size and transfection efficiency. *J Control Release* **2003**, *88*, 159-172.
30. Morimoto, K. *et al.* Molecular weight-dependent gene transfection activity of unmodified and galactosylated polyethyleneimine on hepatoma cells and mouse liver. *Mol Ther* **2003**, *7*, 254-261.
31. Sagara, K.; Kim, S.W. A new synthesis of galactose-poly(ethylene glycol)-polyethylenimine for gene delivery to hepatocytes. *J Control Release* **2002**, *79*, 271-281.
32. Zanta, M.A. *et al.* In vitro gene delivery to hepatocytes with galactosylated polyethylenimine. *Bioconjug Chem* **1997**, *8*, 839-844.
33. Tseng, W.C.; Jong, C.M. Improved stability of polycationic vector by dextran-grafted branched polyethylenimine. *Biomacromolecules* **2003**, *4*, 1277-1284.

34. Ogris, M. *et al.* Melittin enables efficient vesicular escape and enhanced nuclear access of nonviral gene delivery vectors. *J Biol Chem* **2001**, *276*, 47550-47555.
35. Nguyen, H.K. *et al.* Evaluation of polyether-polyethyleneimine graft copolymers as gene transfer agents. *Gene Ther* **2000**, *7*, 126-138.
36. Dincer, S.; Tuncel, A.; Piskin, E. A potential gene delivery vector: N-isopropylacrylamide-ethyleneimine block copolymers. *Macromolecular Chemistry and Physics* **2002**, *203*, 1460-1465.
37. Turk, M. *et al.* In vitro transfection of HeLa cells with temperature sensitive polycationic copolymers. *J Control Release* **2004**, *96*, 325-340.
38. Petersen, H. *et al.* Polyethylenimine-graft-poly(ethylene glycol) copolymers: influence of copolymer block structure on DNA complexation and biological activities as gene delivery system. *Bioconjug Chem* **2002**, *13*, 845-854.
39. Vinogradov, S.V.; Bronich, T.K.; Kabanov, A.V. Self-assembly of poly-amine-poly(ethylene glycol) copolymers with phosphorothioate oligonucleotides. *Bioconjug Chem* **1998**, *9*, 805-812.
40. Tang, G.P. *et al.* Polyethylene glycol modified polyethylenimine for improved CNS gene transfer: effects of PEGylation extent. *Biomaterials* **2003**, *24*, 2351-2362.
41. Petersen, H. *et al.* Synthesis, characterization, and biocompatibility of poly-ethylenimine-*graft*-poly(ethylene glycol) block copolymers. *Macromolecules* **2002**, *35*, 6867-6874.
42. Choi, J.H. *et al.* Effect of poly(ethylene glycol) grafting on polyethylenimine as a gene transfer vector *in vitro*. *Bull Korean Chem Soc* **2001**, *22*, 46-52.
43. Sung, S.-J. *et al.* Effect of polyethylene glycol on gene delivery of polyethylenimine. *Biol Pharm Bull* **2003**, *26*, 492-500.
44. Hildebrandt, I.J. *et al.* Optical imaging of transferrin targeted PEI/DNA complexes in living subjects. *Gene Ther* **2003**, *10*, 758-764.
45. Carlisle, R.C. *et al.* Polymer-coated polyethylenimine/DNA complexes designed for triggered activation by intracellular reduction. *J Gene Med* **2004**, *6*, 337-344.
46. Kircheis, R. *et al.* Tumor-targeted gene delivery: an attractive strategy to use highly active effector molecules in cancer treatment. *Gene Ther* **2002**, *9*, 731-735.
47. Kircheis, R. *et al.* Polycation-based DNA complexes for tumor-targeted gene delivery *in vivo*. *J. Gene Med.* **1999**, *1*, 111-120.
48. Wightman, L. *et al.* Development of transferrin-polycation/DNA based vectors for gene delivery to melanoma cells. *J. Drug Target.* **1999**, *7*, 292-303.

49. Kircheis, R. *et al.* Tumor-targeted gene delivery of tumor necrosis factor-alpha induces tumor necrosis and tumor regression without systemic toxicity. *Cancer Gene Ther* **2002**, *9*, 673-680.
50. Kircheis, R. *et al.* Polyethylenimine/DNA complexes shielded by transferrin target gene expression to tumors after systemic application. *Gene Ther* **2001**, *8*, 28-40.
51. Gharwan, H. *et al.* Nonviral gene transfer into fetal mouse livers (a comparison between the cationic polymer PEI and naked DNA). *Gene Ther* **2003**, *10*, 810-817.
52. Bloomfield, V.A. DNA condensation. *Curr. Opin. Struct. Biol.* **1996**, *6*, 334-341.
53. Fischer, D. *et al.* A novel nonviral vector for DNA delivery based on low molecular weight, branched polyethylenimine: effect of molecular weight on transfection efficiency and cytotoxicity. *Pharm. Res.* **1999**, *16*, 1273-1279.
54. Kunath, K. *et al.* Low-molecular-weight polyethylenimine as a nonviral vector for DNA delivery: comparison of physicochemical properties, transfection efficiency and in vivo distribution with high-molecular-weight polyethylenimine. *J. Control. Release* **2003**, *89*, 113-125.
55. Suh, W. *et al.* An angiogenic, endothelial-cell-targeted polymeric gene carrier. *Mol. Ther.* **2002**, *6*, 664-672.
56. Benns, J.M.; Mahato, R.I.; Kim, S.W. Optimization of factors influencing the transfection efficiency of folate-PEG-folate-graft-polyethylenimine. *J. Control. Release* **2002**, *79*, 255-269.
57. Ogris, M. *et al.* PEGylated DNA/transferrin-PEI complexes: reduced interaction with blood components, extended circulation in blood and potential for systemic gene delivery. *Gene Ther.* **1999**, *6*, 595-605.
58. Sagara, K.; Kim, S.W. A new synthesis of galactose-poly(ethylene glycol)-polyethylenimine for gene delivery to hepatocytes. *J. Control. Release* **2002**, *79*, 271-281.
59. Blessing, T. *et al.* Different strategies for formation of pegylated EGF-conjugated PEI/DNA complexes for targeted gene delivery. *Bioconjug. Chem.* **2001**, *12*, 529-537.
60. Furgeson, D.Y. *et al.* Novel water insoluble lipoparticulates for gene delivery. *Pharm Res* **2002**, *19*, 382-390.
61. Furgeson, D.Y. *et al. Biodistribution of labeled water insoluble lipoparticulates (ISLP) and plasmid DNA after murine tail vein injection. Proc. Controlled Release Society*, Seoul, South Korea, 2002.
62. Furgeson, D.Y. Structural and functional effects of polyethylenimine gene carriers. PhD Dissertation, *Pharmaceutics and Pharmaceutical Chemistry*, University of Utah, Salt Lake City, 2003.
63. Furgeson, D.Y. *et al.* Modified linear PEI-cholesterol conjugates for DNA complexation. *Bioconjug. Chem.* **2003**, *14*, 840-847.

64. Furgeson, D.Y.; Kim, S.W. Linear PEI-cholesterol conjugates for the LDL-R pathway. *Molecular Therapy* **2003**, *7*, S372-S373.
65. Furgeson, D.Y. *et al.* Tumor efficacy and biodistribution of linear polyethylenimine-cholesterol/DNA complexes. *Molecular Therapy* **2004**, *9*, 837-848.
66. Nam, Y.S. *et al.* New micelle-like polymer aggregates made from PEI-PLGA diblock copolymers: micellar characteristics and cellular uptake. *Biomaterials* **2003**, *24*, 2053-2059.
67. Saegusa, T.; Ikeda, H.; Fujii, H. Crystalline polyethyelnimine. *Macromolecules* **1972**, *5*, 108.
68. Chatani, Y. *et al.* Structural studies of poly(ethylenimine). 1. Structures of two hydrates of poly(ethylenimine): sesquihydrate and dihydrate. *Macromolecules* **1981**, *14*, 315-321.
69. Chatani, Y. *et al.* Structural studies of poly(ethylenimine). 2. Double-stranded helical chains in the anhydrate. *Macromolecules* **1982**, *15*, 170-176.
70. Tanaka, R. *et al.* High molecular weight linear poly(ethylenimine) and poly(*N*-methylethylenimine). *Macromolecules* **1983**, *16*, 849-853.
71. Aksenov, S.I. *et al.* NMR spin-echo study of the effect of hydration on the mobility and conformation of linear polyethyleneimine. *Biofizika* **1976**, *21*, 44-49.
72. Brunner, S. *et al.* Overcoming the nuclear barrier: cell cycle independent nonviral gene transfer with linear polyethylenimine or electroporation. *Mol. Ther.* **2002**, *5*, 80-86.
73. Goula, D. *et al.* Rapid crossing of the pulmonary endothelial barrier by polyethylenimine/DNA complexes. *Gene Ther.* **2000**, *7*, 499-504.
74. Brissault, B. *et al.* Synthesis of linear polyethylenimine derivates for DNA transfection. *Bioconjug. Chem.* **2003**, *14*, 581-587.

Chapter 14

Nonviral Gene Therapy with Surfactants

Preeti Yadava, David Buethe and Jeffrey A. Hughes[*]

Department of Pharmaceutics, University of Florida, Gainseville, FL 32610
*Corresponding author: hughes@cop.ufl.edu

With the advantages of nonviral gene delivery vectors over
viral system having been established, the next challenge is to
devise a safe and efficient delivery system. Producing defined
and reproducible particles for gene transfection will be an
important step in this direction. In addition, compacting plasmid DNA (pDNA) into a small size is a fundamental necessity
for efficient nucleic acid delivery. Since plasmid DNA is a
substrate to numerous DNases, the transport vector must be
capable of protecting the plasmid in addition to facilitating the
transport of the plasmid from the administration site to the
nucleus of the desired cell type. Surface-active materials such
as cationic lipids, acting as a vector, can perform both of these
functions. They are also advantageous in terms of toxicity
profiles and manufacturing concerns but often result in a
heterogeneous population of particles after the condensation
event. Cationic detergents with sulfhydryl groups may also be
used as a plasmid condensing and transfection agent, at a
concentration below their critical micelle concentration
(CMC). The condensation reaction is governed by a thermodynamic process resulting in small homogenous pDNA
condensates with unique transfection properties.

Introduction

Gene therapy is an approach for treating or preventing genetic disease by correcting the defective genes responsible for the disease. A genetic disorder is a pathological condition caused by abnormalities in an individual's genetic material. Genetic disorders may be of four types: single-gene, multifactorial, chromosomal, and mitochondrial. Conditions or disorders that arise from mutations in a single gene are the best candidates for gene therapy. Gene therapy also aims at using the human body as a 'Bioreactor' or a biological factory, where the body makes the therapeutic protein at high local concentrations over an extended period of time. Gene therapy may be targeted to repair or replace a defective gene within the germ (gamete-forming cells) or the non-germ cells (somatic cells).

Although gene therapy seems like an elegant way of managing genetic disorders, gene delivery faces its own set of hurdles. One potential risk is that of stimulating the immune system, thus reducing the efficiency of gene therapy. Additionally, the immune response makes administration of multiple doses difficult. Other major concerns include formulation, biological stability, delivery, toxicity, controllability and regulation of expression of the composite delivery system. Another major hurdle has been in the effective delivery of the transgene to the target cell, but of course the problems do not end after the corrective gene or plasmid cross the cell membrane. The acidic environment of the endosome, the presence of nucleases in the cytoplasm and the need to cross the nuclear membrane for expression are just some of the barriers. In order for gene therapy to become a permanent cure for any disease, the therapeutic DNA, introduced into target cells, must remain functional and the cells containing the therapeutic DNA must be long-lived and stable. Integrating therapeutic DNA into the genome has its own concerns and the rapidly dividing nature of most cells leads to a short-lived nature of gene therapy. Finding an effective gene therapy for multifactor or multigene disorders such as arthritis, diabetes, heart disease, high blood pressure and Alzheimer's disease is another major challenge faced by scientists, since these diseases are a mixture of multiple genes and the environment. Two approaches commonly used to address these problems include the use of viral vectors and the delivery of plasmid DNA (nonviral).

Results and Discussion

Nonviral Gene Delivery

Common viruses can be re-engineered to be replicative defective, carry a transgene for a therapeutic purpose *(1)* and have proven to be efficient delivery vectors *(2-6).* In many therapeutic settings, the use of viral vectors is superior to nonviral systems in terms of overall efficiency. Particular conditions requiring transient expression in localized areas or the risk of toxicity create a favorable environment for plasmid based systems.

Unlike viral delivery, in which the DNA/RNA that codes for the protein is actually incorporated into the virus' own genome, nonviral or plasmid-based delivery systems are composed of the complexes of drug carriers (e.g. cationic liposomes) and plasmid DNA. Plasmids are circular forms of DNA found in bacteria and other organisms that are used for the transportation of genes into the host cells. It should be noted that the effectiveness of nonviral gene delivery systems might be improved by alteration of the plasmid. These modifications include optimization of promoters, enhancers, poly-A tail *(7),* introns, and the elimination of bacterial nucleic acid segments *(8,9).* Any one of the listed modifications and others can result in a substantial increase of the transgene activity.

Gene delivery with nonviral vectors faces a number of challenges, including the cell membrane. Along with the physical and chemical obstacles, biological barriers are also present, such as immunogenic responses to the vector as well as immune stimulation by certain DNA sequences *(10,11).* In addition, the use of DNA for nonviral gene expression depends on several factors, namely targeted tissue, extracellular degradation, and intracellular degradation *(10,11).* Thus *in vivo* studies using these vectors will be more challenging.

Cationic Lipids Classifications

Cationic lipid/DNA complexes, also known as cationic lipoplexes *(12),* have become popular for gene transfer. Cationic lipids perform many functions as gene delivery vectors. First, through electrostatic interactions, hydrogen bonding and hydrophobic interactions, the cationic lipid coats and condenses the plasmid to form a complex capable of transfection. Second, the excess of cationic molecules is giving rise to an overall positive charge that leads to enhanced association of the lipoplex with the anionic cell surfaces *(13),* thus increasing cellular uptake via endocytosis *(14,15).* Third, once inside the cell the cationic lipid may play a role in destabilizing the endosomal membrane *(16-18),* thus enabling cytoplasmic delivery of the plasmid.

Cationic lipids (Figure 1) may be classified on the basis of their head group. Quaternary ammonium salts, e.g. cetyl trimethylammonium bromide (CTAB) *(19),* N-[1-(2,3-dioleyloxy) propyl]-N,N,N-tri-methylammoniumchloride (DOT MA) *(20),* and 1,2-diacyl-3-trimethylammonium propane (DOTAP) *(21)* form a major class of cationic lipids used in gene delivery (Figure 2). While all the above mentioned examples are monovalent in nature, polyvalent quaternary ammonium salts such as dioctadecylamidoglicylspermin (DOGS or "transfectam") *(22),* DOSPA *(23),* DPPES *(24),* $(C_8)_2$Gly Sper^{3+} and $(C_{18})_2$Sper^{3+} *(25)*

are also commonly used. Ornithinyl-cysteinyl-tetradecylamide (C_{14}-CO) *(26)* is an example of a divalent cysteine-detergent with two quaternary nitrogens as head group.

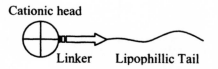

Cationic head

Linker   Lipophillic Tail

Figure 1. Schematic representation of a cationic lipid.

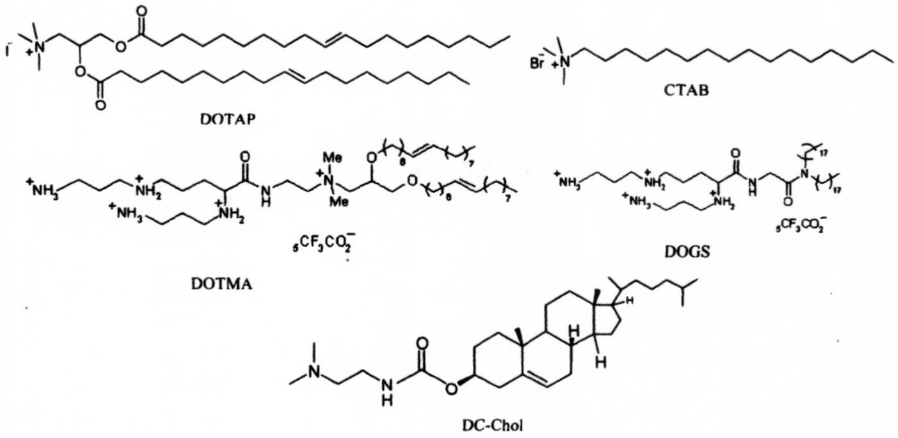

DOTAP

CTAB

DOTMA

DOGS

DC-Chol

Figure 2. Structures of cationic lipids, 1,2-dioleoyloxy-3(trimethylammonio) propane (DOTAP), cetyl trimethylammonium bromide (CTAB), 6-N-[1-(2,3-dioleyloxy)propyl]-N,N,N-trimethylammonium chloride (DOTMA), Dioctadecyl-amidoglicylspermin (DOGS), and α-[N-(N',N'-dimethylaminoethane)-carbamoyl]cholesterol hydrochloride (DC-Chol).

Guanidinium salts represent another commonly encountered class of cationic lipids. The guanidinium moiety has been tested for gene delivery as cyclic guanidines or attached to various alkyl chains by unsaturated glycosides *(27)*. Aminoglycoside and aminoglycoside-derived cationic lipids *(28)* have also been shown to be promising gene transfer agents. Other cationic head groups that have been used include polyamino, polyamidoamine (29,30) and pyridinium (31,32). Cationic lipids may also be grouped on the basis of the tail group they possess. Single-tailed cationic lipids include CTAB and C14-CO, while DOTMA, DOGS and DOTAP are double-tailed. The hydrophobic tail may consist of long carbon chains or cholesterol. Cholesteryl-3β-carboxyamido-ethylenedimethylamine (DC-Chol) (33) is a cholesterol-derived cationic lipid. Other cholesterol derivatives include cholesten-5-yloxy-N-(4-((1-imino-D-thio-galactosyl-ethyl)amino)butyl)formamide (Gal-C4-Chol) and its ethyl formamide and hexylformamide analogues (Gal-C2-Chol, Gal-C6-Chol) (34). The various linkers between the head and the tail group can also be used to classify cationic lipids. Most common linkers include esters (DOTAP), ethers (DOTMA) and amides (DOGS). The linker may also be glycosidic (35), oxyethylene (36,37), diethers, 1,2,4-butanetriol or disulfide (38) in nature. Different types of cationic polymers like poly(ethylenimine) PEI (39), poly(L-lysine) (40-42), protamine (43), chitosan (44) and dendrimers (45,46) have been shown to be a significant part of nonviral gene delivery systems.

Endosome Transport Facilitator Molecules ("Helper Lipids")

It is well known that the chemical structure of cationic lipids can greatly influence the transfection ability of plasmid DNA cationic lipid complexes. It has been proposed that cationic lipids can destabilize lipid bilayers by promoting the formation of nonbilayer structures such as the inverted hexagonal (H_{II}) phase (47-49). In a majority of reported studies, however, cationic liposomes function most efficiently when the cationic lipid is formulated with a helper lipid. Unsaturated phophatidylethanolamines, such as dioleoylphosphatidylethanol-amine (DOPE), are very common. The effectiveness of these helper agents is generally believed to rest on their propensity to form non-bilayer structures that are akin to membrane fusion intermediates. For example, mixtures of the cationic lipid N, N-dioleyl-N,N-dimethylammonium chloride (DODAC) with the anionic lipid cholesteryl hemisuccinate (CHEMS) can form non-bilayer structures such as the hexagonal H_{II} phase under conditions of neutral surface charge. Similarly, mixtures of the cationic lipid 3α-[N-(N',N'-dimethylamino-ethane)-carbamoyl] cholesterol hydrochloride (DC-Chol) and dioleoyl-phosphatidic acid (DOPA) also exhibit H_{II} phase preferences. The ability of cationic lipids to adopt the H_{II} phase structure allows obtaining a measurable parameter to correlate transfection potency of complexes. This parameter is the bilayer-to-H_{II} transition temperature (T_{BH}). The easier the lipid adopts the H_{II} phase, the lower the T_{BH} values. It is interesting to see that cationic lipids with lower T_{BH} values in mixtures with other lipids exhibit more potent transfection properties. In this regard, the transfection potency of cationic lipids has been found to decrease as the saturation of the cationic lipid is increased. The enhanced transfection efficiency observed for unsaturated cationic lipids is thought to be due to their enhanced ability to promote non-bilayer structure in the presence of negatively charged cellular phospholipids (50).

Other endosome transport facilitator molecules such as pH-sensitive peptides (51), e.g. glutamic acid-alanine-leucine-alanine, GALA, the influenza virus hemagglutinin HA2 N-terminal peptide, and pH-sensitive surfactants, e.g. dodecyl-2-(1'-imidazolyl) propionate (DIP) (Figure 3), dodecyl imidazole, and methyl 1-imidazolyl laureate (52) can be included in the transfection complex to increase endosomal escape and transfection. It has been reported that the use of cholesterol as a helper lipid leads to an enhanced transgene expression in animals (53). Besides helper lipids, other agents such as Tween 80 (54) have also been shown to be effective in increasing transgene expression.

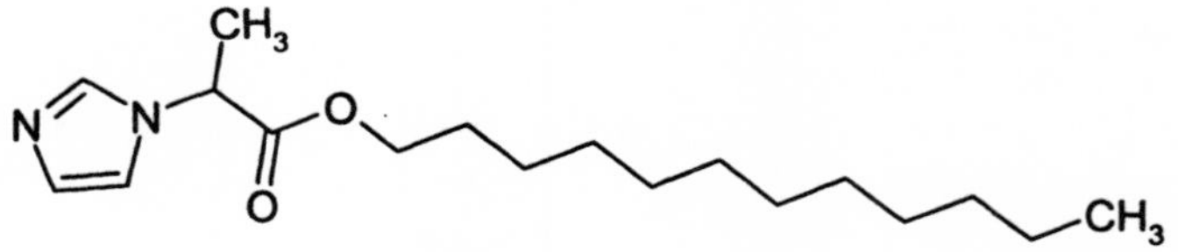

DIP

Figure 3. Dodecyl-2-(1'-imidazolyl) propionate (DIP).

Cationic Lipids Mode of Action

It is generally believed that the mechanism of transfection by synthetic cationic surfactants involves endocytosis (14,55). Xu et al. first proposed a stepwise mechanism for lipoplex entry into a cell via endocytosis, followed by release through destabilization of the endosomal membrane due to the interaction of cationic and anionic lipids (56). Zabner et al. demonstrated that the lipoplex of a 1:1 mixture of N-[1-(2,3-dimyristyloxy)propyl]-N,N-dimethyl-N-(2-hydroxy-ethyl)ammonium bromide (DMRIE) and DOPE crossed the cell membrane by endocytosis (14).

The escape of DNA from the endosome is thought to depend on the ability of some lipids to form a fusogenic inverted hexagonal phase (57,58). Detergents are thought to destabilize the endosomal membrane and induce DNA release into the cytoplasm, in agreement with the good transfection efficiency observed with complexes of DNA bound to vesicles, composed of neutral lipids and cationic detergents (59). The inverted hexagonal phase can be induced by the shape of the cationic detergent (60), the helper lipid (61), or by the interaction of the lipoplex with anionic lipids (50,62). A close association of the protonated surfactant with the phosphate functionalities on the DNA may cause an acid-induced change in morphology of the lipoplex from condensed lamellar to inverted hexagonal. Protonation of the amine groups that get exposed when the bilayer falls apart results in an increase in pH, which occurs along with the phase change *(63)*. Significant protonation of amines causes a large degree of counterion association and increased hydration that brings about an increase in the head group size and favors micelles over bilayers *(64)*. An additional factor that may play an important part in the escape of the DNA from the endosome is the increase in osmotic pressure caused by the increase of dissolved solutes in the endosome.

At a specific concentration, cationic lipids can self-assemble via cooperative hydrophobic intermolecular binding, resulting in the formation of cationic micelles. These micelles can efficiently bind and compact DNA molecules by electrostatic association between the positively charged polar heads of the lipids and the negatively charged phosphate groups of the DNA, thus forming lipoplexes *(65)*. The process is quasi-irreversible and leads to micro-precipitates containing hundreds of DNA molecules per particle. The heterogeneity of the complex makes it difficult to determine what particle species is the most active in transfection. CTAB, a single-tailed cationic lipid (surfactant), has been shown to be an effective transfection reagent *(66-69)*. Although surfactants are believed to have low transfection efficiency and high toxicity in plasmid delivery, they have been shown to function as transfection reagents. Various surfactants e.g. SAINT 2, sugar based cationic Gemini surfactant, C14-CO and GS11 (Figure 4) have been shown to be efficient gene delivery agents.

We synthesized a cationic lipid-oleoyl ornithinate (OLON) and, to decrease cytotoxicity, introduced a biodegradable ester bond in the tail of the lipid to form 6-lauroxyhexyl ornithinate (LHON) (Figure 5). Our data demonstrated that the cytotoxicity of LHON was lower than that of DOTAP or OLON. The transfection efficiency was compared for liposomes produced from (i) double-tailed 1',2'-dioleyl-sn-glycero-3'-succinyl-1, 6-hexanediol ornithine conjugate (DOG SHDO) with an ornithine headgroup, (ii) single-tailed OLON with an ornithine head group, (iii) double-tailed DOTAP with quaternary amine group, and (iv) single-tailed cetyltrimethylammonium bromide (CTAB) with a quaternary amine group. At the optimal ratios OLON/DOPE were shown to result in more than 10-times the transgene expression than other liposomes, even though the DNA uptake was not necessarily greater. This may be explained by the hypothesis that the escape of DNA from endosomes and release of DNA from complexes may be a major barrier of cationic lipid-mediated gene transfection. It was also shown that in the presence of sodium dodecyl sulfate substances a greater fraction of DNA was released from DNA/OLON/DOPE complexes than from DNA/DOTAP/DOPE complexes *(70)*.

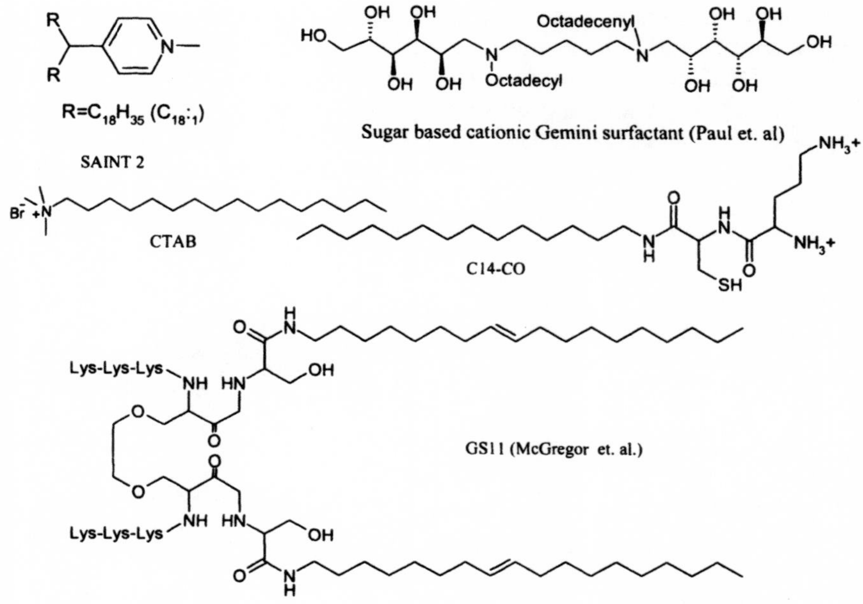

Figure 4. *Structures of SAINT 2, sugar based cationic Gemini surfactant, CTAB, C14-CO and GS11.*

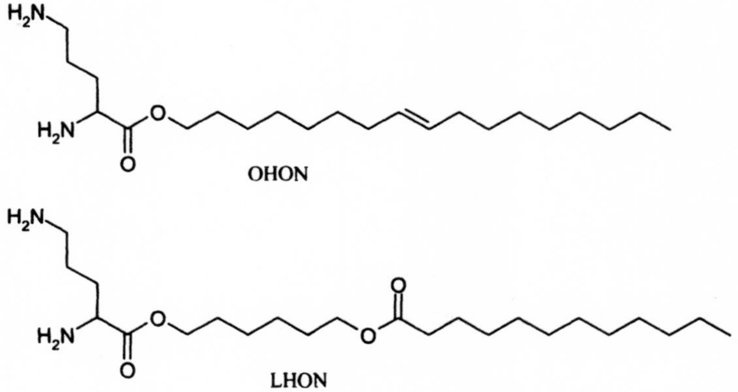

Figure 5. *Cationic lipid-oleoyl ornithinate (OLON) and 6-lauroxyhexyl ornithinate (LHON).*

206

Lipids can induce condensation of DNA in dilute aqueous solution. The polyanion may either be trapped in the core of a liposome, attached to the surface of the lipid bilayer or a combination of both events. DNA condensation by surfactants is greatly dependent on the CMC of the surfactant. As shown in Figure 6, monomolecular collapse of DNA is expected below CMC, while at concentrations above CMC polyanions form much larger complexes with the micelles.

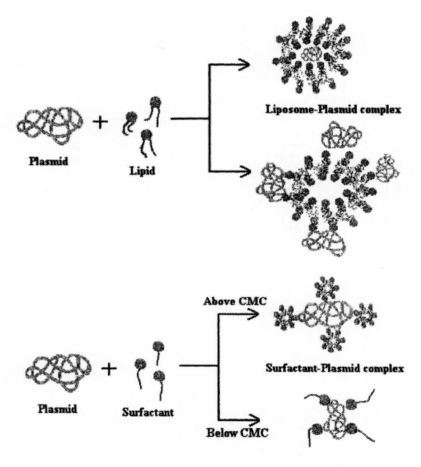

Figure 6. Schematic representation of DNA condensation by lipids and surfactants.

Large complexes (>500 nm) are advantageous for the transfection of cells *in vitro* because they sediment on the cells and carry more plasmid DNA as compared to smaller particles *(71-73)*. However, as expected in animal models their gene-delivery properties are limited, which may correlated to the size-restricted movement and rapid elimination. The size of lipoplexes is mainly dependent on the cationic lipid/DNA charge ratio and not the composition of the lipid or the helper lipid *(74)*. The charge ratio (N/P) is the ratio of the number of nitrogens on the cationic lipid to the number of phosphate groups on the DNA. A neutral charge ratio results in the formation of large aggregates (>1 μm) *(74)*. At a positive charge ratio, the formation of large multilamellar vesicles (LMV) (diameter 300-700 nm) is favored over small unilamellar vesicles (SUV) (diameter 50-200 nm) *(75,76)*. Kennedy *et al.* studied the effect of the order in which DNA and lipid are mixed *(74,77)*. When DNA is added to a lipid, the lipoplex size increases slowly. The opposite order of mixing shows approximately no change in the particle size until the N/P ratio becomes net positive, at which time the particle size increases rapidly *(77)*. Multivalent anions present in the serum or media can facilitate fusion of the lipoplexes, also resulting in an increase in size of the lipoplexes.

It is believed that smaller sized complexes will be resistant to opsonization *(78,79)*, have a longer time in the circulatory system, be able to migrate through the endothelial cells and have better movement when injected into solid tissues. The removal of liposomes from circulation by the reticuloendothilial system (RES) increases with the increase in liposome size *(80)*. Current investigations have used unilamellar vesicles, 50-100 nm in size, for systemic drug delivery applications. Litzinger *et al.* evaluated the biodistribution of three different sizes of liposomes made from dioleoyl-N-(monomethoxy-poly(ethylene glycol) succinyl)-phosphatidylethanolamine (size ranges >300 nm, ~150-200 nm, and <70 nm) and found that the intermediate sized liposomes (~150-200 nm) circulated longest *(81)*. Due to biological constraints, the development of long circulating large (>500 nm) liposomes using steric stabilization methods has not been successful. Hence, consideration of liposome size and its control is vital for efficient drug delivery.

Homogeneous Transfection Complex Size

In light of these factors, designing methods to develop small homogenous transfection particles becomes vital. As mentioned earlier, the formation of transfection particles/complexes results in a heterogeneous mixture of particle sizes *(82,83)* with different abilities to transfect cells. Several methods have been proposed for the production of small monodispersed transfection particles. While rapid mixing can help to produce small particles *(84)*, it might be difficult to develop this approach clinically.

An alternative approach for compacting plasmid DNA into a small size is to condense pDNA by using specific cationic detergents with sulfhydryl groups below their critical micelle concentration (CMC). In the design of surfactant-based transfection reagents it is important to be aware of the CMC. Surfactants with a very low CMC will be difficult to use since they will form micelles that can lead to aggregation when interacting with plasmid DNA. The higher water solubility of detergents over lipids leads to fast DNA release after addition to cells, hence their poor transfection properties. Surfactant-nucleic acid complexes may be stabilized by condensation with a cationic detergent carrying a free sulfhydryl group, which upon oxidation dimerizes to a gemini lipid on the nucleic acid backbone. This procedure combines the advantages of cationic detergents and lipids *(85)*. Disulfide groups also allow release of DNA in the reducing environment inside the cell. The utility of using thermodynamics to produce transfection particles offers advantages over the kinetic approach of bulk mixing of oppositely charged polyelectrolytes.

Since the cell line/tissue types usually influence transfection efficiency, different targets require different vectors. Therefore, expanding the structural diversity of the delivery agents is required. Two detergents, methyl 1-imidazolyl laureate (MIL) and N-dodecyl-2-mercapto imidazole (DMI) (Figure 7), were evaluated for their surface activity by measuring surface tensions of samples under different environmental conditions such as oxidation, reduction and in the presence of a negatively charged molecule, dextran sulfate, which served as a DNA mimic.

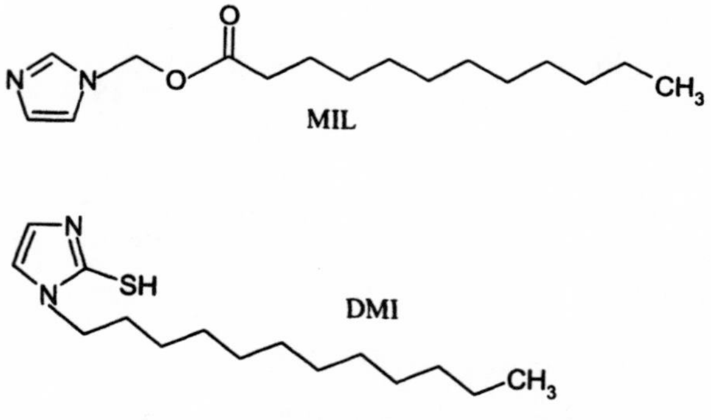

Figure 7. Methyl 1-imidazolyl laureate (MIL) and N-dodecyl-2-mercapto imidazole (DMI).

The CMC values for DMI was found to be around 1 mM, which is 10-times higher than the CMC value of the ornithine-cysteine-based surfactants reported by Behr *et al. (86).* It was also found that DMI forms dimers in the presence of oxygen over time and that both the rate and extent of this interaction can be increased with dextran sulfate. These results may be explained by negatively charged dextran sulfate becoming a template for the oxidative reaction between the two sulfhydryl groups. Thus, the presence of DNA or any DNA-analog would increase the rate of oxidation of cationic surfactants that possess a thiol group. Electron microscopy shows the formation of monodispersed, virus sized particles upon interaction of DMI with pDNA (Figure 8).

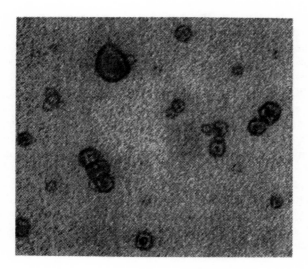

Figure 8. DNA condensation by DMI as seen by electron microscopy (bar = 100 nm).

Conclusions

Surface-active molecules are an attractive method for the transport of nucleic acids inside cells. These versatile systems have been proven to be effective *in vitro.* However, *in vivo* gene transfer is still a challenge and goal of various ongoing studies. Although surface-active agents have been shown to be capable of nuclease protection and facilitating nucleic acid transfer into the cytoplasm, successful clinical use seems a long way. Obvious challenges include toxicity, targeting and efficacy but the versatility of these molecules seems to be the reason that makes their pharmaceutical development appear achievable. Lipoplex-mediated transfection is used in the treatment of cancer, monogenic

210

diseases, especially cystic fibrosis, coronary and peripheral artery diseases, neurological diseases, renal, vision, articular, impairments etc. Biodegradable building blocks, non-toxic counterions, appropriate dose, delivery methods and lipoplex generation techniques are all being investigated in an effort to overcome the problem of cytotoxicity. Studies aiming to correlate the structure of the delivery system with the desired biodistribution of the lipoplex are being undertaken as well.

Although nonviral systems are generally non-infectious and non-hazardous, the major safety issue relates to cytotoxicity *in vivo*. This general toxicity of many cationic lipids is still a major hurdle that needs to be overcome. Lipid toxicity is probably due to the non-natural, non-biodegradable nature of transfection lipids, as well as their ability to disrupt cell membranes. Attempts to counteract cationic lipid toxicity include the use of biodegradable chemical groups, such as pyridinium or ester linkages within the lipid structure. Smaller particles can get into the nucleus of nondividing cells through the nuclear pores, which are 25-50 nm wide, with efficiencies three to four orders of magnitude higher than those seen with noncompacted DNA. In addition, the complexes remain in the circulatory system for a longer period of time because they are resistant to opsonization. This behavior increases the efficiency of the system and reduces the dose required to produce the desired effect, which in turn reduces toxicity.

Small particles are formed with just enough cationic reagents to neutralize the negative charge on the DNA. This results in particles with little or no positive charge. Sometimes the charge may even be negative and, as a result, the particles may not be able to interact with cell membranes. Targeting ligands such as folic acid *(87,88)* may be used to overcome this problem.

Acknowledgment

We would like to thank the NIH (NIH AG 10485) for supporting this work.

References

1. Horellou, P.; Mallet, J. Gene therapy for Parkinson's disease. *Mol Neurobiol.* **1997**, *15*, 241-56.
2. Roth, C.M.; Sundaram, S. Engineering synthetic vectors for improved DNA delivery: insights from intracellular pathways. *Annu Rev Biomed Eng.* **2004**, *6*, 397-426.

3. Tenenbaum, L. et al. Recombinant AAV-mediated gene delivery to the central nervous system. *J Gene Med.* **2004**, *6 Suppl 1*, S212-22.
4. Marti, W.R. et al. Recombinant vaccinia viruses as efficient vectors of biologically active, human B7 costimulation molecules. *Langenbecks Arch Chir Suppl Kongressbd.* **1998**, *115 Suppl I*, p. 131-6.
5. Kootstra, N.A.; Verma, I.M. Gene therapy with viral vectors. *Annu Rev Pharmacol Toxicol.* **2003**, *43*, 413-39.
6. Wilson, D.R. Viral-mediated gene transfer for cancer treatment. *Curr Pharm Biotechnol.* **2002**, *3*, 151-64.
7. Afonina, E.; Stauber, R.; Pavlakis, G.N. The human poly(A)-binding protein 1 shuttles between the nucleus and the cytoplasm. *J Biol Chem.* **1998**, *273*, 13015-21.
8. Tano, K., Foote, R.S.; Mitra, S. High-level expression of the cloned ada gene of Escherichia coli by deletion of its regulatory sequence. *Gene* **1988**, *64*, 305-11.
9. Poppe, C.; Gyles, C.L. Tagging and elimination of plasmids in Salmonella of avian origin. *Vet Microbiol.* **1988**, *18*, 73-87.
10. Ruiz, F.E. et al. A clinical inflammatory syndrome attributable to aerosolized lipid-DNA administration in cystic fibrosis. *Hum Gene Ther.* **2001**, *12*, 751-61.
11. Yew, N.S. et al. Contribution of plasmid DNA to inflammation in the lung after administration of cationic lipid:pDNA complexes. *Hum Gene Ther.* **1999**, *10*, 223-34.
12. Ilies, M.A., Seitz, W.A.; Balaban, A.T. Cationic lipids in gene delivery: principles, vector design and therapeutical applications. *Curr Pharm Des.* **2002**, *8*, 2441-73.
13. Stamatatos, L. et al. Interactions of cationic lipid vesicles with negatively charged phospholipid vesicles and biological membranes. *Biochemistry* **1988**, *27*, 3917-25.
14. Zabner, J. et al. Cellular and molecular barriers to gene transfer by a cationic lipid. *J Biol Chem.* **1995**, *270*, 18997-9007.
15. Wrobel, I.; Collins, D. Fusion of cationic liposomes with mammalian cells occurs after endocytosis. *Biochim Biophys Acta* **1995**, *1235*, 296-304.
16. Zhou, X.; Huang, L. DNA transfection mediated by cationic liposomes containing lipopolylysine: characterization and mechanism of action. *Biochim Biophys Acta* **1994**, *1189*, 195-203.
17. El Ouahabi, A. et al. The role of endosome destabilizing activity in the gene transfer process mediated by cationic lipids. *FEBS Lett.* **1997**, *414*, 187-92.
18. Wattiaux, R. et al. Cationic lipids destabilize lysosomal membrane in vitro. *FEBS Lett.* **1997**, *417*, 199-202.

19. Singh, M. et al. The effect of CTAB concentration in cationic PLG microparticles on DNA adsorption and in vivo performance. *Pharm Res.* **2003**, *20*, 247-51.
20. Felgner, P.L. et al. Lipofection: a highly efficient, lipid-mediated DNA-transfection procedure. *Proc Natl Acad Sci USA* **1987**, *84*, 7413-7.
21. Lardans, V. et al. DNA transfer in a Biomphalaria glabrata embryonic cell line by DOTAP lipofection. *Parasitol Res.* **1996**, *82*, 574-6.
22. Behr, J.P. et al. Efficient gene transfer into mammalian primary endocrine cells with lipopolyamine-coated DNA. *Proc Natl Acad Sci USA* **1989**, *86*, 6982-6.
23. Hofland, H.E.; Shephard, L.; Sullivan, S.M. Formation of stable cationic lipid/DNA complexes for gene transfer. *Proc Natl Acad Sci USA* **1996**, *93*, 7305-9.
24. Mason, J.T.; O'Leary, T.J. Effects of headgroup methylation and acyl chain length on the volume of melting of phosphatidylethanolamines. *Biophys J.* **1990**, *58*, 277-81.
25. Lasic, D.D.; Papahadjopoulos, D. Liposomes revisited. *Science* **1995**, *267*, 1275-6.
26. Dauty, E. et al. Dimerizable cationic detergents with a low cmc condense plasmid DNA into nanometric particles and transfect cells in culture. *J Am Chem Soc.* **2001**, *123*, 9227-34.
27. Herscovici, J. et al. Synthesis of new cationic lipids from an unsaturated glycoside scaffold. *Org Lett.* **2001**, *3*, 1893-6.
28. Belmont, P. et al. Aminoglycoside-derived cationic lipids as efficient vectors for gene transfection in vitro and in vivo. *J Gene Med.* **2002**, *4*, 517-26.
29. Tang, M.X.; Redemann, C.T.; Szoka, Jr., F.C. In vitro gene delivery by degraded polyamidoamine dendrimers. *Bioconjug Chem.* **1996**, *7*, 703-14.
30. Bielinska, A.U. et al. DNA complexing with polyamidoamine dendrimers: implications for transfection. *Bioconjug Chem.* **1999**, *10*, 843-50.
31. Ilies, M.A. et al. Pyridinium cationic lipids in gene delivery: a structure-activity correlation study. *J Med Chem.* **2004**, *47*, 3744-54.
32. van der Woude, I. et al. Novel pyridinium surfactants for efficient, nontoxic in vitro gene delivery. *Proc Natl Acad Sci USA* **1997**, *94*, 1160-5.
33. Yang, K. et al. DC-Chol liposome-mediated gene transfer in rat spinal cord. *Neuroreport* **1997**, *8*, 2355-8.
34. Kawakami, S. et al. Asialoglycoprotein receptor-mediated gene transfer using novel galactosylated cationic liposomes. *Biochem Biophys Res Commun.* **1998**, *252*, 78-83.
35. Bessodes, M. et al. Synthesis and biological properties of new glycosidic cationic lipids for DNA delivery. *Bioorg Med Chem Lett.* **2000**, *10*, 1393-5.

36. Bhattacharya, S.; Dileep, P.V. Cationic oxyethylene lipids. Synthesis, aggregation, and transfection properties. *Bioconjug Chem.* **2004**, *15*, 508-19.
37. Dileep, P.V.; Antony, A.; Bhattacharya, S. Incorporation of oxyethylene units between hydrocarbon chain and pseudoglyceryl backbone in cationic lipid potentiates gene transfection efficiency in the presence of serum. *FEBS Lett.* **2001**, *509*, 327-31.
38. Oupicky, D., Carlisle, R.C.; Seymour, L.W. Triggered intracellular activation of disulfide crosslinked polyelectrolyte gene delivery complexes with extended systemic circulation in vivo. *Gene Ther.* **2001**, *8*, 713-24.
39. Baker, A. et al. Polyethylenimine (PEI) is a simple, inexpensive and effective reagent for condensing and linking plasmid DNA to adenovirus for gene delivery. *Gene Ther.* **1997**, *4*, 773-82.
40. Trubetskoy, V.S. et al. Use of N-terminal modified poly(L-lysine)-antibody conjugate as a carrier for targeted gene delivery in mouse lung endothelial cells. *Bioconjug Chem.* **1992**, *3*, 323-7.
41. Trubetskoy, V.S. et al. Cationic liposomes enhance targeted delivery and expression of exogenous DNA mediated by N-terminal modified poly(L-lysine)-antibody conjugate in mouse lung endothelial cells. *Biochim Biophys Acta* **1992**, *1131*, 311-3.
42. Hashida, M. et al. Targeted delivery of plasmid DNA complexed with galactosylated poly(L-lysine). *J Control Release* **1998**, *53*, 301-10.
43. Sorgi, F.L.; Bhattacharya, S.; Huang, L. Protamine sulfate enhances lipid-mediated gene transfer. *Gene Ther.* **1997**, *4*, 961-8.
44. Erbacher, P. et al. Chitosan-based vector/DNA complexes for gene delivery: biophysical characteristics and transfection ability. *Pharm Res.* **1998**, *15*, 1332-9.
45. Liu, M.; Frechet, J.M.J. Designing dendrimers for drug delivery. *Pharma Sci Technol Today* **1999**, *2*, 393-401.
46. Malik, N. et al. Dendrimers: relationship between structure and biocompatibility in vitro, and preliminary studies on the biodistribution of 125I-labelled polyamidoamine dendrimers in vivo. *J Control Release* **2000**, *65*, 133-48.
47. Hafez, I.M.; Ansell, S.; Cullis, P.R. Tunable pH-sensitive liposomes composed of mixtures of cationic and anionic lipids. *Biophys J.* **2000**, *79*, 1438-46.
48. Jordanova, A.; Lalchev,.; Tenchov, B. Formation of monolayers and bilayer foam films from lamellar, inverted hexagonal and cubic lipid phases. *Eur Biophys J.* **2003**, *31*, 626-32.
49. Rappolt, M. et al. Mechanism of the lamellar/inverse hexagonal phase transition examined by high resolution x-ray diffraction. *Biophys J.* **2003**, *84*, 3111-22.

214

50. Hafez, I.M.; Maurer, N.; Cullis, P.R. On the mechanism whereby cationic lipids promote intracellular delivery of polynucleic acids. *Gene Ther.* **2001**, *8*, 1188-96.
51. Haas, D.H.; Murphy, R.M. Templated assembly of the pH-sensitive membrane-lytic peptide GALA. *J Pept Res.* **2004**, *63*, 451-9.
52. Liang, E. et al. Biodegradable pH-sensitive surfactants (BPS) in liposome-mediated nucleic acid cellular uptake and distribution. *Eur J Pharm Sci.* **2000**, *11*, 199-205.
53. Liu, Y. et al. Factors influencing the efficiency of cationic liposome-mediated intravenous gene delivery. *Nat Biotechnol.* **1997**, *15*, 167-73.
54. Liu, F. et al. New cationic lipid formulations for gene transfer. *Pharm Res.* **1996**, *13*, 1856-60.
55. Remy-Kristensen, A. et al. Role of endocytosis in the transfection of L929 fibroblasts by polyethylenimine/DNA complexes. *Biochim Biophys Acta* **2001**, *1514*, 21-32.
56. Xu, Y.; Szoka, Jr., F.C. Mechanism of DNA release from cationic liposome/DNA complexes used in cell transfection. *Biochemistry* **1996**, *35*, 5616-23.
57. Siegel, D.P.; Epand, R.M. The mechanism of lamellar-to-inverted hexagonal phase transitions in phosphatidylethanolamine: implications for membrane fusion mechanisms. *Biophys J.* **1997**, *73*, 3089-111.
58. Ellens, H. et al. Membrane fusion and inverted phases. *Biochemistry* **1989**, *28*, 3692-703.
59. Pinnaduwage, P.; Schmitt, L.; Huang, L. Use of a quaternary ammonium detergent in liposome mediated DNA transfection of mouse L-cells. *Biochim Biophys Acta* **1989**, *985*, 33-7.
60. Smisterova, J. et al. Molecular shape of the cationic lipid controls the structure of cationic lipid/dioleylphosphatidylethanolamine-DNA complexes and the efficiency of gene delivery. *J Biol Chem.* **2001**, *276*, 47615-22.
61. Koltover, I. et al. An inverted hexagonal phase of cationic liposome-DNA complexes related to DNA release and delivery. *Science* **1998**, *281*, 78-81.
62. Hafez, I.M.; Cullis, P.R. Roles of lipid polymorphism in intracellular delivery. *Adv Drug Deliv Rev.* **2001**, *47*, 139-48.
63. Fielden, M.L. et al. Sugar-based tertiary amino gemini surfactants with a vesicle-to-micelle transition in the endosomal pH range mediate efficient transfection in vitro. *Eur J Biochem* **2001**, *268*, 1269-79.
64. Israelachvili, J.N.; Marcelja, S.; Horn, R.G. Physical principles of membrane organization. *Q Rev Biophys.* **1980**, *13*, 121-200.
65. Felgner, P.L. et al. Nomenclature for synthetic gene delivery systems. *Hum Gene Ther.* **1997**, *8*, 511-2.
66. Denis-Mize, K.S. et al. Plasmid DNA adsorbed onto cationic microparticles mediates target gene expression and antigen presentation by dendritic cells. *Gene Ther.* **2000**, *7*, 2105-12.

67. Lleres, D. et al. Dependence of the cellular internalization and trans-fection efficiency on the structure and physicochemical properties of cationic detergent/DNA/liposomes. *J Gene Med.* **2004**, *6*, 415-28.

68. Petrov, A.I.; Zhdanov, R.I. Complexation of DNA with cationic surfactant CTAB. UV-melting and fluorescent study. in *35th IUPAC Congress* **1995**, 97.

69. Petrov, A.I.; Arslan, A.; Zhdanov, R. DNA interaction with cationic surfactants: DNA-CTAB complex. in *35th IUPAC Congress* **1995**.

70. Tang, F.; Hughes, J.A. Synthesis of a single-tailed cationic lipid and investigation of its transfection. *J Control Release* **1999**, *62*, 345-58.

71. Ogris, M. et al. DNA/polyethylenimine transfection particles: influence of ligands, polymer size, and PEGylation on internalization and gene expression. *AAPS PharmSci.* **2001**, *3*, E21.

72. Jaaskelainen, I. et al. A lipid carrier with a membrane active component and a small complex size are required for efficient cellular delivery of anti-sense phosphorothioate oligonucleotides. *Eur J Pharm Sci.* **2000**, *10*, 187-93.

73. Lee, L.K. et al. Biophysical characterization of an integrin-targeted non-viral vector. *Med Sci Monit.* **2003**, *9*, BR54-61.

74. Xu, Y. et al. Physicochemical characterization and purification of cationic lipoplexes. *Biophys J*, **1999**, *77*, 341-53.

75. Felgner, J.H. et al. Enhanced gene delivery and mechanism studies with a novel series of cationic lipid formulations. *J Biol Chem.* **1994**, *269*, 2550-61.

76. Ross, P.C.; Hui, S.W. Lipoplex size is a major determinant of in vitro lipofection efficiency. *Gene Ther.* **1999**, *6*, 651-9.

77. Kennedy, M.T. et al. Factors governing the assembly of cationic phospho-lipid-DNA complexes. *Biophys J.* **2000**, *78*, 1620-33.

78. Devine, D.V. et al. Liposome-complement interactions in rat serum: implications for liposome survival studies. *Biochim Biophys Acta* **1994**, *1191*, 43-51.

79. Liu, D.; Liu, F.; Song, Y.K. Recognition and clearance of liposomes containing phosphatidylserine are mediated by serum opsonin. *Biochim Biophys Acta* **1995**, *1235*, 140-6.

80. Senior, J.; Crawley, J.C.; Gregoriadis, G. Tissue distribution of liposomes exhibiting long half-lives in the circulation after intravenous injection. *Biochim Biophys Acta* **1985**, *839*, 1-8.

81. Litzinger, D.C. et al. Effect of liposome size on the circulation time and intraorgan distribution of amphipathic poly(ethylene glycol)-containing liposomes. *Biochim Biophys Acta* **1994**, *1190*, 99-107.

82. Niven, R.; Zhang, Y.; Smith, J. Toward development of a non-viral gene therapeutic. *Adv Drug Deliv Rev.* **1997**, *26*, 135-150.

83. Artzner, F.; Zantl, R.; Radler, J.O. Lipid-DNA and lipid-polyelectrolyte mesophases: structure and exchange kinetics. *Cell Mol Biol (Noisy-le-grand)* **2000**, *46*, 967-78.

84. Lee, R.J.; Huang, L. Folate-targeted, anionic liposome-entrapped poly-lysine-condensed DNA for tumor cell-specific gene transfer. *J Biol Chem.* **1996**, *271*, 8481-7.

85. Dauty, E.; Behr, J.P.; Remy, J.S. Development of plasmid and oligo-nucleotide nanometric particles. *Gene Ther.* **2002**, *9*, 743-8.

86. Blessing, T.; Remy, J.S.; Behr, J.P. Monomolecular collapse of plasmid DNA into stable virus-like particles. *Proc Natl Acad Sci USA* **1998**, *95*, 1427-31.

87. Stella, B. et al. Design of folic acid-conjugated nanoparticles for drug targeting. *J Pharm Sci.* **2000**, *89*, 1452-64.

88. Ward, C.M.; Acheson, N.; Seymour, L.W. Folic acid targeting of protein conjugates into ascites tumour cells from ovarian cancer patients. *J Drug Target* **2000**, *8*, 119-23.

Chapter 15

Gene Delivery with Novel Poly(l-tartaramidoamine)s

Yemin Liu[1], Laura Wenning[1], Matthew Lynch[2], and Theresa M. Reineke[1,*]

[1]Department of Chemistry, University of Cincinnati, Cincinnati, OH 45221
[2]Proctor and Gamble Company, Cincinnati, OH 45061
*Corresponding author: reinektm@ucmail.uc.edu

Gene delivery with polymeric vectors has recently developed as a practical alternative to viral delivery systems. In an effort to create nontoxic and highly effective synthetic transfection reagents, we have polymerized a tartarate comonomer with a series of amine comonomers to yield a new family of copolymers for this purpose. Four new poly(L-tartaramidoamine)s (T1-T4) have been designed and studied. Results of gel shift assays indicate that the polymers can bind plasmid DNA (pDNA) at polymer nitrogen to pDNA phosphate (N/P) ratios higher than one. Dynamic light scattering experiments reveal that each polymer compacts pDNA into nanoparticles (polyplexes) in the approximate size range to be endocytosed by cultured cells. The polyplexes formed with TI-T4 and pDNA, containing the firefly luciferase reporter gene, were also examined for their gene expression and toxicity profiles with BHK-21 cells. The poly(L-tartaramidoamine)s exhibit high delivery efficiency without cytotoxic effects, indicating that these polymers show great promise as new gene delivery vehicles.

Introduction

Gene therapy is one of the direct benefits from the completion of the human genome sequencing, and offers exciting promise in treating many genetic disorders (*1*). Cationic polymers are currently being studied as one of the delivery modalities for the intracellular transfer of therapeutic genes due to the serious problems that have surfaced with viral vectors (*2-4*). It has been demonstrated that polycations can readily bind DNA, compact it into nanostructures that facilitate cellular uptake, and efficiently carry pDNA into cultured cells (*2-4*).

Many studies have revealed that structurally different cationic polymers display highly diverse delivery efficiency and toxicity characteristics (*4-7*). For example, chitosan, a common polysaccharide, has minimal toxicity, but the gene delivery profile is low in many cell lines (*8,9*). Conversely, polyethylenimine (PEI), is widely known as the most efficient polymeric vector, unfortunately, it is highly toxic *in vitro* and *in vivo* (*10-12*). In an effort to understand this discrepancy, we have created a series of poly(L-tartaramidoamine)s with structural similarities (containing hydroxyl groups) and secondary amine densities between chitosan (low nitrogen density) and linear polyethylenimine (high nitrogen density). Here, the synthesis, pDNA binding, compaction, delivery efficiency, and toxicity of a new family of poly(Ltartaramidoamine)s (denoted T1-T4) with BHK-21 cells is presented. A comparison of the results of these systems to linear PEI (Jet-PEI) and chitosan is discussed.

Materials and Methods

Synthesis and Purification of Monomers and Copolymers

General

All reagents used in the synthesis, if not specified, were obtained from Aldrich Chemical Co. (Milwaukee, WI) and were used without further purification. Crude pentaethylenehexamine was purchased from Acros (Morris Plains, NJ), and was purified according to a reported procedure (*5*). Chitosan was obtained from Aldrich (low molecular weight, 50-190 kDa), and linear PEI was purchased from Avanti Polar Lipids (Jet-PEI) and used without further purification. All synthetic polymers were purified via dialysis in ultra-pure water using a Spectra-Por (Rancho Dominguez, CA) 1000 MWCO (molecular weight cut-off) membrane prior to transfection. The purified products were lyophilized using a Flexi-dry MP lyophilizer (Kinetics Thermal Systems, Stone Ridge, NY). NMR spectra were collected on a Bruker AV-400 MHz spectrometer.

Polymerization

Each polymer was synthesized through condensation polymerization of an amine comonomer (AA) with dimethyl L-tartarate (BB) in methanol at room temperature *(5,13)* to yield a series of AABBAABB copolymers (Figure 1). All products obtained were of a white color.

Poly(L-tartaramidodiethyleneamine) (T1): Diethylenetriamine (0.29 g, 2.80 mmol) was added to a methanol solution (2.80 mL) of dimethyl L-tartarate (0.50 g, 2.80 mmol) and stirred for 24 hours at room temperature. The mixture was dialyzed against ultra pure water to remove monomer, oligomer, and solvent impurities. Yield: 0.38 g, 1.75 mmol (62.5 %) ^{1}H NMR (D$_2$O): δ 4.46 (s, 2H), 3.31 (broad, 4H), 2.69 (broad, 4H).

Poly(L-tartaramidotriethylenediamine) (T2): Triethylenetetramine hydrate (0.41 g, 2.81 mmol; containing 20.42% H$_2$O) was added and stirred in a methanol solution (2.80 mL) containing dimethyl L-tartarate (0.50 g, 2.80 mmol). After 24 hours, the mixture was dialyzed against ultra pure water. Yield: 0.60 g, 2.31 mmol (82.5 %) ^{1}H NMR (D$_2$O): δ 4.46 (s, 2H), 3.30 (broad, 4H), 2.68 (broad, 8H).

Poly(L-tartaramidotetraethylenetriamine) (T3): Dimethyl L-tartarate (0.50 g, 2.80 mmol) was added to a methanol solution (10.0 mL) containing triethylamine (1.4 g, 14 mmol) and tetraethylenepentamine pentahydrochloride (1.04 g, 2.80 mmol). After 80 hours, the mixture was dialyzed against ultra pure water. Yield: 0.72 g, 2.38 mmol (84.8 %) ^{1}H NMR (D$_2$O): δ 4.45 (s, 2H), 3.31 (broad, 4H), 2.70 (broad, 12H).

Poly(L-tartaramidopentaethylenetetramine) (T4): Dimethyl L-tartarate (0.50 g, 2.80 mmol) was added to a methanol solution (10.0 mL) of triethylamine (1.7 g, 17 mmol) and pentaethylenehexamine hexahydrochloride (1.26 g, 2.80 mmol). After 8 hours, the mixture was dialyzed against ultra pure water. Yield: 0.50 g, 1.44 mmol (51.4 %) ^{1}H NMR (D$_2$O): δ 4.45 (s, 2H), 3.31 (broad, 4H), 2.71 (broad, 16H).

Copolymer and Polyplex Characterization

Gel Permeation Chromatography (GPC) Experiments

The molecular weight, polydispersity and Mark-Houwink-Sakurada (MHS) parameters (α) of the polymers were analyzed by a Viscotek GPCmax Instrument (Houston, TX) equipped with a ViscoGEL GMPW$_{XL}$ column and a

Triple Detection System (static light scattering, viscometry and refractive index). Samples (100 μL, 10-15 mg/mL) were prepared in a sodium acetate buffer (0.5 M, pH 5.0; H_2O/acetonitrile = 4/1, v/v), injected onto the column, and eluted with the buffer at 1.0 mL/min. Detailed characterization results are given in Table 1.

Gel Shift Assays

The DNA-binding ability of T1-T4 was examined via gel electrophoresis shift experiments. Plasmid DNA containing the firefly luciferase reporter gene (gWizLuc) was purchased from Aldevron (Fargo, ND). All of the polymer and pDNA solutions were prepared in DNAse and RNAse free water purchased from Gibco BRL (Gaithersburg, MD). Plasmid DNA (10 μL, 0.1 μg/μL) was mixed with an equal volume of polymer solution at N/P ratios between 0 and 50. Each solution was incubated for 30 min to allow polymer-DNA binding and polyplex formation. Aliquots (10 μL) of the polyplex solutions were run in a 0.6% agarose gel containing 6 μg of ethidium bromide/l00 mL TAE buffer (40 mM Tris-acetate, 1 mM EDTA). Results are detailed in Figure 2.

Dynamic Light Scattering

Particle (polyplex) size was measured on a Zetapals dynamic light scattering instrument (Brookhaven Instruments Corporation, Holtsville, NY) at 662.0 nm. All reagents were prepared in DNAse and RNAse free water. Plasmid DNA (150 ~tL, 0.02 ~lglpL) was complexed with each polymer at NIP ratio 5 and 30 (Table I P_5 and P_{30} respectively) and then allowed to stand for one hour before diluting to 0.7 mL with DNAse, RNAse free water.

Cell Culture Experiments (5)

Cell Transfection and Luciferase Assay

Media and supplements were purchased from Gibco BRL (Gaithersburg, MD). BHK-21 cells were purchased from ATCC (Rockville, MD) and maintained in Dulbecco's Modified Eagle Medium (DMEM) supplemented with 10% FBS, 100 units/mg penicillin, 100 μg/mL streptomycin, and 0.25 μg/mL amphotericin at 37 °C and 5% CO_2. Chitosan and Jet-PEI were used as positive controls in these experiments. Untransfected cells and naked pDNA were used as negative controls.

Polymer (T1-T4, chitosan, and Jet-PEI) and pDNA (gWiz-Luc) solutions were prepared in DNA and RNA-free water. The polyplexes were prepared immediately before transfection by combining solutions of each polymer (150 µL) with pDNA (150 µL, 0.02 µg/µL) at N/P ratios of 5, 10, 20, and 30. The mixtures were allowed to incubate for one hour and diluted to 900 µL with reduced serum media (pH 7.2). BHK-21 cells were plated at 50,000 cells per well in 24-well plates and incubated for 24 hours prior to transfection. The cells were transfected with 300 µL of polyplexes solution or with naked pDNA in reduced serum media in triplicate. After 4 hours, 800 µL of DMEM was added to each well. 24 hours after the transfection, the media was replaced with 1 mL of DMEM. 47 hours after the transfection, cell lysates were analyzed for luciferase activity with Promega's luciferase assay reagent (Madison, WI) and for cell viability as described below. For each sample, light units were integrated over 10 s in duplicate with a luminometer (GENios Pro, TECAN US, Research Triangle Park, NC), and the average relative light unit (RLU) value was obtained as shown in Figure 3.

Cell Viability

The toxicity of the polyplexes formed with T1-T4, chitosan, and Jet-PEI was studied by measuring cell viability. This was measured by the amount of protein in the cell lysates obtained 47 hours after transfection with polyplexes formed with each of the polymers. The protein level of cells transfected with each type of polyplex was determined by Bio-Rad's DC protein assay (Hercules, CA) against a protein standard curve of various concentrations (0.26 mg/mL-2.37 mg/mL) of bovine serum.albumin (98%, Sigma, St. Louis, MO) in cell culture lysis buffer and normalized with protein levels of untransfected cells. Results are detailed in Figure 4.

Results and Discussion

Polymer Synthesis and Characterization

Polymers (Figure 1) were synthesized by polycondensation of each diamine (diethybenetriamine, triethylenetetramine, tetraethylenepentamine or pentaethylenehexamine) and dimethyl L-tartarate at room temperature (*5*). The similar degrees of polymerization for all of the poly(L-tartaramidoamine)s were obtained by optimizing the reaction concentrations and times. The MHS α values, molecular weights, polydispersity indices, and degrees of polymerization were determined via static light scattering and viscometry. The results of these analyses are detailed in Table I.

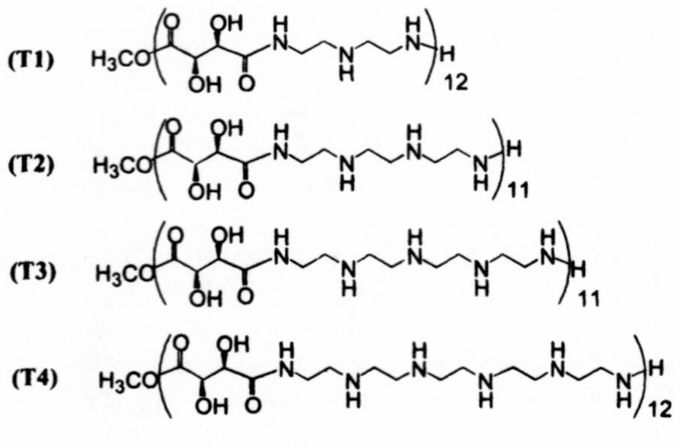

222

Figure 1.The new poly(L-tartaramidoamine)s T1-T4.

Table I. Polymer and Polyplex Characterization Data

Polymer	α	M_w (kDa)	Mw/M_n	n	P_5 (nm)	P_{30} (nm)
T1	0.733	2.7	1.3	12	94	328
T2	0.761	2.9	1.3	11	139	256
T3	0.772	3.2	1.3	11	67	90
T4	0.826	4.3	1.3	12	74	85

The pobydispersity values obtained for this family of condensation polymers was quite low (most likely due to elimination of smaller oligomers during dialysis). NMR studies as well as the a values of Mark-Houwink-Sakurada equation suggest that these polymers are mostly linear, with a low degree of branching off the secondary amines (5). The similarity in all of the physical characteristics of T1-T4 assured that an accurate comparison of the chemical structure-biological property relationships was being made and that the differences in the biological parameters was not due to molecular weight variations.

pDNA Binding Studies and Polyplex Characterization

The polymers were complexed with pDNA at several N/P ratios, and the points of charge neutralization were determined by agarose gel electrophoresis (Figure 2). Poly(L-tartaramidoamine)s T1-T4 bound pDNA at N/P ratios 2, 2, 2, 1 respectively, which are similar to the N/P ratios that both Jet-PEI and chitosan bind pDNA (*10,14*).

0 0.5 1 2 3 4 5 10 20 30 40 50   0 0.5 1 2 3 4 5 10 20 30 40 50

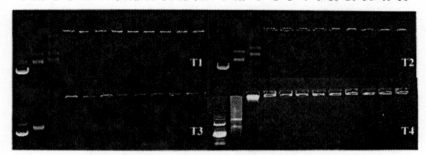

Figure 2. Gel electrophoresis shift assay of the
poly(L-tartaramidoamine,)s T1- T4.

Polyplex sizes were determined via dynamic light scattering by combining each polymer with pDNA at N/P ratios of 5 and 30. The results of these samples are reported in Table 1. As shown, as the charge ratio is increased, the polyplex sizes increased. The results indicate that T1-T4 can compact pDNA into nanoparticles in a similar size range to both Jet-PEI and chitosan, demonstrating that the polyplexes are in the correct size range to be taken up into cells through the endocytotic pathway (*10,14*).

Transfection and Toxicity of the Polyplexes *In Vitro*

The transfection efficiency of poly(L-tartaramidoamine)s T1-T4 was determined by assaying for luciferase protein activity with BHK-21 cells. The results are reported in relative light units (RLUs) per mg of protein. As shown in Figure 3, T1-T4 all deliver pDNA into BHK-21 cells with varying degrees of efficiency. It is noticed that as the number of secondary amine increases between the tartarate comonomers along the polymer backbone, gene expression generally increases (*5*).

It should be noted that T1 and T2 revealed very similar gene delivery profiles even though the secondary amine density of T2 is slightly higher. Additionally, it was noticed that for each synthetic polymer, as the N/P ratio increased, gene expression was enhanced. This effect has been noted with other polycationic systems and may be related to the lower molecular weight (*6,15*). These data show that the new vectors all display enhanced gene delivery efficiency over chitosan, however, the gene expression values were still slightly lower than Jet-PEI. These results suggest that the low delivery efficiency that is characteristic of chitosan is most likely due to a significant decrease in the amine density with this polymer structure as compared to linear PEI.

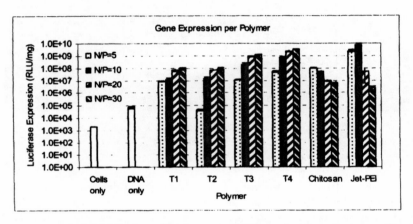

Figure 3. Luciferase gene expression with BHK-21 cells of polyplexes formed with polymers T1-T4, chitosan, and Jet -PEI Delivery efficiency data for N/P ratios of 5, 10, 20, and 30 are displayed. Untransfected cells and naked pDNA were the controls in this experiment.

Cell viability profiles of BHK-21 cells exposed to the polyplexes at various N/P ratios were determined by measuring the total protein concentration in the cell lysates after 47 hours following initial transfection and normalizing the data to protein levels from untransfected cells. As shown in Figure 4, polymers T1-T4 exhibit the nontoxic properties of chitosan and display significantly lower cytotoxicity than Jet-PEI. This result indicates that the tartarate comonomer with this PEI-like backbone significantly reduces the toxicity associated with synthetic delivery systems. In addition, cell viability for T1-T4 was greater than

80%, even at the point of maximum gene expression (N/P = 30). As shown for Jet-PEI, at maximum gene expression (N/P = 5), this vector displays only 57% cell viability. These results indicate that T4 is a nontoxic yet highly efficient gene delivery vector and shows great promise for further *in vitro* and *in vivo* investigations.

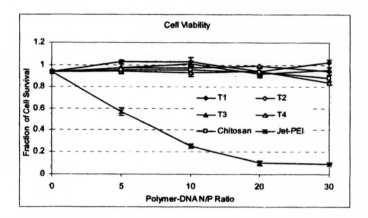

Figure 4. Viability of BHK-21 cells exposed to polyplexes formed with T1-T4, chitosan, and Jet-PEI at N/P ratios of 0 (DNA only), 5, 10, 20, and 30. The protein assay data was normalized to results obtained for the untransfected cellular control.

Conclusion

In this study, we have synthesized and characterized a new family of poly(L-tartaramidoamine)s (T1-T4). Results indicate that these structures reveal low cytotoxicity (equivalent to chitosan) and T4 promotes high gene delivery efficiency (similar to Jet-PEI), suggesting these structures are promising for further *in vitro* and *in vivo* studies. It was also noticed that as the number of amine units between the tartarate comonomers increased, gene expression was enhanced, even though no comparable increase in cytotoxicity was observed with the synthetic structures. Future experiments are aimed at elucidating the source of toxicity reduction. It is currently not understood if this reduction is due to a decrease in the amine density or if the low toxicity is an effect of the presence of hydroxy groups within the structures.

226

References

1. Rubanyi, G.M. The future of human gene therapy. *Mol. Asp. Med.* **2001,** *22,* 113-142.
2. Davis, M.E. Non-viral gene delivery systems. *Curr. Opin. Biotech.* **2002,** *13,* 128-131.
3. Ma, H.; Diamond, S.L. Nonviral gene therapy and its delivery systems. *Curr. Pharm. Biotech.* **2001,** *2,* 1-17.
4. Han, S.; Mahato, R.I.; Sung, Y.K.; Kim, S.W. Development of biomaterials for gene therapy. *Mol. Ther.* **2000,** *2, (4),* 302-317.
5. Liu, Y.; Wenning, L.; Lynch, M.; Reineke, T.M. New poly(D-glucaramido-amine)s induce DNA nanoparticle formation and efficient delivery efficiency into mammalian cells. *J. Am. Chem. Soc.* **2004,** *126, (24),* 7422-7423.
6. Reineke, T.M.; Davis, M.E. Structural effects of carbohydrate-containing polycations on gene delivery. 1. Carbohydrate size and its distance from charge centers. *Bioconjugate Chem.* **2003,** *14, (1),* 247-254.
7. Reineke, T.M.; Davis, M.E. Structural effects of carbohydrate-containing polycations on gene delivery. 2. Charge center type. *Bioconjugate Chem.* **2003,** *14, (1),* 255-261.
8. MacLaughlin, F.C.; Mumper, R.J.; Wang, J.; Tagliaferri, J.M.; Gill, I.; Hinchcliffe, M.; Rolland, A.P. Chitosan and depolymerized chitosan oligomers as condensing carriers for in vivo plasmid delivery. *J. Controlled Release* **1998,** *56,* 259-272.
9. Erbacher, P.; Zou, S.; Bettinger, T.; Steffan, A.-M.; Remy, J.-S. Chitosan-based vector/DNA complexes for gene delivery: biophysical and characteristics and transfection ability. *Pharm. Res.* **1998,** *15, (9),* 1332-1339.
10. Boussif, O.; Lezoualch, F.; Zanta, M.A.; Mergny, M.D.; Scherman, D.; Demeneix, B.; Behr, J.-P. A versatile vector for gene and oligonucleotide transfer into cells in culture and in vivo: Polyethylenimine. *Proc. Natl. Acad. Sci. U.S.A* **1995,** *92,* 7297-7301.
11. Wightman, L.; Kircheis, R.; Rossler, V.; Carotta, S.; Ruzicka, R.; Kursa, M.; Wagner, E. Different behavior of branched and linear polyethylenimine for gene delivery in vitro and in vivo. *J. Gene Med.* **2001,** *3,* 362-372.
12. Chollet, P.; Favrot, M.C.; Hurbin, A.; Coll, J.-L. Side-effects of a systematic injection of linear polyethylenimine-DNA complexes. *J. Gene Med.* **2002,** *4,* 84-91
13. Ogata, N.; Hosoda, Y. Synthesis of hydrophilic polyamide from L-tartrate and diamines by active polycondensation. *J. Polym. Sci., Polym. Chem. Ed.* **1975,** *13,* 1793-1801.

14. Borchard, G. Chitosans for gene delivery. *Adv. Drug Deliv. Rev.* **2001**, *52*, 145-150.
15. Plank, C.; Tang, M.X.; Wolf, A.R.; Szoka, Jr., F.C. Branched cationic peptides for gene delivery: role of type and number of cationic residues in formation and *in vitro* activity of DNA polyplexes. *Hum. Gene Ther.* **1999**, *10*, 319-332.

Chapter 16

Plasmid DNA Encapsulation Using an Improved Double Emulsion Process

Edmund J. Niedzinski*, Yadong Liu, and Eric Y. Sheu

Genteric, Inc., 50 Woodside Plaza, Suite 102, Redwood City, CA 94061
*Corresponding author: eniedzinski@genteric.com

The route of administration for an active pharmaceutical ingre-dient (API) often requires the development of novel formu-lations to facilitate effective delivery. Many API, for example most proteins, polypeptides, oligonucleotides and plasmid DNA, degrade under extreme pH or in the presence of digestive enzymes. In order to successfully deliver these API orally, formulations that are able to protect and deliver them to the targeted sites are mandated. One common strategy for protecting API is encapsulation, using biocompatible poly-meric materials. The encapsulation materials and the methods used determine the stability of the formulation and the bio-availability of the API. We have developed a novel method for the encapsulation of plasmid DNA into a compatible poly-mer. This method produces particles that are smaller than those made by conventional methods, encapsulate plasmid DNA with greater than 90% efficiency, protect plasmid DNA from degradation, and have release characteristics similar to convential nonviral delivery methods. These cumulative data suggest that this delivery system may be suitable for delivery of DNA into harsh environments.

Introduction

The effective delivery of an active pharmaceutical ingredient (API) requires a thorough understanding of the properties of the API and the obstacles that the API encounters during its application. Since many API are vulnerable to chemical and biological obstacles, advanced formulations are required for effective delivery. For example, conventional oral delivery is not possible for most proteins, polypeptides, oligonucleotides and plasmid DNA due to the low pH in the stomach and enzymatic degradation along the enteric route. In order to orally adminster sensitive API, formulations that protect and deliver them to the targeted sites are mandated. The encapsulation of an API into a biocompatible polymeric material, such as poly(lactide-co-glycolide), is a common strategy to protect and deliver sensitive API.

As the field of gene therapy grows, novel formulations for plasmid DNA are needed (1,2). A successful gene therapy product will require a delivery system that can remain stable and successfully deliver a genetic payload to the target cells. Although a number of nonviral gene therapy formulations are in clinical trials (3), many of these formulations may not tolerate harsh conditions. Therefore, a more robust system would extend the applicability of gene therapy. To meet this goal, we have developed a polymeric encapsulation strategy that would withstand the harsh chemical and biological obstacles that are encountered in many common routes of delivery.

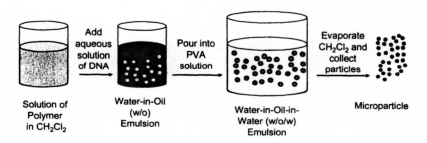

Figure 1. Overview of Double Emulsion Formulation Process.

The double emulsion process shown in Figure 1 is a convenient method to generate particles for drug delivery (4). While this method was originally designed to encapsulate water-soluble polypeptides and small molecule drugs (5), this method can also be used to encapsulated DNA. Unfortunately, the use of double emulsions to encapsulate hydrophilic bio-therapeutics such as proteins and plasmid DNA results in relatively low encapsulation efficiency and high particle size polydispersity (6). Although this particle size range may be suitable for delivery to antigen presenting cells, the uptake of particles by these cells limits the clinical application of this technology. Through the incorporation of cationic lipids, we have modified the double emulsion process to produce submicron particles with greater than 90% enapsulation efficiency. Herein, we present the development of this technology.

Materials and Methods

Materials

50:50 poly(D,L-lactic-*co*-glycolic) acid (PLGA) was purchased from Birmingham Polymers. Zwittergent 3-14 was purchased from Aldrich. Poly-(vinyl alcohol) was purchased from Sigma (30-70kDa), Aldrich (13-23 kDa), Fluka (31 kDa) and Alfa Aesar (22-26 kDa). DNA was prepared as an endotoxin-free reduced, supercoiled plasmid using anion exchange resins (Qiagen) at Genteric, Inc. Cationic lipids were purchased from Avanti Polar Lipids (Birmingham, AL).

Representative Formulation Procedure

An aqueous DNA solution (0.5 mg of plasmid DNA in 0.75 mL TE buffer) is added to a solution of polymer (50 mg of 50:50 poly(lactic-*co*-glycolic) acid) in CH_2Cl_2 (1 mL) to form a water-in-oil (w/o) emulsion. This mixture is emulsified by vortexing at 2500 rpm for 15 sec. A proper quantity of ABM is added and the emulsion is further mixed by vortexing (2500 rpm/15 sec.). The resulting emulsion is added to an aqueous solution (8% PVA, 25 mL) to form a water-in-oil-in-water (w/o/w) emulsion. The solution is allowed to stir (1500 rpm) until the CH_2Cl_2 evaporates, resulting in a solid particle. The particles are collected by centrifuging (4000 rpm, 15 min.). The supernatant is decanted and the particles are washed with water (25 mL). This process is repeated three times and the microparticles are lyophilized. The particles are then collected and stored at 0°C. The supernatant and wash solutions are collected to be analyzed for DNA concentration.

Particle Size

The particle size was determined by examining the microparticles under 400x magnification on a Micromaster optical microscope (Fisher Scientific), equipped with a Kodak MDS290 documentation system.

Encapsulation Efficiency

The relative encapsulation efficiency of the microparticles was assessed by measuring the amount of DNA in the supernatant and washes of the microparticle preparation. This solution was diluted with 1% Zwittergent 3-14 in TAE buffer and then analyzed using the PicoGreen® reagent (Molecular Probes). The assay was conducted according to the manufacturer's instructions, using a Gemini XS Fluorescence Microplate Reader (Molecular Devices).

Preparation and Purification of Reporter Genes

The plasmid pBATSEAP contains the secreted alkaline phosphatase (SEAP), operably linked to human cytomegalovirus major immediate early enhancer/promoter, which is positioned upstream of the first intron of human β-globin. pBATLuc is an analogous plasmid, in which the SEAP gene was replaced with the luciferase gene. The plasmid vectors were produced by bacterial fermentation and purified with an anion exchange resin (Qiagen, Santa Clarita, CA) to yield an endotoxin-reduced, supercoiled plasmid, containing less than 100 E.U./mg DNA as measured by clot LAL assay (Charles River Endosafe). Stock DNA solutions were prepared using sterile water.

Analysis of Encapsulated Plasmid DNA

A microparticle sample (2-5 mg) in a 4-mL glass vial was treated with CH_2Cl_2 (1 mL) and gently mixed for 16 hours. The suspension was treated with a solution of 1% Zwittergent in TAE Buffer (pH 7.4) and the resultant emulsion was gently mixed for 4 hours. The sample was allowed to stand until the two phases separate, and the aqueous layer was tested for DNA concentration using the PicoGreen® reagent (Molecular Probes). The assay was conducted according to the manufacturer's instructions, using a Gemini XS Fluorescence Microplate Reader (Molecular Devices).

Cell culture

CHO cells from the American Type Culture Collection (ATCC, Rockville, MD) were cultured in 75 cm^2 cell culture flasks with HAMS F12 media containing 10% bovine calf serum. NIH3T3 cells were grown in Dulbecco's modified Eagle's medium with 10% fetal calf serum. The cells were maintained at 37°C in a 5% CO_2 environment. Cells were split into 24-well plates at 60% confluence 24 hours before transfection.

Transfection of Cultured Cells

Growth media was removed from the CHO cells and replaced with 100μL of Dulbecco's modified Eagle's medium with or without 10% fetal calf serum. The formulations were diluted with water to obtain a DNA concentration of 1mg/100mL and 100μL was administered to each well. Liposome formulations were prepared as previously described. After 2 hours, the formulation solution was removed, frozen for DNA analysis, and replaced with 500μL of Dulbecco's modified Eagle's medium with or without 10% fetal calf serum. After 24 hours, the medium was removed, frozen for SEAP analysis, and replaced with 500μL of fresh growth medium. This process was repeated at 48 hours and 120 hours.

Secreted Alkaline Phosphatase Assay

Tissue supernatants, prepared in luciferase lysis buffer, were analyzed for SEAP concentration using the chemiluminescent SEAP Reporter Gene Assay available from Roche Diagnostics (Indianapolis, IN). Briefly, the samples were diluted 1:50 with Dilution Buffer then incubated in a water bath for 30 min at 65°C. The samples were then centrifuged for 1 min at 13,000 rpm and 50 μL of the resulting supernatant was transferred to a microtiter plate. 50 μL of the provided Inactivation Buffer was added, followed by a 5-minute incubation at room temperature. The Substrate Reagent (50 μL) was added prior to a 10-minute incubation at room temperature. Light emissions from the samples were measured using an L-max plate luminometer from Molecular Devices (Sunnyvale, CA). The relative light units were converted to mass of SEAP protein, based on a standard curve run contemporaneously with the samples. Plasma concentrations of SEAP were measured using the same assay, except that the samples were diluted 1:7 with the Dilution Buffer.

Results and Discussion

The addition of a cationic lipid (DSTAP) into a double emulsion significantly increases the efficiency of plasmid DNA encapsulation (Figure 2). The efficiency increases as more cationic lipid is added to the double emulsion, with a three-fold increase for a formulation containing DSTAP:DNA in a 4:1 ratio compared to a formulation without DSTAP.

Besides increasing the encapsulation efficiency, the addition of a cationic lipid to a double emulsion has a profound effect on the size of PLGA microparticles. As shown in Figure 3, increasing the DSTAP:DNA ratio resulted in a population of particles that were smaller and less polydisperse. The microparticles with DSTAP were approximately 1-3 μm in size, while the particles without DSTAP were noticeably larger (5-10 μm), as determined by light microscopy. However, increasing the charge ratio beyond 4:1 did not result in an increase in encapsulation efficiency or decrease in particle size (data not shown).

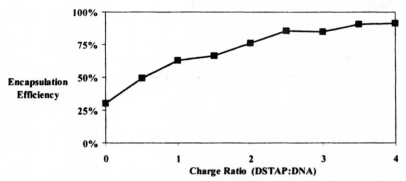

Figure 2. Effect of cationic lipid concentration on DNA encapsulation efficiency. Particles were generated using different concentrations of DSTAP. The amount of DNA contained in the microparticles was measured using Pico-Green®. The encapsulation efficiency was calculated as a percentage of DNA found in the supernatant samples relative to amounts of DNA that was added to the formulation.

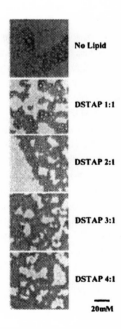

Figure 3. Effect of cationic lipid concentration on particle size. Particles were generated using different concentrations of DSTAP, and particle size was analyzed using light microscopy.

The encapsulation efficiency of this formulation appeared to be influenced by the cationic lipid structure (Table I). Specifically, adding cationic lipids with the same cationic head group and longer hydrophobic domains slightly increase the encapsulation efficiency. Interestingly, the inclusion of cationic lipids with different structures had no observable effect on the particle size (data not shown).

Table I. Effect of Cationic Lipid Structure on DNA Encapsulation Efficiency.[a]

Cationic Lipid	Encapsulation Efficiency
None	30.38%
DMTAP (C14:0)	90.38%
DPTAP (C16:0)	91.92%
DSTAP (C18:0)	92.69%

[a]The encapsulation efficiency was calculated as a percentage of the DNA found in the supernatant samples relative to amounts of DNA that was added to the formulation. The particles were formulated at 4:1 charge ratio.

Although the double emulsion process is conducted under relatively mild conditions, high stir rates and temperatures can degrade sensitive API. To determine if any decomposition occurred during formulation, the integrity of the encapsulated plasmid DNA was analyzed using agarose gel electrophoresis. As shown in Figure 4, plasmid DNA that was extracted from the microparticles appears to be slightly nicked in comparison to control plasmid, a phenomona that has been observed in other studies (7). Fortunately, subsequent studies revealed that the encapsulated plasmid retained full biological activity in cell culture.

To test the ability of these formulations to protect DNA under degradative conditions, the formulations were challenged by incubating with DNase type I (Figure 5). Control plasmid DNA and formulated plasmid DNA was incubated for 20 minutes in buffer and various concentrations of DNase type I. After incubation, the plasmid DNA was extracted from the particles and the structural integrity of the plasmid DNA was analyzed using gel electrophoresis. After incubating for 20 minutes with 1.6 units/mL of DNase I, the encapsulated plasmid remained intact while the control plasmids was completely degraded.

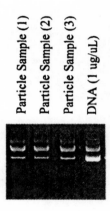

Figure 4. Agarose gel electrophoretic analysis of plasmid DNA. To remove the polymeric coating, three identical microparticle samples (1, 2, and 3) were incubated with CH₂Cl₂ for 16 hours. A 1% solution of zwittergent in TE buffer was used to extract the plasmid DNA and disrupt any plasmid DNA cationic lipid association. An aliquot of this aqueous layer was then analyzed using a 0.8% agarose gel. Samples were concurrently analyzed with control plasmid.

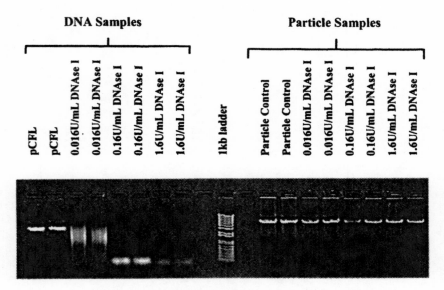

Figure 5. Analysis of DNA stability following incubation with DNase type I. Particle formulated with DSTAP at 4:1 charge ratio were incubated with various concentrations of DNase I for 20 minutes at 40 °C. Control samples were incubated in TAE buffer. The DNA was extracted from the particles and analyzed using gel electrophoresis.

The gene transfer efficiency of the particles was assessed by treating cultured Chinese Hamster Ovary (CHO) cells with particles containing secreted alkaline phosphates (SEAP). Encapsulated plasmid DNA and control conditions were administered to CHO cells in 24-well tissue culture plates. First, DNA uptake was analyzed by measuring the amount of DNA that remained in the supernatant 2 hours after administration (Figure 6). The highest concentration of plasmid DNA was found in the supernatant of cells that were treated with DNA in water. Significantly less DNA was detected in the other solutions *(8)*. After 24 hours of incubation, the supernatant was collected and measured for SEAP expression (Figure 7). This process was repeated at 48 and 72 hours after administion. Microscopic analysis of the cell population did not reveal any observable toxicity. Interestingly, the levels of SEAP expression from cell treated with DSTAP liposomes are similar to the cells treated with the microparticle formulation.

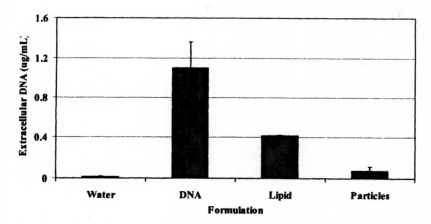

Figure 6. Analysis of extracellular DNA concentration. Formulations were tested by administering 100 µL to each well containing CHO cells in 100 µL of serum positive (+) media. After 2hours, the media/formulation was removed and frozen for analysis. The samples were then thawed and tested for DNA concentration using Pico-Green® (samples diluted 1/10 in 1% zwittergent in TAE buffer). Each data point represents the average of three wells +/- SD. Assay sensitivity <100pg/mL.

Although current double emulsion procedures are mild enough to encapsulate sensitive molecules, such as antibodies *(9)* and DNA *(10)*, these processes often result in poor encapsulation efficiency, which may be attributed to the diffusion of the active pharmaceutical ingredient from the inner aqueous phase to the outer aqueous phase as the oil phase is evaporating. Since this interstitial layer is typically an organic solvent, this problem is exacerbated as the active ingredient decreases in polarity. To circumvent this problem, we have been investigating methods to increase the encapsulation efficiency of plasmid DNA into a microparticle by adding cationic lipids to the formulation process.

Cationic lipids are being actively investigated as adjuvants for gene delivery, usually in the form of cationic liposome:plasmid DNA complexes *(11)*. These formulations have primarily been used to compact plasmid DNA for delivery into cells *(12)*. Plasmid DNA is compacted by cationic lipids due to the ionic interaction of their cationic ammonium salts with the anionic phosphate salts of plasmid DNA. *In vivo*, these adjuvants have been successful in certain routes of administration, such as intracranial administration *(13)*, but the dynamic nature of these complexes make them sensitive to the chemical and enzymatic barriers that are encountered when these complexes are administered through routes such as oral or intravenous. Cationic particles, generated by adding plasmid DNA to pre-formed cationic lipid-PLGA particles, have been investigated as a DNA-vaccine delivery method *(14)*. While this may overcome the issues of formulation stability, plasmid DNA is still exposed to degradation. Rather than adding DNA after the particle has been formed, we believe that a

238

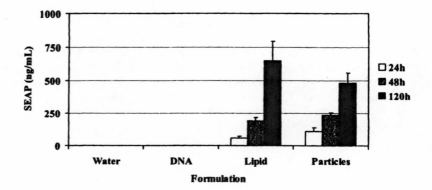

Figure 7. Analysis of microparticle and lipid transfection efficiency in CHO Cells. Formulation were tested by administering 100 μL (1μg of DNA) to each well containing CHO cells in 100 μL of serum positive (+) media. After 2 hours, the media/formulation was removed and replaced with 500 μL of serum positive media. At 24 (white), 48 (grey), and 120 (black) hours, the media were removed, immediately frozen until analysis, and replaced with fresh media. Each data point represents the average of three wells +/- SD.

formulation that utilizes plasmid DNA as an intergral part of the particle would have an even greater applicability.

In an emulsion formulation, the interaction of a cationic lipid with plasmid DNA appears to compact the aqueous interior of the double emsulsion. As the organic solvent evaporates from the formulation, a smaller particle is formed. The interaction of the plasmid DNA and the cationic lipid may also control the diffusion of DNA through the interstitial organic phase in the double emulsion. When bound to plasmid DNA, the larger alkyl chains of the lipid provides a larger hydrophobic barrier that prevents plasmid DNA from crossing the organic phase. This argument is supported by the apparent dependency of carbon chain length on encapsulation efficiency (Table 1).

In addition to increasing the encapsulation efficiency and controlling the particle size, the encapsulation process must be gentle enough to encapsulate the plasmid DNA without comprising its structural integrity. Based upon electro-phoretic analysis (Figure 4), our process did not cause signifigant damage to the plasmid. In addition, the formulated plasmid DNA could sustain exposure to degradative conditions (Figure 5). These two results are paramount in the development of a formulation technology that is designed to accommodate a chemically sensitive active ingredient.

Based upon previous publications, we anticipated that these particles would release the encapsulated plasmid DNA in a sustained manner. In fact, in initial release rate studies, which were conducted by incubating the particles in TRIS buffers, less than 10% of the plasmid DNA was released over a seven day period. However, when these particles were administered to cultured cells, the kinetics of SEAP expression from the cells treated with the particles was very similar to the kinetics of SEAP expression from the cells treated with lipid:DNA complexes (Figure 7). Since only a small fraction of the DNA was detected in the extracellular solution of cells treated with particles (Figure 6), the DNA is either being degraded or released into the cell. These results suggest that the release of the DNA is triggered by an intracellular phenomenon. This observation is beneficial for *in vivo* applications. While a sustained DNA release formulation would have many applications, a formulation that releases DNA rapidly after cell uptake would be more beneficial. This form of controlled release would circumvent many of the obstacles that are encountered when DNA is administered in a dynamic formulation, such as a cationic lipid:DNA complex. Currently, further studies are being conducted to evaluate the *in vivo* applicability of this technology.

Conclusion

The goal of this project was to develop a formulation protocol that could efficiently encapsulate DNA. By carefully adjusting the parameters of the formulation and introducing cationic lipids, we were able to successfully produce particles with greater than 90% DNA encapsulation efficiency and small particle size (approximately 1 μm). Currently, we are testing the ability of this formulation to encapsulate other API and to deliver plasmid DNA to the gastrointestinal system.

Acknowledgements

We would like to thank Kathleen Galligan-Lail for preparation and purification of the plasmid DNA, Will Frazier for assistance with the cell culture experiments, and Hosna Mujadidi for conducting the protein assays.

References

1. For example, see: Zhuang, F.F.; Liang, R.; Zou, C.T.; Ma, H.; Zheng, C.X.; Duan, M.X.J. High efficient encapsulation of plasmid DNA in PLGA microparticles by organic phase self-emulsification. *Biochem Biophys Methods* **2002**, *52*, 169-178.

2. Perez, C.; Sanchez, A.; Putnam, D.; Ting, D.; Langer, R.; Alonso, M.J. Poly(lactic acid)-poly(ethylene glycol) nanoparticles as new carriers for the delivery of plasmid DNA. *J Control Release* **2001**, *75*, 211-224.

3. For an update of clinical trials in gene therapy, see: http://www4.od.nih.gov/oba/rac/clinicaltrial.htm.

4. Fattal, E.; Roques, B.; Puisieux, F.; Blanco-Prieto, M.J.; Couvreur, P. Multiple emulsion technology for the design of microspheres containing peptides and oligopeptides. *Adv Drug Del Rev* **1997**, *28*, 85-96.

5. Yamamoto,; M.; Takada, S.; Ogawa; Y. Method for producing microcapsule. U.S. Patent Number 4,954,298.

6. Tinsley-Bown, A.M.; Fretwell, R.; Dowsett, A.B.; Davis, S.L.; Farrar, G.H. Formulation of poly(D,L-lactic-co-glycolic acid) microparticles for rapid plasmid DNA delivery. *J Control Release* **2000**, *66*, 229-241.

7. Jones, D.H.; Corris, S.; McDonald, S.; Clegg, J.C.S.; Farrar, G.H. Poly (DL-lactide-co-glycolide)-encapsulated plasmid DNA elicits systemic and mucosal antibody responses to encoded protein after oral administration. *Vaccine* **1997**, *15*, 814-817.

8. Note: Since the DNA assay does not detect encapsulated plasmid DNA, the microparticle solution may contain undetected plasmid DNA.

9. Wang, J.; Chua, K.M.; Wang, C.H. Stabilization and encapsulation of human immunoglobulin G into biodegradable microspheres. *J Colloid Interface Sci.* **2004**, *271*, 92-101.

10. Hao, T.; McKeever, U.; Hedley, M.L. Biological potency of microsphere encapsulated plasmid DNA. *J Control Release* **2000**, *69*, 249-259.

11. Gershon, H.; Ghirlando, R.; Guttman, S.B.; Minsky, A. Mode of formation and structural features of DNA-cationic liposome complexes used for transfection. *Biochemistry* **1993**, *32*, 7143-7151.

12. Niidome, T.; Huang, L. Gene therapy progress and prospects: nonviral vectors. *Gene Ther* **2002**, *9*, 1647-1652.

13. Hecker, J.G.; Hall, L.L.; Irion, V.R. Nonviral gene delivery to the lateral ventricles in rat brain: initial evidence for widespread distribution and expression in the central nervous system. *Mol Ther* **2001**, *3*, 375-384.

14. Singh, M.; Ott, G.; Kazzaz, J.; Ugozzoli, M.; Briones, M.; Donnelly, J.; O'Hagan, D.T. Cationic microparticles are an effective delivery system for immune stimulatory cpG DNA. *Pharm Res* **2001**, *18*, 1476-1479.

Chapter 17

A New Microencapsulation Technique Based on the Solvent Exchange Method

Yoon Yeo and Kinam Park[*]

Departments of Pharmaceutics and Biomedical Engineering, Purdue
University, West Lafayette, IN 47907
[*]Corresponding author: kpark@purdue.edu

Protein microencapsulation is difficult because of the sensitivity of
proteins to various stresses encountered during the encapsulation
process and release period. In an attempt to overcome such difficuities,
a new microencapsulation method has been developed based on an
interfacial phenomenon between a polymer solution and an aqueous
solution, which we call "solvent exchange." A dual microdispenser
system consisting of two ink-jet nozzles was employed to test the
concept. This article describes the concept and advantages of the
solvent exchange method.

Introduction

Controlled drug delivery can influence the performance of a drug by manipulating its concentration, location, and duration *(1)*. Since the emergence of
early controlled release products in late 1960s, controlled drug delivery systems
have evolved to such an extent that drug release can be modulated in various
manners to comply with needs of the body. Potential advantages of controlled
drug delivery systems include: (i) improving patient compliance by reducing the
number of drug administrations; (ii) reducing side effects by maintaining the

blood level of the drug within the therapeutic range; (iii) improving drug efficacy by extending duration of drug concentration in an effective level; and (iv) providing opportunities for targeted drug delivery (2). Controlled release technology has been highly successful with low molecular weight drugs; however, controlled delivery of high molecular weight drugs, such as genes, peptides, and proteins has been difficult.

Peptides and proteins constitute an important group of therapeutic compounds. The high specificity and potency are major advantages of protein drugs as compared to traditional low molecular weight drugs (2). Advances in biotechnology have made recombinant peptide and protein drugs available in large quantities. Moreover, recent completion of the human genome project is expected to bring discovery of new protein drugs with superb bioactivities. Currently, most protein drugs are administered via invasive parenteral routes on a regular basis because the oral delivery of a protein drug is not a viable option at present. In this regard, use of biodegradable polymeric microparticles that can release a drug at a controlled rate for a specified period has been considered as an attractive alternative to the frequent parenteral administration method.

A number of microencapsulation methods have been developed during the past few decades. Advances in drug delivery technologies led to successful launch of commercial products on the market, such as Lupron Depot® (Leuprolide acetate, TAP Pharmaceuticals Inc.), Zoladex® Depot (Goserelin acetate, AstraZeneca), Sandostatin LAR® Depot (Octreotide acetate, Norvatis), and Trelstar™ Depot (Triptorelin pamoate, Pfizer). However, such successes are mostly limited to low molecular weight drugs or oligopeptides. Despite more than 20 years of effort to develop protein-encapsulated microparticle systems, only one product has reached the market: Nutropin Depot® (Human growth hormone, Genentech Inc.), which was approved by the U.S. Food and Drug Administration in 1999. Major challenges in protein microencapsulation come from difficulties in preserving structural and functional integrity of the encapsulated protein throughout the lifetime of the product, which result in undesirable release profiles as well as protein instability problems (3,4).

Numereous studies discovered that proteins are sensitive to mechanical and chemical stresses and can easily be destroyed during the microencapsulation process and the prolonged release period. The most widely recognized problem in the contemporary microencapsulation techniques is that they can generate stressful conditions during the fabrication process and/or the release period such as (i) extensive exposure of proteins to a large water/organic solvent (w/o) inter-facial area during microencapsulation process (5,6), (ii) mechanical stresses such as emulsification or homogenization (7), and (iii) extended contact with hydrophobic polymers and their degradation products (8). Proteins exposed to such damaging environments during fabrication of microparticles and/or the release period tend to undergo various structural modifications, leading to loss of their biological functions or making them unavailable for release. As a result, a successful control of the release profile using the microparticle systems has seldom been achieved.

Our interest in developing a new microencapsulation system is based on the hypothesis that protein stability and the release profiles can be improved by minimizing those damaging conditions. The objective of this research is to develop a simple and efficient microencapsulation method that is distinguished from contemporary techniques. This new approach minimizes the formation of w/o interface, employs a mild instrumental system that is highly compatible with stress-sensitive drugs, and generates reservoir-type microcapsules that reduce the contact between encapsulated drugs and the potentially damaging environments.

Solvent Exchange Method

The new microencapsulation technique, which we call the "solvent exchange method," is based on a hypothesis that interfacial mass transfer bet-ween two contacting liquids can be utilized for making reservoir-type microcapsules. Figure 1 describes one method of making microcapsules, using two separate droplets based on the solvent exchange method. Microcapsules can be made as a layer of polymer solution encapsulates an aqueous droplets and then leaves a polymer membrane on the aqueous surface. Due to the surface tension gradient, most organic solvents that dissolve water-insoluble polymers can spread on the aqueous surface. Solid membrane occurs when interfacial mass transfer results in decrease in the solvent quality. In order to ensure this interfacial phenomenon, mutual solubilities of the two liquids are necessary. In order to provide a condition that allows contact between the aqueous droplets and the layer of polymer solution, we have developed a dual microdispenser system that consists of two ink-jet nozzles. This article presents how the new microencapsulation technique has been developed and the dual microdispenser system was used in implementing the new concept of micro-encapsulation.

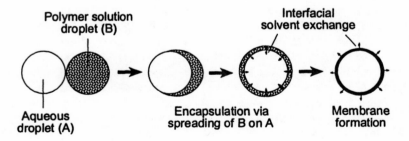

Figure 1. Microencapsulation based on the solvent exchange method.

Materials and Methods

Screening of Organic Solvents

Organic solvents that allow easy formation of a polymer membrane on an aqueous surface were screened as described previously *(9)*. Briefly, poly(lactic acid-*co*-glycolic acid) (PLGA) was used as the encapsulating polymer, and organic solvents having a Hildebrand solubility parameter of 16–24 MPa$^{1/2}$ were tested for (i) capability of solubilizing the PLGA polymer, (ii) diameter and (iii) turbidity of the polymer membranes that PLGA solutions of the solvents left on a 0.5% agarose gel.

Microcapsule Preparation by Dual Microdispenser System

The dual microdispenser system consisted of two ink-jet nozzles (Figure 2). Microcapsules were produced as described previously *(9)*. Briefly, a 2% PLGA-ethyl acetate solution and an aqueous solution containing a model drug and/or an excipient of choice were fed through each nozzle at a controlled flow rate. For confocal microscopy, FITC-dextran and Nile Red were added to the aqueous solution and the PLGA-ethyl acetate solution, respectively. The liquid streams were perturbed by a frequency generator (Hewlett-Packard model 33120A) to produce a series of droplets of uniform size. The trajectories of the two jets were precisely controlled to ensure collisions between every pair of droplets. The collision behavior was observed using a video camera under stroboscopic illumination. The microcapsules thereby formed were collected in a water bath. For comparison, microcapsules were also produced using the double emulsion-solvent evaporation method described in the literature *(10)*, with slight modification.

Particle Size Control Using an Ink-Jet Nozzle

In order to understand the effects of instrumental parameters on size of the droplets formed by an ink-jet nozzle, nozzle orifice diameter d, forcing frequency f, and volumetric liquid flow rate Q were varied while introducing distilled water into an ink-jet nozzle as described previously *(11)*. The drop sizes were determined from stroboscopic images. Apparently, the drops were highly homogeneous in size; thus, representative ones were taken to determine the size generated under the specific condition. In determining sizes of micro-capsules, light microscopic pictures of the collected microcapsules were used. Reported values were averages of 30-50 microcapsules.

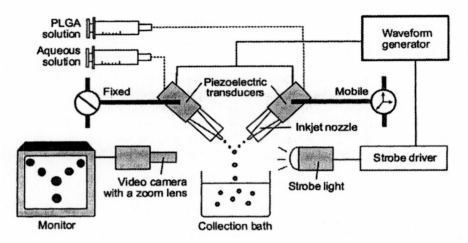

Figure 2. Schematic description of a dual microdispenser system.

Microscopic Observation of Microcapsules

Nascent microcapsules were observed using a bright-field microscope. A droplet of the microcapsule suspension was placed on a glass cover slip and the microcapsules were observed with a Nikon Labophot 2 microscope. The internal structure of the microcapsules was imaged using an MRC-1024 Laser Scanning Confocal Imaging System (Bio-Rad) equipped with a krypton/argon laser and a Nikon Diaphot 300 inverted microscope.

Results and Discussion

Solvent Selection

The new encapsulation method depends on the formation of polymer membranes on aqueous microdroplets; thus, it is important to find solvents which can dissolve the polymer (PLGA in this study) but form solid membrane upon contact with aqueous media. Thus, sixty organic solvents having solubility parameters similar to that of PLGA polymers (i.e., 16-24 $MPa^{1/2}$) were tested for their capability of dissolving the PLGA polymer. Dried PLGA (125 mg) was added to glass vials containing 5 ml of the test organic solvent. The vials were agitated overnight at room temperature. Solvents were classified into four groups: good solvents that formed clear polymer solutions; intermediately good solvents that formed turbid polymer solutions upon heating; intermediately poor solvents that were marginally able to swell the polymer; and poor solvents in which the polymer remained intact. The result is summarized in Figure 3, which was drawn after the method of Teas *(12)*. It should be noted that only half of the screened organic solvents were able to dissolve PLGA. The rest were only able to marginally swell the polymer or were non-solvents. It is also noticeable that the good solvents form a reasonably well-defined area on the Teas graph. This result indicates that solubility of a polymer in a given solvent relies on the nature of intermolecular interactions as well as the closeness of the solubility parameters. Application of the Teas graph in the solubility prediction was described in detail in reference *(9)*.

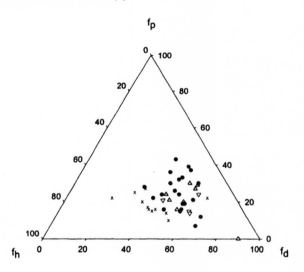

Figure 3. Comparison of Hansen's solubility parameters for various solvents: contributions of dispersion forces (f_d), polar interactions (f_p), and hydrogen bonding (f_h). (●) Good solvent; (▲) intermediately good solvent; (▽) intermediately poor solvent; and (✕) poor solvent. (Copyright 2003 Elsevier.)

Selected solvents for PLGA were further refined by their spreading capability and the quality of the formed membranes. Solutions of polymer in different solvents were placed on a layer of agarose gel. The polymer films thereby formed were evaluated with respect to their diameters and optical densities, which were used to estimate the degree of spreading of each solution and the quality of the polymer membrane, respectively. That a polymer solution formed a membrane of a relatively large diameter implied that the solution would spread easily over an aqueous droplet in the encapsulation process. The solutions forming membranes which displayed relatively high turbidities were discarded, since it meant formation of a rough and discontinuous precipitate that would not be able to control the drug release. Therefore, solvents making polymer solutions which formed membranes of large diameters and low turbidities were selected as desirable solvents, as indicated in Figure 4. Among the candidate solvents, ethyl acetate was used in following studies.

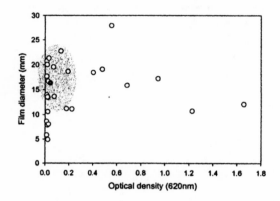

Figure 4. Evaluation of organic solvents based on their spreading over an agarose gel surface and subsequent formation of a polymer membrane. Preferable solvents lie in the shaded portion of the plot. Gray point indicates ethyl aceate. (Copyright 2003 Elsevier.)

Formation of Microcapsules by the Solvent Exchange Method

Figure 5 shows the formation of microcapsules in air from collision and merger of two microdroplets generated by two ink-jet nozzles. The merged droplets were subsequently collected in a water bath to leave reservoir-type microcapsules. The geometry of the microcapsules, consisting of a single aqueous core and a polymeric membrane surrounding the core, was demonstrated by confocal microscopy and clearly contrasted with those generated by the double emulsion method, as shown in Figure 6.

It is noteworthy that the reservoir-type microcapsules having a continuous polymer membrane formed only when the solvents in the candidate area (shown

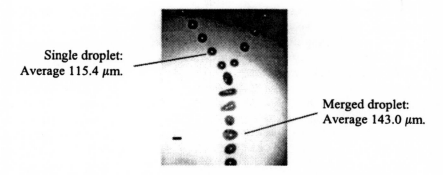

Single droplet:
Average 115.4 µm.

Merged droplet:
Average 143.0 µm.

Figure 5. Stroboscopic image of the microcapsule formation via midair collision between two component liquids. Here, the left stream is 0.25% alginate solution, and the right stream is 4% PLGA solution. Nozzle orifice diameter d = 60 µm; volumetric flow rate Q = 0.6 ml/min; Forcing frequency f = 10.6 kHz. Scale bar = 100 µm.

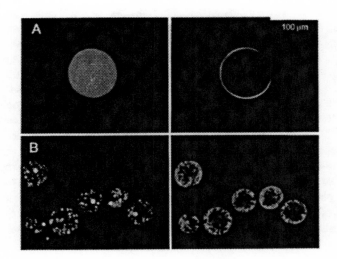

Figure 6. Confocal laser microscopic images of microparticles produced by (A) the solvent exchange method and (B) the double emulsion-solvent evaporation method. Left and right panels indicate aqueous phases labeled with FITC-dextran and polymer phase labeled with Nile Red, respectively.

in Figure 4) were used. Benzyl alcohol and acetic acid, for example, which lie outside the candidate area, were not able to form the membrane around the aqueous core or formed discontinuous membrane, respectively. Therefore, this result indicates that the criteria used in the solvent selection were valid guides for selection of appropriate solvents.

Another interesting finding is that the size of microcapsules was relatively homogeneous, determined by the interplay among different instrumental variables *(11)*. The size of microdroplets produced by an ink-jet nozzle is a function of three variables: diameter d of the orifice, linear velocity of the jetted solution V (or volumetric flow rate Q), and forcing frequency f. In theory, the diameter d_d of the microdroplets can be calculated by equating the volume of a fraction of the liquid jet ($\pi(d/2)^2\lambda$) emerging from the ink-jet nozzle and that of resulting spheres ($(1/6)\pi d_d^3$):

$$d_d = (\frac{3d^2V}{2f})^{\frac{1}{3}} = (\frac{6Q}{\pi \cdot f})^{\frac{1}{3}}$$

In our previous study it was shown that the experimental values precisely agreed with the calculation *(11)*. That is, the droplet size was primarily determined by the nozzle orifice, decreased with increasing frequency at a fixed flow rate, and increased with increasing flow rate at all tested levels of frequency. As for the size of microcapsules, which formed as a result of merging two equal-sized microdroplets, it was expected that the microcapsule diameter would be 1.26 times of the single droplets, assuming that there was no loss of material upon their collision. The sizes measured from the stroboscopic pictures satisfied these expectations for both single and merged droplets as shown in Figure 5. On the other hand, the majority of microcapsules collected in the water bath were close to single droplets in size, and the membrane existed only as a thin membrane. First, it is possible that the polymer layer shrank as the solvent that constituted 95-98% of the polymer phase was extracted into the aqueous phases by the solvent exchange. Alternativerly, considering the existence of occasional satellites, it is also possible that portions of the polymer layer separated during stirring of the bath, leading to reduction of the membrane thickness.

Conclusions

The solvent exchange method has been developed to address the traditional difficulties in protein microencapsulation. Reservoir-type microcapsules were produced, using a dual microdispenser system based on midair collision bet-ween component materials, followed by interfacial phase separation of the polymeric membrane. From the mild nature of the encapsulation process and the unique geometry of the microcapsules, it is expected that this method will provide several advantages over contemporary methods, especially in encapsulation of proteins or peptides. First, the process does not include potentially damaging conditions such as an emulsification step, which often exerts unfavorable influences on the stability of encapsulated drugs by exposing them to the w/o interface and excessive physical stress. Second, in the mononuclear microcapsules, undesirable interactions between protein and organic solvent or polymer matrix are limited to the interface at the surface of the core. However, whether these potential advantages will be reflected through enhanced release profiles and stability of the encapsulated proteins remains to be seen. Third, the organic solvent for polymers can be chosen with more flexibility than in conventional methods; hence, toxicity concerns over residual solvents, in particular methylene chloride, can be avoided. Fourth, the use of ink-jet nozzles allows for a precise control over the particle size.

Acknowledgments

This study was supported in part by the National Institutes of Health through grant GM67044, Samyang Corporation, Purdue Research Foundation, and NSF Industry/University Center for Pharmaceutical Processing Research.

References

1. Robinson, J.R. Controlled drug delivery: past, present, and future. In *Controlled Drug Delivery Challenges and Strategies*; Park, K., Ed.; ACS Professional Reference Book: Washington DC, **1997**, 1-7.
2. van de Weert, M. Structural Integrity of Pharmaceutical Proteins in Polymeric Matrices. Utrecht University (Thesis), The Netherlands, **2001**.
3. Schwendeman, S.P. Recent advances in the stabilization of proteins encapsulated in injectable PLGA delivery systems. *Crit. Rev. Ther. Drug Carrier Syst.* **2002**, *19*, 73-98.
4. van de Weert, M.; Hennink, W. E.; Jiskoot, W. Protein instability in PLGA microparticles. *Pharm. Res.* **2000**, *17*, 1159-1167.
5. Sah, H. Protein instability toward organic solvent/water emulsification: implications for protein microencapsulation into microspheres. *PDA J. Pharm. Sci. Technol.* **1999**, *53*, 3-10.
6. Perez, C.; Castellanos, I.J.; Costantino, H.R.; Al-Azzam, W.; Griebenow, K. Recent trends in stabilizing protein structure upon encapsulation and release from bioerodible polymers. *J. Pharm. Pharmacol.* **2002**, *54*, 301-313.

7. Morlock, M.; Koll, H.; Winter, G.; Kissel, T. Microencapsulation of rh-erythropoietin, using biodegradable PLGA: protein stability and the effects of stabilizing excipients. *Eur. J. Pharm. Biopharm.* **1997**, *43*, 29-36.
8. Kim, H.K.; Park, T.G. Microencapsulation of human growth hormone within biodegradable polyester microspheres: protein aggregation stability and incomplete release mechanism. *Biotechnol. Bioeng.* **1999**, *65*, 659-667.
9. Yeo, Y.; Basaran, O.A.; Park, K. A new process for making reservoir-type microcapsules using ink-jet technology and interfacial phase separation. *J. Controlled Rel.* **2003**, *93*, 161-173.
10. Cohen, S.; Yoshioka, T.; Lucarelli, M.; Hwang, L.H.; Langer, R. Controlled delivery systems for proteins based on PLGA microspheres. *Pharm. Res.* **1991**, *8*, 713-720.
11. Yeo, Y.; Chen, A.U.; Basaran, O.A.; Park, K. Solvent exchange method: a novel microencapsulation technique using dual microdispensers. *Pharm. Res.* **2004**, In press.
12. Teas, J.P., Graphic analysis of resin solubilities. *J. Paint Technol.* **1968**, *40*, 19-25.

Chapter 18

Hollow Microcapsules for Drug Delivery by Self-Assembly and Cross-Linking of Amphiphilic Graft Copolymers

Kurt Breitenkamp[1], Denise Junge[2], and Todd Emrick[1,*]

[1]Department of Polymer Science and Engineering and [2]Conte Center for Polymer Research, University of Massachusetts at Amherst, 120 Governors Drive, Amherst, MA 01003
*Corresponding author: tsemrick@mail.pse.umass.edu

The synthesis of hollow, micron-sized capsules is described, using an oil-water interfacial assembly, and subsequent cross-linking, of amphiphilic graft copolymers. Poly(cyclooctene)-g-poly(ethylene glycol) (PEG) copolymers, synthesized by ring-opening metathesis copolymerization of PEG-functionalized cyclooctene macromonomers with other cyclooctene derivatives, are observed to segregate to the toluene-water interface. Covalent crosslinking by ring-opening cross-metathesis with a bis-cyclooctene PEG derivative imparts mechanical integrity to these hollow capsules. Fluorescently-labeled crosslinked capsules are examined in solution using laser scanning confocal microscopy (LSCM). This strategy can be utilized for the preparation of encapsulants and carriers for hydrophobic molecules, such as the potent anti-cancer drug doxorubicin (DOX). These novel capsules may be well suited for controlled release therapies, where the transport of drugs can be regulated by factors such as crosslink density, hydrolytic stability, and enzymatic stability of the polymer.

Introduction

Advances in modern medicine for the treatment of human illness continue to improve the length and quality of human life. While new therapeutic agents play a major role in these advances, drug delivery strategies also represent key components of successful treatments. Synthetic polymers are important for implementing delivery strategies, and hold significant promise for enhancing considerably the therapeutic benefits of current and future drugs. While polymer-based drug delivery systems provide a means to improve the bioavailability, safety, and efficacy of drug therapy, several important problems are currently under investigation. For example, a number of strategies are being developed to improve the *controlled* nature of therapeutic delivery to include long time periods, predetermined time intervals, and response to physiological changes. In addition, functional polymeric materials are being designed to provide *targeted* drug delivery mechanisms with a specificity that simultaneously promotes healing and diminishes unwanted side effects.

Polymer-Drug Conjugation

New synthetic polymers are critically important to addressing complex issues in the drug delivery field. Since the introduction of polymer-based drug delivery principles by Ringsdorf and coworkers, *(2)* polymer therapeutics has expanded considerably in scope and is now poised to revolutionize medical treatment protocols *(3)*. The covalent attachment of drugs to polymeric backbones (i.e., polymer-drug conjugation) has been explored extensively in efforts to improve blood solubility, circulation time, and efficacy of small molecule drugs. This is evidenced by a growing number of poly(ethylene glycol) (PEG)-drug conjugates, both reported in the literature and entering clinical practice. Such PEGylation provides an aqueous solubility to hydrophobic drugs, and simultaneously slows metabolism and renal clearance of the drug *(4)*. The key features of PEGylation extend to delivery of protein- and gene-based drugs that are normally susceptible to rapid *in vivo* degradation.

New polymeric materials and conjugation techniques, appropriately tailored for drug delivery, will drive further growth in polymer therapeutics. This is exemplified by a number of new conjugate systems, based on well-defined polymer architectures, which promise to enhance drug delivery systems by cell targeting. For example, poly(N-(2-hydroxypropyl)methacrylamide) (HPMA) drug conjugates (Figure 1), reported by Duncan and coworkers, have shown good promise in clinical trials. The covalent coupling of chemotherapy drugs such as doxorubicin (DOX) and paclitaxel to these HPMA copolymers resulted in increased uptake into tumor cells relative to non-conjugated DOX, attributed to the enhanced permeability and retention (EPR) effect *(5)*. The EPR effect results from the disorganized nature of the tumor vasculature, which leads to an increased permeability to polymer therapeutics. Poor lymphatic drainage in the tissue allows drug retention at the tumor site. The EPR effect is a unique feature of polymer-based therapeutics, which arises from increased blood circulation times of the polymer-drug conjugate relative to their non-conjugated analogues.

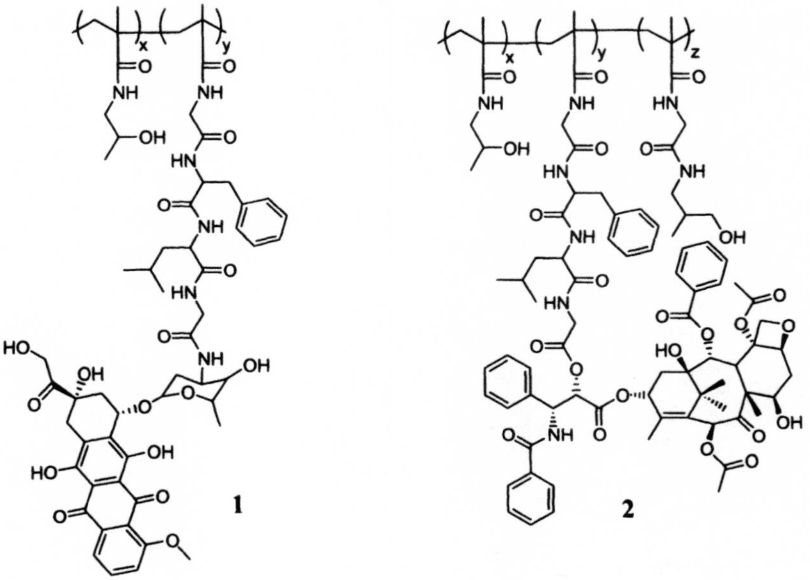

Figure 1. Chemical structures of HPMA-drug conjugates used in clinical trials: doxorubicin-HPMA copolymer (1) and paclitaxel-HPMA copolymer (2).

The manipulation and control of polymer architecture offers exciting opportunities in drug delivery. Conjugates based on branched polymer architectures are particularly interesting, as these polymers possess (1) multiple chain-ends for drug attachment, and (2) the potential for engineering degradation sites and cell targeting moieties into the structure. In this context, dendritic polymers are quite unique as their very highly branched structure leads to a large number of chain-ends for drug attachment. In addition, the uniform molecular weight distribution of dendrimers, reminiscent of proteins, makes them quite special relative to typical polymers and provides a well-characterized molecular plat-form that carries appeal to the medical community. Fréchet and coworkers are among the leaders in designing dendrimers for drug delivery and have reported several examples of water soluble, polyester-based dendrimer-DOX conjugates *(6)*. These conjugates are modular by design and can be equipped with higher drug payloads than linear polymer-drug conjugates, as well as chain-end solubi-lizing groups and cell-targeting moieties. The utility of dendrimer-DOX conju-gates was demonstrated through *in vitro* cell models, which show dramatically increased serum half-life compared to free DOX, as well anti-proliferation effects in cancer cells. In preliminary *in vivo* studies, the dendrimer-drug conjugates were well-tolerated upon intravenous injection into mice, and showed no accumulation in liver, heart, or lung tissue *(7)*.

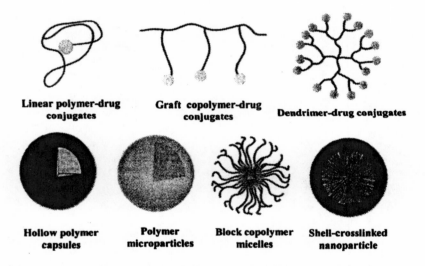

Linear polymer-drug conjugates   **Graft copolymer-drug conjugates**   **Dendrimer-drug conjugates**

Hollow polymer capsules   **Polymer microparticles**   **Block copolymer micelles**   **Shell-crosslinked nanoparticle**

Figure 2. Polymer architectures and encapsulation techniques for drug delivery. (See page 3 of color inserts.)

Drug Encapsulation

An emerging effort in polymer therapeutics centers on drug encapsulation using polymer assemblies. Both block copolymers and functional polymers of complex architectures can be used as precursors to micro- and nanoscale drug delivery vehicles. For example, amphiphilic triblock copolymers known as Pluronics®, composed of poly(ethylene oxide)-*b*-poly(propylene oxide)-*b*-poly (ethylene oxide), are the polymers of choice for a number of researchers wor-king in the polymer therapeutics field. In water, these amphiphilic copolymers assemble by multimolecular micellization when present in solution above the critical micelle concentration (CMC). The hydrophobic PPO block collapses to form the micellar core, while the hydrophilic PEG blocks form a peripheral corona (Figure 3). These core-shell polymer micelles have been used to encapsulate molecules, including small molecule drugs, polypeptides, and polynucleotides *(8)*.

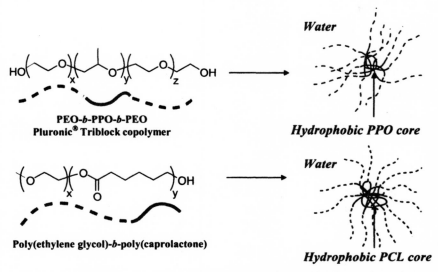

Figure 3. Schematic representation of amphiphilic triblock and diblock copolymer assembly in water.

A variety of amphiphilic diblock copolymers can self-assemble in aqueous solution to form nano- and micron-sized structures, with applications from drug encapsulation to artificial viruses and cells *(9)*. While PEG is most often used as the hydrophilic segment, a variety of hydrophobic core-forming blocks can be utilized, providing a means to tune such features as drug loading, stability, and release characteristics *(10)*. Figure 3 (bottom) illustrates a PEG-*b*-poly(ε-caprolactone) (PEG-*b*-PCL) diblock copolymer, where the PCL core encapsulates and solubilizes hydrophobic compounds such as indomethacin *(11)*. Recently, Eisenberg and coworkers used PEG-*b*-PCL copolymers to demonstrate *in vitro* delivery of organic dyes to intracellular compartments *(12)*. Issues to address for present and future development of block copolymer drug delivery vehicles include (1) improved stability of the assembly upon dilution in the bloodstream, (2) improved drug payload, and (3) improved targeting by integration of cell-seeking functionality. Wooley and coworkers are among the research groups addressing these issues through novel synthesis, for example by covalently crosslinking a poly(acrylic acid) (PAA) micellar corona around a

258

poly(ε-caprolactone) core to form shell-crosslinked nanoparticles (SCN) *(13)*. The cross-linking imparts stability to the nanostructures, while removal of the polyester core by hydrolysis significantly increases the encapsulation volume.

Amphiphilic Polyolefin-g-PEG Copolymers

We are investigating a new series of amphiphilic graft copolymers for drug encapsulation, specifically polyolefins with pendant hydrophilic PEG grafts. These copolymers are synthesized by ring-opening metathesis polymerization (ROMP) of PEG-functionalized cyclic olefin macromonomers to afford graft copolymers with a number of tunable features, such as graft density and length, crystallinity, and amphiphilicity *(14)*. The amphiphilic nature of these graft copolymers can be exploited for assembly, and the unsaturation present in the poly(cyclooctene) backbone can be used for covalent crosslinking of the polymers at the oil-water interface to afford hollow capsules *(15)*. This covalent crosslinking results in a thin, mechanically robust, polymer membrane, with features that can be tuned to control drug release. The copolymerization of PEG-functional cyclic olefins with those containing oligopeptides and other bio-functionality may lead to particularly effective use of these novel capsules in delivery applications as well as other bio-related applications of synthetic polymers.

Materials and Methods

Cyclooctene, cyclooctadiene, succinic anhydride (99%), and lithium aluminum hydride (95%) were obtained from Alfa Aesar. *m*-Chloroperoxybenzoic acid (MCPBA) (77%), 4-dimethylaminopyridine (DMAP) (99%), 1,3-dicyclohexylcarbodiimide (DCC) (99%), ethyl vinyl ether (99%), and 1,3-bis-(2,4,6-trimethylphenyl)-2-(imidazolidinylidene)dichloro (phenylmethylene) (tricyclohexylphosphine) ruthenium (Grubbs' Generation II catalyst) were purchased from Aldrich. Poly(ethylene glycol) monomethyl ether (mPEG) 750 was purchased from Polysciences, Inc., and purified by column chromatography (in CH$_2$Cl$_2$/Acetone/MeOH mixtures) before use. mPEG 1000 was purchased from Shearwater, Inc. and used without further purification.

Molecular weights and polydispersity indices (PDIs) were measured using gel permeation chromatography (GPC) in DMF (0.01 M LiCl, 0.5 mL/min), and referenced against linear polystyrene standards. The chromatographic system utilized three Polymer Laboratories Mixed-D columns, an HP 1048 gradient pump, and a refractive index detector (HP R4010).

NMR spectra were collected on a Bruker DPX 300 spectrometer (referenced to CDCl$_3$): ^{1}H at 300 MHz and ^{13}C at 75 MHz. UV data was obtained on a Hitachi U-3010 spectrophotometer at a scan rate of 60 nm/min. Fluorescence confocal microscope images were obtained using an inverted microscope with a TCS SP2 confocal system (Leica). Atomic force microscope (AFM) images were obtained using a Digital Instruments Dimension 3100 AFM.

Results and Discussion

Amphiphilic polymers are particularly useful for mediating interfaces, as demonstrated by their history of applications as polymeric surfactants. While amphiphilic block copolymers and polyelectrolytes have been well-studied, amphiphilic graft copolymers have received comparatively little attention (16-18). The novel amphiphilic graft polymers synthesized and studied in our labs contain a hydrophobic polyolefin backbone with hydrophilic poly(ethylene glycol) (PEG) pendant chains. We have used ruthenium benzylidene catalysts such as bis(tricyclohexylphosphine)benzylidene ruthenium (IV) dichloride (Grubbs' Generation I catalyst), and the mono-1,3-dimesitylimidazolidine-2-ylidene derivative (Grubbs' Generation II catalyst) to prepare amphiphilic graft copolymers by ring-opening metathesis polymerization (ROMP) of PEG-functional macromonomers, as illustrated in Figure 4. These novel functional polyolefins cannot be made by classic olefin polymerization techniques, but rather by newer ring-opening methods that are tolerant of polar functionality (19). In addition to PEG grafts, metathesis polymerization provides an effective method for the introduction of a variety of functionality onto the polymer backbone through copolymerization of X-substituted cyclic olefins. Cyclooctene derivatives containing alcohol and carboxylic acid functionality, fluorescent labels, and pendant polymer chains such as PEG and oligopeptides were prepared and polymerized in these studies. Utilizing the macromonomer approach is important, as this precludes the need for post-polymerization grafting that often presents difficulties in terms of yields and the grafting of multiple components. In addition, using the macromonomer approach, copolymers can be tuned considerably in terms of backbone and graft molecular weights. By varying the concentration of the macromonomer relative to unsubstituted cyclooctene, properties such as water-solubility, amphiphilicity, and crystallinity can be altered.

Polymers 6-11 in Table I were prepared by copolymerizations of macromonomer 1 and cyclooctene in equimolar ratios using either Grubbs' Generation I or Generation II catalyst. Catalyst choice and stoichiometries are key factors in determining macromonomer conversion and copolymer properties. Co-polymers 6-8 were synthesized using Grubbs' Generation I catalyst, where 60-70% of the macromonomer became incorporated into the polymer product, as determined by ^{1}H NMR spectroscopy. Copolymers 9-11 represent similar studies, using the more active Grubbs' Generation II catalyst, where nearly complete conversion of macromonomer 3 was obtained. To demonstrate the tunability of polymer molecular weight, copolymers 9-11 were synthesized using different ratios of monomer to Grubbs' Generation II catalyst. Monomer-to-catalyst ratios of 500:1 (copolymer 9) and 100:1 (copolymer 11) gave products with M_n 540,000 and 220,000 g/mol, respectively. The amphiphilic polycyclooctene-*g*-PEG copolymers are soluble in a range of organic solvents (MeOH, acetone, CH_2Cl_2, $CHCl_3$, THF, DMF, and toluene).

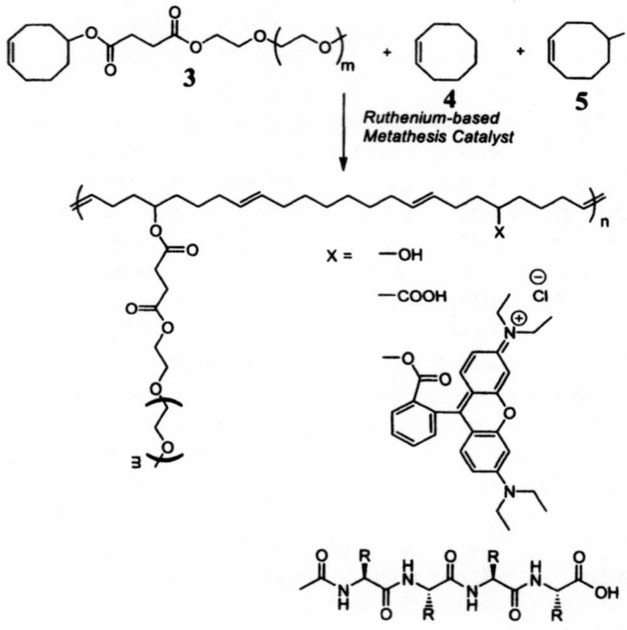

Figure 4. Synthesis of functional, amphiphilic graft copolymers by ring-opening metathesis polymerization of PEG-functional cyclooctene macromonomers with other cyclooctene derivatives.

The amphiphilic graft copolymers described above dissolve in water and stabilize oil-in-water emulsions for hours to days or more, depending on macro-monomer content. To enhance the structural stability of these assemblies, we have investigated covalent crosslinking strategies at the oil-water interface, utilizing the unsaturated polyolefin backbone of the graft copolymer. For this we prepared difunctional, bis-cyclooctene PEG amphiphiles, such as compound 13, for co-assembly at the oil-water interface and subsequent crosslinking by ring-opening cross-metathesis. This is depicted in Figure 5 for the generation of hollow microcapsules, containing the desired crosslinked graft copolymer membrane. Experimentally, a graft copolymer sample of type 12 is dissolved in toluene with the *bis*-cyclooctene cross-linker. Grubbs' Generation II catalyst is added to the solution, and the mixture is added to water. Gentle shaking for several minutes gives cross-linked capsules that can be isolated by filtration or by drying onto a solid substrate. The benefits of capsule cross-linking for delivery systems are two-fold. First, crosslinking makes the capsule amenable to delivery applications *in vivo*, where bloodstream stability is paramount. Second, the crosslinked polymer membrane can be utilized and tuned to control diffusion and release of encapsulated therapeutics. The ester-containing cross-links hydrolyze in aqueous solution, and the extent of crosslinking, combined with variation of the PEG linker chemistry, provides opportunities for controlling drug diffusion across the membrane. We are studying these topics presently.

Table I. Macromonomer Incorporation and Molecular Weights of Polycyclooctene-*g*-PEG Copolymers

Entry[a]	M:Cat I[b]	M:Cat II[c]	[M] (mol/L)	f_{macro}[d]	F_{macro}[e]	10^{-3} M_n[f]	PDI
1	500	-	2.0	0.5	0.35	330	1.61
2	250	-	2.0	0.5	0.4	220	1.59
3	100	-	2.0	0.5	0.4	140	1.54
4	-	500	2.0	0.5	0.49	540	2.2
5	-	250	1.0	0.5	0.5	330	1.56
6	-	100	1.0	0.5	0.5	220	1.6

[a]Polymerizations performed at 40°C in CH_2Cl_2. [b]Cat I = Grubbs' Generation I catalyst. [c]Cat II = Grubbs' Generation II catalyst. [d]f_{macro} = feed ratio of macromonomer relative to cyclooctene, determined by ^{1}H NMR. [e]F_{macro} = fraction of macromonomer incorporated into polymer, determined by ^{1}H NMR. [f]Determined by GPC in DMF versus linear polystyrene standards

Integration of fluorescent moieties into the polymer backbone, such as the rhodamine B labeled cyclooctene shown in Figure 4, allows for the observation of interfacial segregation using laser scanning confocal microscopy. Several such confocal images are shown in Figure 6. Figure 6a shows a cross-section of an oil-filled microcapsule in water (fluorescence emission at 556 nm), and illustrates the strong interfacial segregation of the graft copolymer. Figure 6b is a three-dimensional projection image of the capsules, which shows their spherical nature and confirms droplet coverage by the graft copolymer. To demonstrate the effectiveness of the interfacial crosslinking, we replaced the oil-water medium with ethanol as a single solvent. Rather than dissolve away, the capsule membrane collapses, as shown in Figure 6c. The collapsed capsules can be isolated by drying onto a solid substrate and subsequently redispersed in a solvent such as water or ethanol.

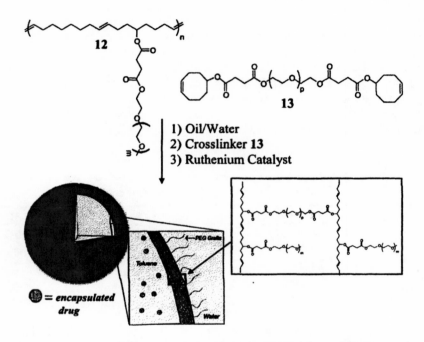

Figure 5. Schematic illustration of interfacial assembly of polycyclooctene-g-PEG and crosslinking by ring-opening cross-metathesis with a bis-cyclooctene crosslinker 13. (See page 4 of color inserts.)

The ease with which these oil-filled capsules can be prepared suggests a practicality for many types of encapsulation, where the encapsulant is either dissolved in an oil phase or is itself a hydrophobic liquid. In many cases, clinical use of attractive drug candidates has been impeded by their poor aqueous solubility. For the chemotherapy drug doxorubicin (DOX) (shown in Figure 6), water solubility is achieved by protonation of the secondary amine. Drug modifications of this type can alter pharmacokinetic profiles and in some cases increase renal clearance and thus shorten circulation lifetimes. Using the oil-in-water assembly method described here, we have encapsulated DOX in its neutral form in the hollow capsule interior. The inherent fluorescence of DOX allows its visualization in the capsule by confocal microscopy. Figure 6d shows a three-dimensional projection image of these DOX-filled capsules (fluore-scence at 580 nm). Relative to lipid emulsions or polymer micelles, drug encapsulation using these hollow crosslinked capsules allows for higher drug loadings and is limited only by the solubility of the drug in the oil medium. We are now developing new crosslinking strategies that permit drying and redispersion of the drug filled capsules in water or ethanol (data not shown). Diffusion studies of DOX across the polymer membrane, as a function of crosslink density, are also in progress.

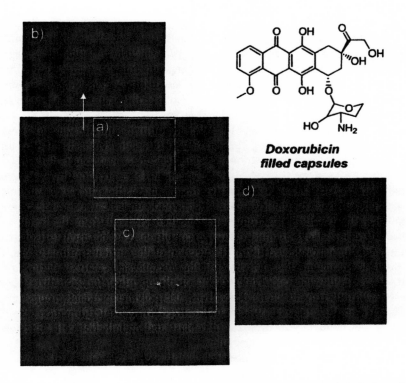

Doxorubicin
filled capsules

Figure 6. Laser scanning confocal microscopy of fluorescent-labeled graft copolymer capsules. (a) cross-sectional slice; (b) 3-D projection image of polymer covered capsules; (c) collapsed capsule after introduction of ethanol; (d) projection image of DOX-filled capsules. (See page 5 of color inserts.)

While the EPR effect provides a passive method for cell targeting with polymer-drug conjugates and polymer assemblies, researchers are now investigating delivery systems that contain ligands designed to interact preferentially with particular cell types. Such targeted delivery techniques are expected to enhance drug localization at the disease site and increase the efficacy of the drug, while effectively lowering the required dosage. This is particularly important for

264

the delivery of highly toxic drugs such as many chemotherapy drugs, where reducing the risk of side effects on healthy cells and tissue is critical. Functionality of interest for cell targeting includes folic acid, oligopeptides, and monoclonal antibodies. Conjugation of such targeting groups to polymeric materials has been reported on amphiphilic diblock copolymers, shell-cross-linked nanoparticles, and a variety of PEG-drug conjugates *(20-22)*. Despite the advances contained in these reports, considerable efforts lie ahead for researchers to develop practical, effective, and disease-targeting polymer therapeutics.

An attractive feature of capsules formed by interfacial assembly of amphiphilic graft copolymers is the relative ease with which highly functional capsule surfaces can be achieved. Tedious end-group functionalization is typically required to integrate functionality to the surface of block copolymer assemblies. In the case of amphiphilic graft polymers, functionality can be easily integrated into the copolymer structure, and thus the capsule surface, through the polymerization of functional monomers. We are utilizing this feature to functionalize capsules with cell-targeting moieties through synthesis and copolymerization of oligopeptide functional cyclooctene macromonomers for an RGD-substituted cyclooctene (Figure 7). Oligopeptide 15 is prepared by solid phase coupling chemistry on a 2-chlorotrityl resin, and carboxylic acid substituted cyclooctene 14 is then coupled to the oligopeptide under carbodiimide coupling conditions; the resulting macromonomer is then cleaved from the resin. The choice of resin is key, as mild cleavage conditions can be used to isolate the oligopeptide in its protected form. Protection of the arginine primary amine is required to maintain the activity of the metathesis catalyst during polymerization. Following copolymerization of oligopeptide-functionalized monomer 16 with other cyclooctene derivatives, the polymer pendant groups can be deprotected in the presence of acid and utilized for capsule formation. Studies are currently underway to evaluate fibroblast and endothelial cell adhesion to RGD-functionalized polyolefin capsules.

The defining quality of the crosslinked polyolefin-g-PEG capsules presented here is the versatility of the chemistry by which they are synthesized. Through controlled variation of features such as PEG density and chain length, crosslink density, and exterior functionality, the capsules can be tuned in terms of size, drug release profile, and cell-targeting capabilities. Future work on these novel drug delivery systems will focus on optimizing many of these parameters and evaluating polymer and capsule structure-property relationships. We are also exploring the effects of PEG-chain length on capsule size, parti-cularly towards

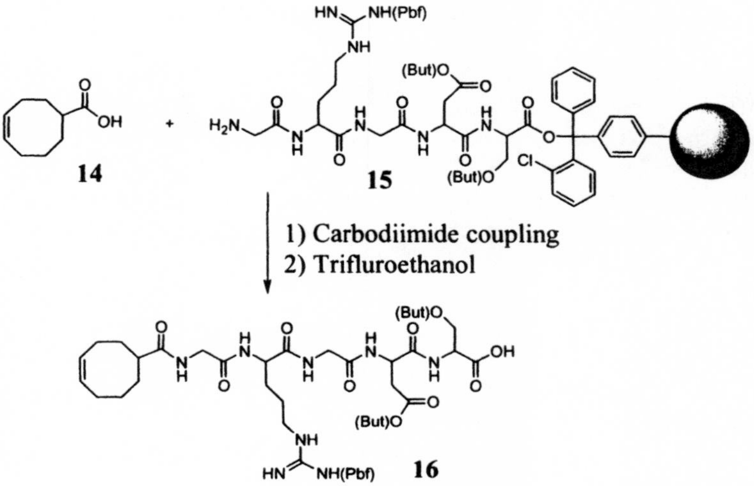

Figure 7. Synthesis of protected RGD-containing oligopeptide macromonomer.

materials in the sub-micron range. One example currently under investigation uses an oligopeptide-functional crosslinker, which is preferentially cleaved in the presence of prostate-specific antigen (PSA), an enzyme that is overproduced in prostate cancer cells.

Conclusions

In summary, we have described the synthesis, interfacial assembly and co-valent crosslinking of novel, amphiphilic graft copolymers at the oil-water interface to generate hollow capsules with mechanical integrity and tunable porosity. By integration of fluorescent monomers into the polymer backbone, the capsule can be viewed and characterized in solution by laser scanning confocal microscopy. This encapsulation method is particularly applicable to hydrophobic drugs such as doxorubicin, and the synthetic approach provides a number of opportunities for integration of cell-targeting and other types of functionality onto the capsule periphery. Future reports will quantify drug diffusion across the polymer membrane and employ crosslinking strategies that provide a controlled drug release over long and defined time periods.

266

Acknowledgements

The authors acknowledge financial support from the Center for University-Industry Research on Polymers at UMass Amherst (Exploratory Research Award), the National Science Foundation Research Site for Educators in Chemistry (RSEC) program at UMass Amherst, the Office of Naval Research (N00014-03-1-1000), and the National Science Foundation Materials Research Science & Engineering Center (MRSEC) at UMASS Amherst.

References

1. Address: Department of Chemistry, Keene State College, 229 Main Street, Keene, NH 03435
2. Ringsdorf, H. Structure and Properties of Pharmacologically Active Polymers. *J. Polym. Sci. Pol. Sym.* **1975**, 135-153.
3. Duncan, R. The dawning era of polymer therapeutics. *Nat. Rev. Drug Discov.* **2003**, *2*, 347-360.
4. Greenwald, R.B. PEG drugs: an overview. *J. Control. Release* **2001**, *74*, 159-171.
5. Duncan, R.; Gac-Breton, S.; Keane, R.; Musila, R.; Sat, Y.N.; Satchi, R.; Searle, F. Polymer-drug conjugates, PDEPT and PELT: basic principles for design and transfer from the laboratory to clinic. *J. Control. Release* **2001**, *74*, 135-146.
6. Ihre, H.R.; De Jesus, O.L.P.; Szoka, F.C.; Fréchet, J.M.J. Polyester dendritic systems for drug delivery applications: Design, synthesis, and characterization. *Bioconjugate Chem.* **2002**, *13*, 443-452.
7. De Jesus, O.L.P.; Ihre, H.R.; Gagne, L.; Fréchet, J.M.J.; Szoka, F.C. Polyester dendritic systems for drug delivery applications: In vitro and in vivo evaluation. *Bioconjugate Chem.* **2002**, *13*, 453-461.
8. Kabanov, A.V.; Batrakova, E.V.; Alakhov, V.Y. Pluronic® block copolymers as novel polymer therapeutics for drug and gene delivery. *J. Control. Release* **2002**, *82*, 189-212.
9. Discher, D.E.; Eisenberg, A. Polymer vesicles. *Science* **2002**, *297*, 967-973.
10. Adams, M.L.; Lavasanifar, A.; Kwon, G.S. Amphiphilic block copolymers for drug delivery. *J. Pharm. Sci.-US* **2003**, *92*, 1343-1355.
11. Kim, S.Y.; Shin, I.G.; Lee, Y.M.; Cho, C.S.; Sung, Y.K. Methoxy poly(ethylene glycol) and epsilon-caprolactone amphiphilic block copolymeric micelle containing indomethacin. II. Micelle formation and drug release behaviours. *J. Control. Release* **1998**, *51*, 13-22.
12. Savic, R.; Luo, L.B.; Eisenberg, A.; Maysinger, D. Micellar nanocontainers distribute to defined cytoplasmic organelles. *Science* **2003**, *300*, 615-618.

13. Zhang, Q.; Remsen, E.E.; Wooley, K.L. Shell cross-linked nanoparticles containing hydrolytically degradable, crystalline core domains. *J Am Chem Soc* **2000**, *122*, 3642-3651.
14. Breitenkamp, K.; Simeone, J.; Jin, E.; Emrick, T. Novel amphiphilic graft copolymers prepared by ring-opening metathesis polymerization of poly(ethylene glycol)-substituted cyclooctene macromonomers. *Macromolecules* **2002**, *35*, 9249-9252.
15. Breitenkamp, K.; Emrick, T. Novel polymer capsules from amphiphilic graft copolymers and cross-metathesis. *J. Am. Chem. Soc.* **2003**, *125*, 12070-12071.
16. O'Donnell, P.M.; Brzezinska, K.; Powell, D.; Wagener, K.B. "Perfect comb" ADMET graft copolymers. *Macromolecules* **2001**, *34*, 6845-6849.
17. Hester, J.F.; Banerjee, P.; Mayes, A.M. Preparation of protein-resistant surfaces on poly(vinylidene fluoride) membranes via surface segregation. *Macromolecules* **1999**, *32*, 1643-1650.
18. Heroguez, V.; Breunig, S.; Gnanou, Y.; Fontanille, M. Synthesis of alpha-norbornenylpoly(ethylene oxide) macromonomers and their ring-opening metathesis polymerization. *Macromolecules* **1996**, *29*, 4459-4464.
19. Scholl, M.; Ding, S.; Lee, C.W.; Grubbs, R.H. Synthesis and activity of a new generation of ruthenium-based olefin metathesis catalysts coordinated with 1,3-dimesityl-4,5-dihydroimidazol-2-ylidene ligands. *Org. Lett.* **1999**, *1*, 953-956.
20. Pan, D.; Turner, J.L.; Wooley, K.L. Folic acid-conjugated nanostructured materials designed for cancer cell targeting. *Chem. Commun.* **2003**, 2400-2401.
21. Yamamoto, Y.; Nagasaki, Y.; Kato, M.; Kataoka, K. Surface charge modulation of poly(ethylene glycol)-poly(D, L-lactide) block copolymer micelles: conjugation of charged peptides. *Colloid Surface B* **1999**, *16*, 135-146.
22. Jones, S.D.; Marasco, W.A. Antibodies for targeted gene therapy: extracellular gene targeting and intracellular expression. *Adv. Drug Del. Rev.* **1998**, *31*, 153-170.

Chapter 19

Microencapsulation in Yeast Cells and Applications in Drug Delivery

G. Nelson[*], S. C. Duckham, and M. E. D. Crothers

Micap plc, No.1, The Parks, Lodge Lane, Newton-Le-Willows,
WA12 0JQ, United Kingdom
*Corresponding author: gnelson@micap.co.uk

Yeast cells as preformed microcapsules can be used to improve the bioavailability of poorly soluble drugs in the gastrointestinal tract. Microorganisms have been recognised as
potential preformed natural microcapsules since the early
1070s, when Swift and Co., USA, patented a technique using
specifically prepared yeast containing high concentrations of
lipid, greater than 40% by weight. Using commercially
available yeast strains, such as *Saccharomyces cerevisiae*,
from the baking and brewing industry, this paper describes the
encapsulation of peppermint, fenofibrate and econazole nitrate
and explores the release of peppermint in the mouth of human
volunteers and fenofibrate in the duodenum of beagle dogs.

Introduction

Microorganisms have been recognised as potential preformed natural
microcapsules since the early 1070s, when Swift and Co., USA, patented a
technique using specifically prepared yeast, containing high concentrations of
lipid, greater than 40% by weight *(1)*. Synthetic polymeric carrier materials and
polymers extracted from natural materials have been investigated for drug

delivery applications over the last few decades *(2)*. However, as an alternative, in recent years microorganisms have been considered as novel vectors for the delivery of recombinant vaccines and other bioactive proteins and peptides via the gastrointestinal tract *(3-5)*. Working particlualry with live *lactobacillus* species and common baker's yeast (*Saccharomyces cerevisiae*), researchers have promoted the use of these organisms for targeting site-specific areas throughout the digestive tract. The live organism would potentially manufacture the beneficial compounds *in-situ*, either expressing on the surface or secreting the compound in response to the digestive environment or as a result of cell lysis *(6-8)*. However, many would consider the use of live organisms for drug delivery difficult to accept and have approached drug delivery using non-viable bacteria *(9)*, yeast and yeast cell walls, where much of the site-specific targeting remains possible without problematic issues in controlling the activity of the microorganism. This paper will concentrate on the roll of non-viable yeast cells as a drug delivery vehicle.

Yeast are unicellular fungi and must be considered as one of the most commercially important classes of microorganism with applications in brewing, (beer and wine production), ethanol production, baking, and production of recombinant proteins and biopharmaceuticals *(10)*. The cell mass is used to produce yeast extract for food and fermentation media applications, and whole yeast cells are incorporated into human food and animal feed *(11)*. In the health food industry, yeast cells and the vitamins they contain are utilized widely *(12)*.

Although there are many species of yeast which reproduce by budding or fission, by far the most commercially important belong to the genus *Saccharomyces*, probably the best-characterized of the yeast genera. The ascomycetous yeast *Saccharomyces cerevisiae* is a key microorganism in traditional biotechnological processes as well as a tool of choice to understand the eukaryotic cell.

An average *Saccharomyces* cell is approximately 5 micrometers in diameter, although other species of yeast can be as much as 20 micrometers in diameter. The ovoid shape is defined by the cell wall, a physically rigid structure which primarily protects the yeast internal membrane and organelles from the environment *(13)*. The cell wall is composed of complex and highly cross-linked glucan, mannan and, to a lesser degree, chitin which is associated with the bud scars remaining after a daughter cell has budded from the mother cell. The wall is approximately 100-200 nm thick, comprising 15-25% of the dry mass of the cell. The cell wall surrounds the much thinner (<10 micrometers) plasma membrane, a typical bilayer unit membrane comprising phospholipids, sterols and neutral lipids, represented mainly by triacyl glycerols and sterol esters.

For molecules to enter yeast, they must first pass through the cell wall and then across the plasma membrane, the major permeability barrier. In live yeast, transport can be by passive or facilitated diffusion and active transport. In microencapsulation processes, the encapsulation take place in dead as well as live cells, indicating that encapsulation takes place by simple diffusion, a process which follows Overton's Rule, stating that permeability coefficients correlate well with oil/water partition coefficients *(14)*. The cell wall also limits permeation, based on size and shape. Although many results have been obtained for the permeation limits, typical results are for molecular weight 620-740 Da and for molecular radius 0.81nm – 0.89nm *(15)*.

Microorganisms as Preformed Microcapsules

As an alternative to traditional microencapsulation processes, the use of preformed natural microorganisms as microcapsules was first considered in the 1970s. It was observed that yeast cells (*Saccharomyces cerevisiae*), when treated with a plamolyser, could be used to encapsulate water-soluble substances for use in medical, cosmetic and food products (16). The process involved removal of the cytoplasmic contents from the yeast, followed by absorbtion of water-soluble extracts such as onion juice.

Yeast was considered as a cost-effective source of microcapsules suitable for low cost high volume applications as waste yeast cells were available as a biproduct of the fermentation industry. When grown in fermenters, the microbial capsules reach a uniform size-distribution, and their physical make-up can be modified simply by altering the nutrient balance within the fermentation medium.

Later in the 1970s, Swift and Co utilized specific yeast able to accumulate high concentrations (>40%) of lipid material when grown on specific high-nitrogen based media *(1)*. The yeast strains *Torulopsis lipofera* and *Endomyces vernalis* were particularly useful and were able to accumulate lipophilic compounds such as dyes, vitamins and drugs by dissolving the active ingredients within lipid globules found in the cytoplasm. The product was particularly designed for carbonless copy paper, where physical forces were used to release the encapsulated dye. For drug delivery, although acetyl salicylic acid (Aspirin) was encapsulated successfully, release from the strongly resistant yeast cell wall proved difficult if not impossible.

The use of microorganisms for microencapsulation progressed to a more practical proposition when a small company, AD2, based in Birmingham, UK, found that it was possible to encapsulate lipophilic substances in yeast cells with less than 5% lipid content *(17)*. This patented process allowed food-grade yeast strains, which were already produced widely by the brewing, ethanol and baking industries, to be utilized. Yeast strains employed included *Saccharomyces* species, *Kluyveromyces fragilis* and *Candida utilis*. The encapsulation process was fairly simple, whereby the desired water-insoluble liquid core material was added with agitation to a yeast suspension at temperatures generally above 35°C.

With this process, payloads of up to 70% by weight were claimed, and examples of material encapsulated included dyes, flavors and agrochemicals. The release of actives from the capsules was considered to be by simple diffusion, by physical pressure, chemical activity or by biodegradation.

Further application of the AD2 technology were developed by the British Textile Technology Group, who were trying to develop a durable fragrant and biocidal finish on cotton and woolen textiles *(18)*. However, the food-grade quality of the yeast cells and their ability to encapsulate high concentration of essential oil-based fragrances suggested that application for delivery of food flavors was more practical. In a more detailed examination of the properties of yeast cells containing essential oil it was found that the rate of permeation of essential oils into the yeast cell varied considerably due to the variation in terpene chemistry *(19)*. The rate of permeation was increased at elevated temperature, up to a limit of 60°C, and molecular size and shape was also demonstrated to be important in success of encapsulation.

The yeast microencapsulation process and application continue to be exploited by Micap plc (formerly Fluid Technologies plc) in many industrial sectors, including the food, agrochemical, speciality chemical and pharmaceutical industries *(20)*. Prior work has demonstrated that yeast cells can be used to encapsulate a variety of chemical entities, however, active ingredient release was directed to industrial processes where great physical pressure or the presence of solvent initiated release. This paper explores the release of poorly soluble model drug compounds within the GI tract and determines whether yeast cells can be used to target delivery of active ingredients to the lining of the duodenum with a resultant improvement in bioavailability.

Materials and Methods

Yeast Strains

Yeast (*Saccharomyces cerevisiae* and *Saccharomyces boulardi*) fresh, active dried and spray dried were obtained from a number of commercial suppliers including Lesaffre International (Marcq-en-Barceoul, France), Lallemand Inc.(Montreal, Canada), DHW-OHLY (Hamburg, Germany), Quest International (Menstrie, Scotland) and Agrano (Riegel, Freiburg, Germany). Peppermint oil was obtained from Firmenich, S.A. (Geneva).

Preparation of Microcapsules

The conditions used for encapsulation of peppermint oils in yeast were based on those outlined in original work by AD2 *(17)*. The yeast was suspended in water to a final concentration of 33% (w/v). Peppermint oil was added to a final concentration equal to 50% of the weight of the yeast. The mixture was stirred at 40°C for 4 hours in a water-jacketed vessel, using a Teflon coated paddle. The yeast cells containing encapsulated peppermint oil were harvested by centrifugation (2000x g), and twice washed with water, to remove residual unencapsulated flavor. The washed yeast capsules containing peppermint oil were diluted to 35% dry solids with water and spray dried (Büchi Mini Spray Dryer, B-290).

For fenofibrate or econazole nitrate encapsulation, an aliquot of solvent (usually benzyl alcohol) was added to a previously calculated amount of the drug. The drug/solvent mixture was then gently warmed until all drug had dissolved before being allowed to cool to room temperature. Evidence of drug precipitation was observed by visual inspection and monitored over 48 hours. If no precipitation occurred, more drug was added to the solution and the above process repeated until a solution was obtained containing a maximum concentration of drug, with no precipitation occurring on cooling. Yeast (*Saccharomyces cerevisiae*) and freshly distilled deionized water were mixed in a reaction vessel, placed into a water bath (Clifton NE4-P)) and stirred continuously using an overhead stirrer (Stuart Scientific SS10) for 30 minutes, until a homogenous mixture was obtained. A solution of drug and solvent (obtained using the method described above) was added to the vessel and the mixture allowed to stir continuously at low shear for 5 hours. Once completed, the mixture was removed from the reaction vessel and washed with deionized distilled water and the aqueous component removed using an ALC PK130R centrifuge operating at 4000 rpm, 5°C for 20 minutes. The supernatant was decanted, the remaining pellet suspended in deionized distilled water (20% solids) and then spray dried using a Büchi B290 spray dryer, operating at 180°C, 100% Aspiration and 30% pump rate.

Preparation of chewing gums containing peppermint oil and yeast micro-encapsulated peppermint oil

Two chewing gum samples were prepared, one containing 0.520% peppermint oil (the control gum) and the other 1.926% yeast-encapsulated peppermint oil, which (with a flavour loading of 27%, w/w) also contained 0.520% of peppermint oil (the sample gum.)

Their formulations were as follows:

Ingredient %	Control	Sample
Gum base	30.1	30.1
Powdered sucrose	37.4	36.0

Corn syrup	32.0	32.0
Peppermint oil	0.520	0
Yeast-encapsulated peppermint	0	1.926
Equivalent peppermint oil	n/a	0.520

The method of their production was as follows:

- Melt the gum over a pan of hot water in a single-use plastic dish.
- Add the flavour and corn syrup to the gum and mix well.
- Sprinkle the powdered sucrose onto a clean, dry work surface.
- Spoon the gum mixture onto the sucrose and knead into a ball.
- Flatten ball with a rolling pin and pass through a pasta machine's rollers to form a thin sheet of gum.
- Cut sheet into strips and wrap in foil.

The chewing gums were informally tasted by five volunteers to ensure they were acceptable for release. The gum made with the encapsulated peppermint was perceived to have a more intense and longer-lasting flavor than the gum made with the peppermint oil. The samples were used for both mechanical chewing and real time breath-by breath analysis. In the mechanical chewing device artificial saliva was used, incubated to 37°C and, where appropriate, SDS was added to 1% (w/v). The menthol content was monitored using standard gas chromatography methods.

Extraction and Estimation of Encapsulated Peppermint Oil

The extraction of yeast encapsulated peppermint oil was based on the method of Anddaraman and Reineccius *(21)*, with the exception that ethanol was used as the solvent due to extraction efficiencies

Extraction and Estimation of Encapsulated Drug

The dry yeast encapsulates were hydrated with an aliquot of distilled/ acidified water, mixed using a rotamixer (Hook and Tucker Instruments) and left for 30 minutes. Excess water-miscible and drug-soluble solvent was then added and the mixture left for a further 30 minutes before being centrifuged for 20 minutes, using a HERMLE Z-160-M centrifuge operating at ambient temperature and 13,000 rpm. The supernatant was removed, filtered (Gelman, 0.45μm) and assayed for drug loading, using an Agilent 1100 series HPLC system connected to a diode array detector. Carrier-loading was determined using an Agilent 6890 GC system with a 7683 automatic injector and a flame ionisation detector.

Particle Size Determination and Particle Analysis

Particle size distribution of the dry yeast capsules was determined at ambient temperature, using a Malvern Sirrocco 2000 particle sizer and analysed using the MIE model. The moisture content of all samples was obtained using an Ohaus MB45 moisture balance by placing an aliquot of the yeast sample onto an aluminium pan and heating the dry yeast encapsulates to 85°C for 10 minutes.

Pharmacokinetic Analysis

Male purebred beagle dogs (9.0-12.0 kg, n=12) were fasted overnight (>18 hours) prior to the administration of the test formulations. All studies were conducted at a fixed dose of 30 mg/kg and food was allowed 2 hours post dose administration. The animals in group 1 (n=2) received a single oral admini-stration of Tricor® capsules (Abbott Laboratories) and the animals in groups 2-3 (n=5 each group), which received the yeast capsule formulations (dose volume 1 mL/kg), were anesthetized with Telazol (5mg/kg, 1M). Isoflurane delivered via a oxygen vaporiser with a nose cone was used as needed. An endotracheal tube was placed and general anaesthesia maintained with isoflurane in oxygen. A flexible endoscope was passed down the oesophagus through the stomach and into the duodenum. A catheter was passed down the working channel and the test formulations delivered. The endoscope and catheter were withdrawn and the endotracheal tube was removed when the animal regained its swallowing reflex. The animals were monitored for normal recovery from anaesthesia. Blood samples were collected (2mL whole blood) via the jugular vein or other appropriate vessels, placed in tubes containing EDTA, and stored on ice until centrifuged. Sample intervals were prior to dosing and approximately 0.5, 1, 2, 3, 4, 6, 8 and 24 hours after dosing. After centrifugation, 100μL of the plasma sample was transferred into a 2.0-mL microcentrifuge tube and 20 mL methanol added. The sample was centrifuged for 10 minutes, and the supernatant was transferred directly into an HPLC vial. Plasma concentrations of fenofibric acid were determined using a Waters Series HPLC system consisting of a Waters 2695 pump, Waters 2695 autosampler, diode array UV detector (287nm), Phenomenex Luna C-18 column and Waters Millennium data collection software (ver. 4.0). The mobile phase used was 55:45 acetonitrile:0.2% phosphoric acid solution (v/v) with a flow rate of 1.2 mL per minute and an injection volume of 20 µL.

Two model low-molecular weight lipophilic drug entities, econazole nitrate (drug A) and fenofibrate (drug B), were chosen for trial encapsulation in yeast cell capsules, based on availability of the drugs, molecular size and partition coefficient, the latter being properties known to effect successful loading of lipophilic flavor molecules in yeast capsules. Three yeast varieties were used to determine whether the encapsulation was yeast strain specific. The encapsulations proved successful for both drugs in all three yeast varieties. The loadings of each drug with yeast variety are detailed in Figure 3. The particle size of the spray-dried products, approximately 30 micrometers in diameter, resulted from agglomerates of yeast cell capsules (Figure 4). The drug loading levels were found to be in the range 100-200 mg/g yeast (10-20% w/w). Further development work on the encapsulation process suggests that this is typical for a wide range of pharmaceutical actives. Successful encapsulation can be achieved with low molecular weight drugs with masses in the range 200-1000 g/mol and octanol-water partition coefficients of greater than 2.0 – 6.0.

The yeast variety influences the loading level for each drug. For example, with fenofibrate the loading varies between 100 and 180mg/g yeast between the three yeast varieties. The difference in loading levels observed for different yeast types may be attributed to a number of structural characteristics of the yeast cells such as the membrane and cell wall composition or to the pre-processing and manufacturing history of the yeast strain used, for example whether the yeast had been produced in a batch or continuous process or whether the yeast had been dried, or even the type of drying used (e.g. roller drying, fluidized-bed drying or spray drying).

Results and Discussion

Flavor release in the mouth was initiated by contact with the mucous membrane surface, when a peppermint-flavored chewing gum was prepared and the flavor release monitored using real time breath-by-breath analysis (Figure 1).

However, when placed in a mechanical chewing device, the peppermint flavor did not release from the chewing gum by mechanical action or by the presence of saliva, unless SDS was present, as monitored by gas chromatography (Figure 2). On analysis by light microscopy, the yeast cell capsules were seen to be intact.

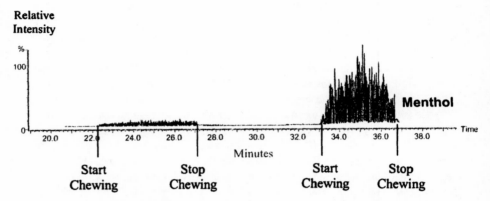

Figure 1. Real time breath-by-breath analysis of menthol, the main component of peppermint, during chewing of chewing gum samples containing peppermint oil (left) and yeast encapsulated peppermint oil (right).

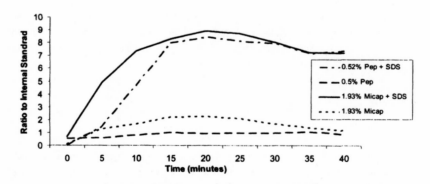

Figure 2. The release of menthol as an indication of peppermint flavor released from a chewing gum during mechanical chewing (peppermint was added to 0.52% w/w), and from yeast capsules (1.93% w/w, the equivalent flavor loading). SDS was added to the dissolution phase at 1% w/v.

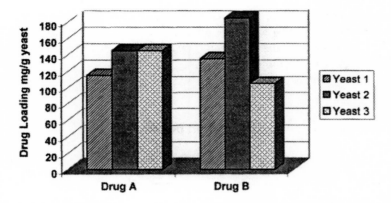

Figure 3. Plot showing encapsulation levels of two model drugs in three different yeast types (drug A – econazole nitrate), drug B – fenofibrate).

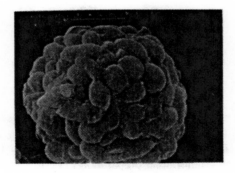

Figure 4. A typical agglomerate of yeast capsules produced by spray-drying after encapsulation of fenofibrate.

It is well known from previous work with lipophilic flavor molecules in yeast cell capsules that the active ingredients, once encapsulated, remain inside the yeast cells in the presence of water and are not released unless the cell membrane is disrupted, for example by the use of surface active agents.

This phenomenon was also found with encapsulated drugs (Figure 5). The dissolution profile indicates that SDS must be present above the critical micelle concentration for release to take place. With flavor release on the surface of the tongue, the direct contact with the cell surface also initiated release. The pharmacokinetic profiling work was designed to determine if a lipophilic active could be released from the yeast cell without the use of surface active agents, and ultimately enter the bloodstream.

Dissolution profile

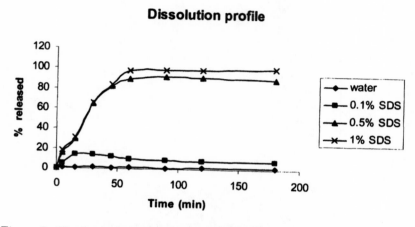

Figure 5. The in-vitro release profile of fenofibrate in the presence of sodium dodecyl sulphate.

The two yeast varieties containing fenofibrate were administered directly to the duodenum of 5 beagle dogs, and the level of fenofibrate acid in the plasma quantified over a 24-hour period using HPLC analysis. For comparison, a commercially available fenofibrate preparation (Tricor®) was administered to 2 beagle dogs orally. The results, detailed in Figure 5, indicate that the fenofibrate was successfully released from the yeast cell capsules into the systemic circulation. Both yeast cell capsule formulations and Tricor® reached the maximum drug concentration in the systemic circulation within the same time period, although the yeast formulation 2 reached its maximum concentration slightly earlier than the other two formulations. All three formulations have a similar absorption profile, with the profile of yeast formulation 1 being most similar to that of the commercial formulation, although yeast formulation 1 shows a much more controlled release of fenofibrate from the formulation than Tricor®, and the initial burst of drug into the systemic circulation was reduced. There appears to be more fenofibrate absorbed from the commercial Tricor® formulation. However, the area under the curve provides a better indication of total amount of drug absorbed from the formulations, which in turn is a better indication of the potential bioavailability. The areas under the curve calculated from Figure 6 above are plotted in Figure 7. From this plot it is clear that most drug was absorbed into the systemic circulation from yeast formulation 1, compared to the commercial product and yeast formulation 2. It therefore appears that with careful choice of yeast variety and successful encapsulation, the bioavailability of lipophilic drugs can be improved. The potential improvement may be large, considered that no formulation development had taken place during the yeast cell capsule preparations.

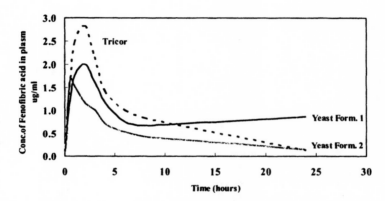

Figure 6. Time-dependent concentration of fenofibric acid in plasma for a commercial formulation, Trico®r and two yeast formulations (yeast 1 – Saccharomyces cerevisiae and yeast 2 –Saccharomyces boulardi).

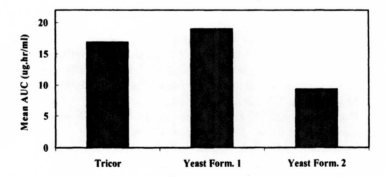

Figure 7. The mean area under the curve (AUC) (µg hr/ml) for yeast encapsulated fenofibrate compared to the commercial formulation Tricor® (yeast 1 – Saccharomyces cerevisiae and yeast 2 –Saccharomyces boulardi).

Conclusions

Yeast cells can be utilized as microcapsules for the encapsulation of lipophilic drug molecules. The drug remains stable within the capsule until release is initiated by addition of a surfactant or by contact with a mucous membrane. When administered directly into the duodenum, the lipophilic drug is released from the cell and enters the blood stream with a reduced burst effect and prolonged release profile.

Acknowledgements

The authors would like to acknowledge the help and advice of Juerg Lange of SkyPharma, Muttenz, Switzerland, in preparing the protocols for this development.

References

1. *Swift and Company*; US patent specification 40001480, **1977**.
2. Langer R. Drug delivery and targeting. *Nature*, **1998**, *392*, Suppl. 5-10.
3. Fichetti, V.A.; Medaglini, D.; Pozzi, G. Gram-positive commensal bacteria for mucosal vaccine delivery. *Curr Opin Biotechnol*; **1996**, *7*, 659-666.
4. Blanquet, S., Antonelli, R., Laforet, L., Denis, S., Marol-Bonnin, S., Alric, M. Living recombinant Saccharomyces cerevisiae secreting proteins or peptides as a new drug delivery system in the gut. *J. Biotechnol.*, **2004**, *110*, 37-49.
5. Stubbs, A.C., Martin, K. S., Coeshott, C., Skaates, S.V., Kuritzkes, D. R., Bellgrau, D., Franzusoff, A., Duke, R.C., Wilson, C.C. Whole recombinant yeast vaccine activates dendritic cells and elicits protective cell-mediated immunity. *Nature Medicine*, **2001**, *7*, 625-629.
6. Drouault, S., Corthier, G., Ehrlich, S.D., Renault, P. Survival, physiology, and lysis of Lactococcus lactis in the digestive tract. *Appl. Environ Microbiol*, **1999**, *65*, 4881-4886.
7. Schreuder, M.P., Deen, C., Boersma, W.J.A., Pouwels, P.H., Klis, F.M. Yeast expressing hepatitis B virus surface antigen determinants on its surface: Implications for a possible oral vaccine. *Vaccine*, **1996**, *14*, 383-388.
8. Steidler, L., Hans, W., Schotte, L., Neirynck, S., Obermeier, F., Falk, W., Fiers, W., Remaut, E. Treatment of murine colitis by Lactococcus lactis secreting interleukin-10. *Science*, **2000**, *289*, 1352-1355.

9. Paukner, S., Kohl, G., Jlava, K., Lubitz, W. Sealed bacterial ghosts - Novel targeting vehicles for advanced drug delivery of water-soluble substances. *J. Drug Targeting*, **2003**, *111*, 151-161.

10. *Biology of Microorganisms*, Brock, T.D.; Madigan, M.T.; Prentice-Hall, Englewood Cliffs, New Jersey, **1988**, 381.

11. Dziezak, J.D. Yeasts and yeast derivatives-Definitions, characteristics, and processing. *Food Technology*, **1987**, 104-108.

12. *Biotechnology in Food Processing*; Trivedi, N.; Noyes Publications, Park Ridge, **1986**, 115.

13. *Yeast, Physiology and Biotechnology*; Walker, G.M.; Wiley: Chichester, **1998**, 17.

14. Overton, E. *Naturforsch. Ges. Zuerich*, **1899**, *44*, 88.

15. Scherrer, R.; Louden, L.; Gerhardt, P. Porosity of yeast-cell wall and membrane. *J Bacteriol*, **1974**, *118*, 534-540.

16. *Serozym Laboratories*; French patent specification 2179528, **1973**.

17. *AD2 (now owned by Micap plc)*; European patent specification, 0242135, **1987**.

18. *British Textile Technology Group*; European patent specification, 0511258, **1991**

19. Bishop, J.R.P.; Nelson, G.; Lamb, J. Microencapsulation in yeast cells. *J. Microencapsulation* **1998**, 15, 761-773.

20. Nelson, G. Application of microencapsulation in textiles. *Int. J. Pharm.* **2002**, *242*, 55-62.

21. Anandaraman, S.; Reinceccius, G.A. *Perfumer and Flavorist*, **1987**, *12*, 33.

Chapter 20

Chelants for Delivery of Metal Ions

R. Keith Frank*, Philip S. Athey, Gyongyi Gulyas, Garry E. Kiefer, Kenneth McMillan, and Jaime Simón

The Dow Chemical Company, Freeport, TX 77541
*Corresponding author: rkeithfrank@dow.com

Chelating agents play an important role in the delivery of metal ions for medical applications. These applications include both diagnosis and therapy. In diagnostic applications, the metal ions can be radioactive (gamma-emitter), paramagnetic, or fluorescent. Therapeutic applications generally require metals that are either chemically active on their own or emit ionizing radiation (beta or alpha-emitter). The ability of the chelant to form a stable *in vivo* complex is critical to its role in delivering the metal ion effectively and safely to its target. Examples using chelating agents for metal ion delivery in medical applications include therapeutic bone agents and bifunctional chelating agents.

Introduction

Chelating agents play an important role in the delivery of metal ions for medical applications. These applications include both diagnosis and therapy. In diagnostic applications, the metal ions can be radioactive, paramagnetic, or fluorescent.

Radioactive metals such as technetium-99m (^{99m}Tc) are used in nuclear medicine procedures to evaluate physiologic conditions such as bone cancer, renal function, or heart disease. In these applications, γ-emitting isotopes are used in conjunction with either simple scintillation cameras or more complex imagers such as single-photon emission computed tomography (SPECT) and positron emission tomography (PET). Very precise anatomical information can be ascertained by magnetic resonance imaging (MRI). A contrast agent consisting of chelated paramagnetic metal ions such as gadolinium (Gd) are typically used to enhance images in MRI.

Therapeutic applications generally require metals that are either chemically active on their own or emit ionizing radiation. Ionizing radiation used in therapy is typically particulate (β or α), but photon (γ) emissions can be used as well. Therapeutic effect is achieved with the cell-killing caused by hydroxyl radicals formed by the interaction of ionizing radiation with water.

Occasionally, metal ions can be administered as a simple salt, as is the case for Metastron® (^{89}Sr chloride) for bone pain due to metastatic bone cancer. More commonly, however, a chelating agent is required to control the biodistribution and pharmacokinetics of the metal ion. Sometimes a metal ion is conjugated to a carrier moiety using a bifunctional chelant, depicted as the claw in Figure 1. The carrier may be a polymer or a biological targeting molecule such as a monoclonal antibody (mAb) used in radioimmunotherapy (RIT).

Antibody, protein, peptide, polymer, etc

Bifunctional chelating agent binds metal to carrier

Metal ion

Figure 1. Roll of a bifunctional chelant in a conjugate.
(See page 5 of color inserts.)

The ability of the chelant to form a stable metal ion complex is critical to its role in delivering the metal effectively and safely to its target. Therapeutic bone agents and bifunctional chelants are two example chelating agent applications for metal ion delivery that will be discussed.

Results and Discussion

Therapeutic Bone Agents

Quadramet® (Samarium Sm-153 Lexidronam Injection)

The bone-seeking chelating agent Lexidronam, also known as ethylene-diaminetetramethylenephosphonic acid (EDTMP), has been used to deliver the β-emitting radioisotope ^{153}Sm to sites of bone lesions. The therapeutic bone agent Quadramet, shown in Figure 2 (without cations and waters of hydration), was approved in 1997 for the treatment of bone pain due to metastatic bone cancer.

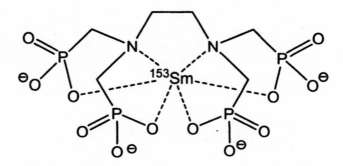

Figure 2. ^{153}Sm-EDTMP (Quadramet®) for bone cancer. (Cations and waters of hydration not shown.)

^{153}Sm is a moderate-energy β emitter (702 KeV – 44.1%, 632 KeV – 34.1%, 805 KeV – 21%) but also has an imagable γ photon (103 KeV – 28.3%), which allows viewing the biolocalization of the ^{153}Sm using a scintillation camera, as shown in Figure 3. There is a lack of any accumulation in soft tissue with only skeletal uptake evident. In general, about half of the injected radioactivity accumulates in the bone with the remainder being excreted very rapidly via the kidneys (less than two hours in rodents).

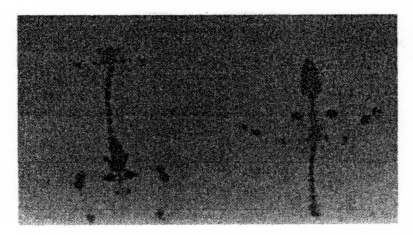

Figure 3. ^{153}Sm-EDTMP scintillation scan 3 hours post injection in a rabbit.
(See page 6 of color inserts.)

One of the benefits of bone-seeking radiopharmaceuticals such as Quadramet is that they have higher uptake in lesions or rapidly growing bone than in normal bone. This behavior is illustrated in the rabbit drill-hole model shown in Figure 4. In this model, small holes are drilled in the femur to simulate a bone tumor. After injection of ^{153}Sm-EDTMP, the rabbit is imaged using a scintillation camera, as shown on the left. On the right is an image of the lesion, showing a 20-fold excess of radioactivity in the lesion versus the surrounding normal bone.

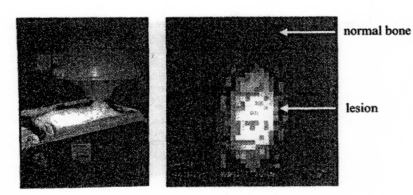

Figure 4. Rabbit drill-hole model. Left - scintillation camera. Right - image
shows selective uptake of ^{153}Sm-EDTMP in drill-hole lesion.
(See page 6 of color inserts.)

Figure 5 shows an image from an early human trial of ^{153}Sm-EDTMP. Sites of metastatic bone disease are clearly visualized by the selective uptake of the radiopharmaceutical.

Sites of
metastatic
disease

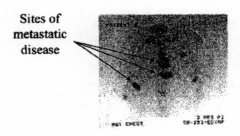

Figure 5. Image from human trial of ^{153}Sm-EDTMP showing sites of metastatic bone tumors. (See page 6 of color inserts.)

EDTMP forms a moderately labile complex with Sm *in vivo*. A large excess of EDTMP over Sm (nearly 300-fold) is required in order to control the delivery of the metal ion and prevent uptake in non-target tissues. At much lower chelant-to-metal ratios, significant amounts of radiometal accumulate in tissues such as the liver. For example, as can be seen from the data in Table , the uptake (percent injected dose) in the liver increases as the EDTMP to Sm ratio decreases *(1)*.

Table I. Liver uptake of ^{153}Sm in rats at 2 hours with varying EDTMP to Sm ratios.

EDTMP : Sm ratio	Liver (%ID)
38	0.79
76	0.59
115	0.37
153	0.33
268	0.25

Skeletal Targeted Radiotherapy (^166Ho-DOTMP)

A developmental spin-off of ^{153}Sm-EDTMP is Skeletal Targeted Radio-therapy (STR^{TM}). STR is currently in late stage clinical evaluation for bone marrow ablation as part of a bone marrow transplantation protocol for the treatment of multiple myeloma. For this application, the bone seeking chelant, 1,4,7,10-tetraazadodecane-1,4,7,10-tetramethylenephosphonic acid (DOTMP), is being used to deliver the β-emitter ^{166}Ho (Figure 6) to bone.

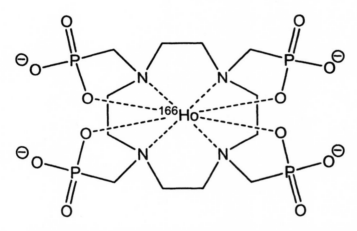

Figure 6. ^{166}Ho-DOTMP (STR) for bone marrow ablation. (Cations and waters of hydration not shown.)

^{166}Ho was selected because of its high β energy (1854 KeV – 51%, 1774 KeV – 48%), which allows complete ablation of the bone marrow to be achieved. Because of its short half life (26.8 hours), the actual transplant can occur within a few days. This is important because patients having undergone bone marrow ablation are immune-compromised and prone to infection. ^{166}Ho also has an imagable γ photon (81 KeV – 6.2%), which allows for visualization.

The macrocyclic DOTMP chelant forms very stable lanthanide complexes. Due to this high stability, a large excess of chelant is not required. This is important in this application because, in order to achieve complete ablation, a large amount of radioactivity is required. It is undesirable to inject a large amount of free chelating agent due to potential toxicity concerns.

Bifunctional Chelating Agents

Bifunctional chelants (BFC) are molecules that have the capacity to sequester a metal ion and, in addition, have the facility to become attached to another entity. The three parts of a BFC molecule are illustrated below in Figure 7. Shown is the BFC α-[2-(4-nitrophenyl)ethyl]-1,4,7,10-tetraazacyclodo-decane-1,4,7,10-tetraacetic acid (PA-DOTA). There is a chelant portion, a linker portion, and a spacer connecting the two. The chelant portion of the BFC binds the metal ion and, for (+3)-metals such as ^{90}Y, ^{111}In, or ^{177}Lu, is typically either an acyclic diethylenetriaminetetraacetic acid (DTPA) analog or a macro-cyclic 1,4,7,10-tetraazacyclodocane-1,4,7,10-tetraacetic acid (DOTA) analog (shown). The linker is a reactive group that can be attached to a carrier. Shown is a typical isothiocyanate, which can be used to attach to lysine ε–amines on a protein. Various spacing groups have been used to alter the biological proper-ties of the final conjugate.

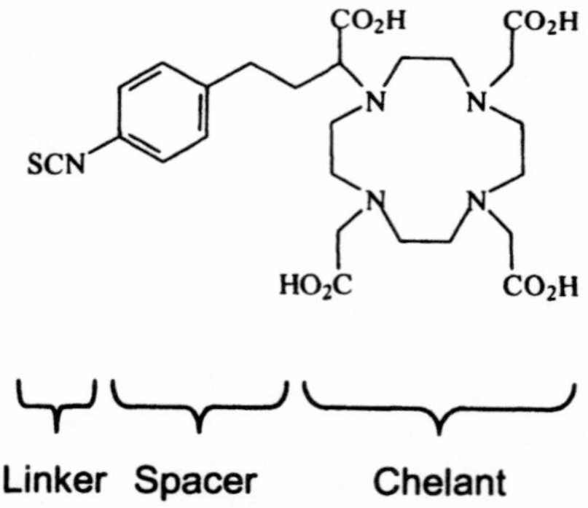

Linker Spacer   Chelant

Figure 7. PA-DOTA bifunctional chelant showing the three parts of the BFC.

Biological Targeting Molecules

Biological targeting molecules such as monoclonal antibodies (mAb) can be conjugated to BFCs to deliver metal ions to specific sites in the body. This has both diagnostic and therapeutic applications. For example, a cancer-seeking mAb can be labeled with a radiometal to visualize sites of metastatic disease. In radioimmunotherapy, this radiolabeling combines the cytotoxic nature of ionizing radiation with the targeting ability of mAbs to destroy cancer cells. An example of such a biotargeted radiopharmaceutical is Zevalin™, developed by IDEC and recently approved as a treatment for non-Hodgkin's lymphoma.

It is particularly important in therapeutic applications that a stable BFC-metal ion complex be maintained *in vivo*. Failure to achieve suitable stability will allow the radioactive metal to be released and thus accumulate in non-target tissue. This can be illustrated by an experiment comparing the three bioconjugates shown in Figure 8. These identical CC49 F(ab')$_2$ fragments were conjugated to three different BFCs, Bz-DTPA, MX-DTPA, and PA-DOTA. They were all labeled with ^{177}Lu and injected into BALB/C mice (n=5). The percent injected dose (%ID) in the liver over time post-injection (PI) is shown in Figure 9.

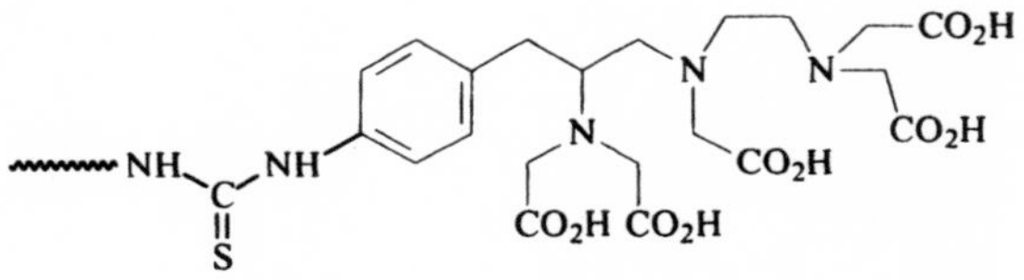

Bz-DTPA

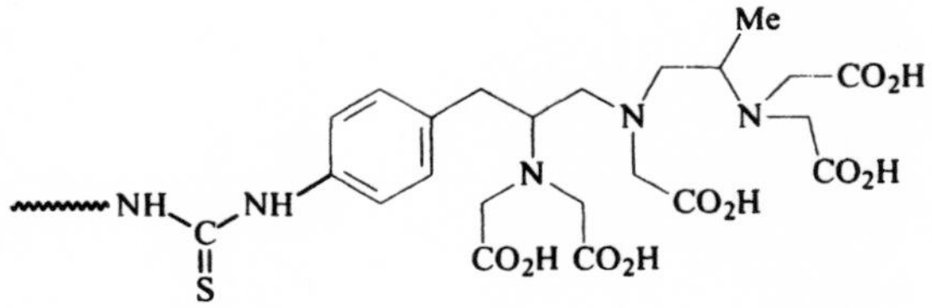

MX-DTPA

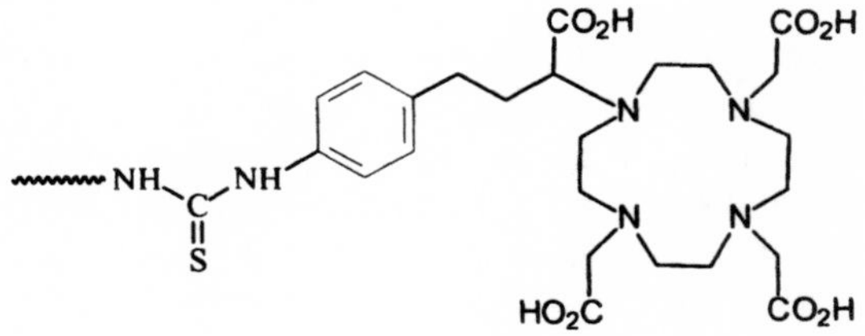

PA-DOTA

Figure 8. Three CC49 F(ab')$_2$ bioconjugates based on Bz-DTPA, MX-DTPA, and PA-DOTA.

The three bioconjugates show significantly different uptakes of ^{177}Lu in the liver. These are inversely related to the stability of the metal complexes (i.e.,

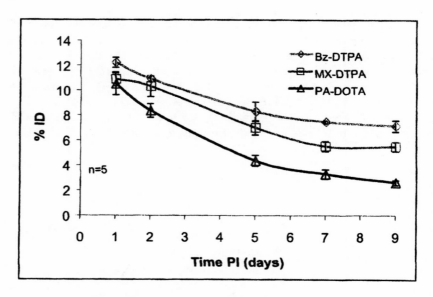

Figure 9. Biodistribution study of three ^{177}Lu-labeled bioconjugates.
(See page 7 of color inserts.)

Bz-DTPA forms the least stable Lu-complex and results in the highest Lu liver uptake, whereas PA-DOTA, which forms the most stable Lu-complex gives the lowest Lu liver uptake).

Similar biodistribution differences were seen in the bone (not shown). The increased localization of ^{177}Lu in these tissues for less stable complexes can be explained by radiometal "leaking" out of the chelant. It is well know that "free" lanthanides such as Lu will accumulate in the liver and bone. Free metal ions form colloids which go to the liver, and lanthanides are calcium ion mimics in the bone.

Calcium mimicking by lanthanides is clearly undesirable, as accumulation in non-target tissues such as liver or bone can result in undesired radiotoxicity. New BFCs such as α-(5-isothiocyanato-2-methoxyphenyl)-1,4,7,10-tetraaza-cyclododecane-1,4,7,10-tetraacetic acid (MeO-DOTA), shown in Figure 10, have been specifically designed to maintain stable *in vivo* complexes, particularly for lanthanide metal ions.

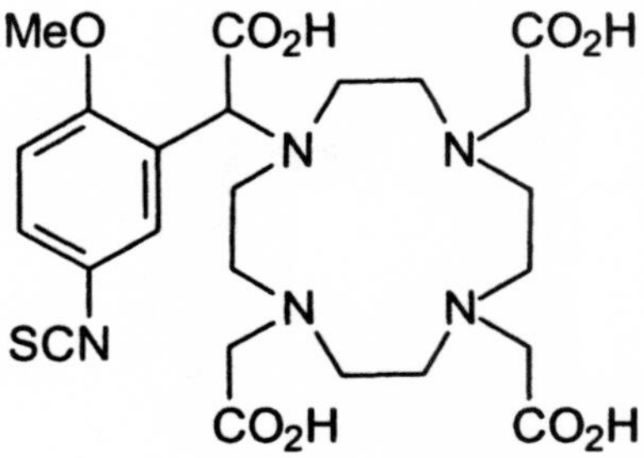

Figure 10. MeO-DOTA bifunctional chelating agent.

Nanoparticles

Nanoparticles are of great interest because they have the potential to carry a large payload. BFCs are being used to attach metal ions, particularly Gd, to nanoparticles. Gd, being a paramagnetic metal, is important as a contrast agent in magnetic resonance imaging (MRI). Unlike radioisotopes, however, a relatively large amount of Gd is required for imaging. Whereas a single isotope attached to a biological targeting molecule (e.g. an ^{111}In-labeled mAb such as ProstaScint®) could make an effective imaging agent, this is not the case with Gd. Nanoparticles allow a large number of Gd ions to be localized at target sites.

Carbohydrate Hydrogel Nanoparticles

Carbohydrate hydrogel nanoparticles are one example of this kind of system. Targeting peptides such as integrins, EGFR, and VEGR-2 can be attached to the surface in order to target the nanoparticle to particular sites *(2)*. The Gd-chelate of MeO-DOTA, shown on the left in Figure 11, has been used to bind multiple Gd ions to the nanoparticle (not shown in the figure is the coordinated water molecule essential for MRI contrast). The isothiocyanate is used to react with amine groups on the particle surface to form a covalent linkage. A depiction of the targeted nanoparticle is shown in Figure 11 on the right.

292

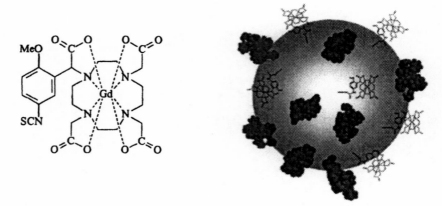

Figure 11. Gd-MeO-DOTA bifunctional chelate (left, shown without water of hydration) and depiction of its use in a targeting nanoparticle.

Lipid Nanoparticles

Another type of particle being developed is lipid-covered perfluorocarbon nanoparticles containing targeting molecules on their surface. The targeting molecules enable the particles to accumulate in certain sites such as arterial plaque. One goal of this work is the targeted delivery of a contrast agent for MRI *(3)*.

BFCs have been modified to contain lipophilic tails, designed to incorporate themselves into these lipid nanoparticles. Compounds such as DTPA-PE, shown in Figure 12, are being evaluated as a means to attach a large number of Gd-ions to targeted lipid nanoparticles *(4)*.

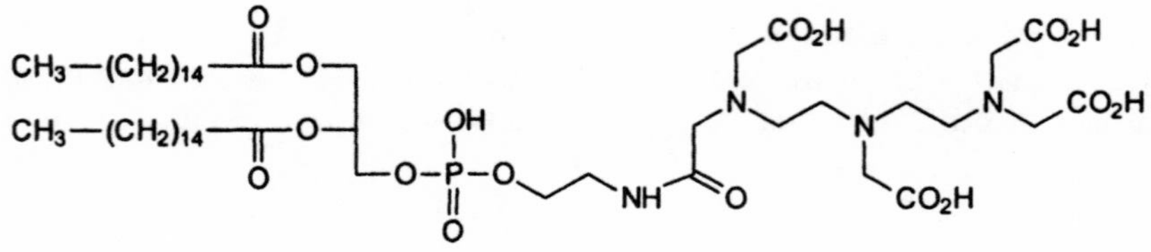

Figure 12. Chemical structure of DTPA-PE.

An example of a targeting lipid nanoparticle is represented in Figure 13. The BFC shown is a derivative of MeO-DOTA and has a lipophilic tail, which allows it to be attach to the lipid monolayer. If the metal ion "M" in the figure is Gd, one can make an MRI contrast agent. A radiopharmaceutical can be produced if "M" is a radiometal. This could be diagnostic, such as ^{111}In or a therapeutic such as ^{177}Lu or [90]Y.

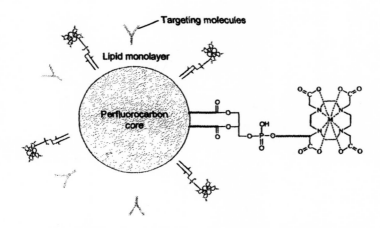

Figure 13. Targeting lipid nanoparticle. (See page 7 of color inserts.)

Conclusions

In summary, chelating agents can be used to effectively control the destiny of metal ions in the body. The chelating agent is important, and not all chelants are created equal.

Looking toward the future, medical chelants for metal ion delivery will continue to evolve. Next generation chelants will have varying charge and/or lipophilicity, which can provide improved pharmacokinetics of small molecules such as targeting peptides. A variety of linking groups and methodologies will provide conjugation under very mild conditions and will provide chelants suitable for use in solid-phase peptide synthesis. Novel chelant designs will allow for faster metal complexation kinetics while maintaining *in vivo* stability.

294

References

1. Unpublished data from The Dow Chemical Company
2. Unpublished information from Alnis BioSciences, Inc.
3. Wickline, S.A.; Lanza, G.M. Nanotechnology for molecular imaging and targeted therapy. *Circulation* **2003**, *107*, 1092-1095.
4. Winter, P.M.; Caruthers, S.D.; Yu, X.; Song, S.K.; Chen, J.; Miller, B.; Bulte, J.W.M.; Robertson, J.D.; Gaffney, P.J.; Wickline, S.A.; Lanza, G.M. Improved molecular imaging contrast agent for detection of human thrombus. *Mag. Reson. Medicine* **2003**, *50*, 411-416.

Chapter 21

Molecular Imaging and Therapy: New Paradigms for 21st Century Medicine

G. M. Lanza[1,*], P. M. Winter[1], M. S. Hughes[1], S. D. Caruthers[1,2], J. N. Marsh[1], A. M. Morawski[1], A. H. Schmieder[1], M. J. Scott[1], R. W. Fuhrhop[1], H. Zhang[1], G. Hu[1], E. K. Lacy[1], J. S. Allen[1], and S. A. Wickline[1]

[1]Division of Cardiology, Washington University Medical School, 660 South Euclid Boulevard, St. Louis, MO 63110
[2]Philips Medical Systems, 600 Beta Drive, Cleveland, OH 44143
*Corresponding author: greg@cvu.wustl.edu

Molecular imaging agents are extending the potential of non-invasive medical diagnosis from basic gross anatomical descriptions to complicated phenotypic characterizations, based upon the recognition of unique cell-surface biochemical signatures. Originally the purview of nuclear medicine, molecular imaging is now a prominent feature of most clinically relevant imaging modalities, in particular magnetic resonance imaging and ultrasound. Nanoparticulate molecular imaging agents afford the opportunity not only for targeted diagnostic studies but also for image-monitored site-specific therapeutic delivery. Combining imaging with drug delivery permits verification and quantification of treatment and offers new clinical strategies to many diseases, e.g., cardiovascular, oncologic, and rheumatologic pathology. Ligand-targeted perfluorocarbon nanoparticles represent a novel platform technology in this emerging field, providing molecular imaging for all relevant imaging modalities and vascular-targeted drug delivery across a broad disease spectrum.

Actively-targeted molecular imaging agents provide contrast to tissues by recognizing unique biochemical signatures and providing a highly amplified signal that is subsequently detectable with a clinical imaging modality. In general, assessment of molecular information requires target-specific probes, a robust signal amplification strategy and a sensitive high-resolution imaging modality *(1)*. This concept, which is routinely applied *in vitro* by polymerase chain reactions or immunohistochemistry, is now extended to noninvasive *in vivo* applications to facilitate the detection and visualization of important molecular or cellular moieties present in nano- and picomolar concentrations. A successful targeted contrast agent must have (i) prolonged circulation half-life (ideally greater than 60 minutes), (ii) persistence at the targeted site, (iii) selective binding to epitopes of interest, (iv) prominent contrast-to-noise enhancement, and (v) an acceptable toxicity profile. These macrocomposite systems must be practical to manufacture and convenient to use in a clinical setting.

We have developed a multidimensional targeted nanoparticle platform, that addresses many of the issues that have led most targeted imaging and drug delivery approaches to fail (Figure 1) *(2-4)*. This novel agent is a ligand-targeted, lipid-encapsulated, *nongaseous* perfluorocarbon emulsion, produced through microfluidization techniques. Perfluorocarbon nanoparticles have poor inherent acoustic reflectivity *(5,6)* until concentrated upon the surfaces of tissues or membranes, for example clots, endothelial cells, smooth muscle cells, synthetic membranes etc *(2)*. This feature provides marked improvement in contrast signal without increasing the background level. Moreover, unlike perfluorocarbon-based microbubbles, liquid perfluorocarbon nanoparticles exhibit a prolonged half-life and are stable to high incident acoustic pressure *(7)*.

The nanoparticle platform may be modified for robust MRI applications by including enormous payloads (i.e., >50,000) of paramagnetic chelates into the outer surfactant layer for T1-weighted contrast or by taking advantage of the inherent high [19]Fluorine content provided by the perfluorocarbon cores for imaging or spectroscopy. Active targeting to vascular biomarkers is typically accomplished by covalent crosslinking of homing ligands, e.g., monoclonal antibodies, peptides, peptidomimetics, and others, and their natural steric exclusion of >200nm particles from extravascular locations. The emulsion nanoparticles have long circulatory half-lives due to their size (i.e., >90 minutes), without further modification of their outer lipid surfaces with polyethylene glycol or utilization of polymerized lipids.

From a regulatory perspective, perfluorocarbon emulsions have been approved for parenteral commercial use as artificial blood substitutes. Using USP defined materials and well characterized functionalized lipids for homing and metal chelation, the nanoparticles may be produced through a common, self-assembly emulsification process at large-scale volumes with excellent reproducibility. The final product is amenable to heat sterilization and we and others have demonstrated shelf stability in excess of 12-month at room temperature.

SITE-SPECIFIC NANOPARTICLES

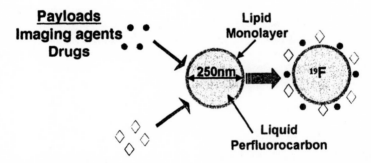

Figure 1. Platform paradigm for ligand-targeted, perfluorocarbon nanoparticles (250nm diameter nominal) with combination capability for imaging and therapy.

Results and Discussion

Ultrasonic Molecular Imaging

Intravascular and intracardiac thromboses are important etiologies for stroke and infarction with profound adverse health impacts in western societies. In the first report of ultrasonic molecular imaging, marked acoustic enhance-ment of thrombi created by perfluorocarbon nanoparticles, targeted systemically *in vivo* in a canine model, illustrated the concept of contrast-based tissue characterization (Figure 2) *(2)*. In that report, we demonstrated that perfluoro-carbon nanoparticles, although too small to be detected in circulation at the low molecular imaging dosage employed, were able to persist in blood, specifically bind, and acoustically enhance acute vascular thrombi. In this seminal work, homing of the nanoparticles to thrombus utilized an avidin-biotin based targe-ting scheme, similar to those popular at that time in nuclear medicine. But, in subsequent *in vivo* studies *(8-11)*, homing ligands have been chemically conju-gated to the surfactant layer. This change avoided time-consuming multi-step methods, which are untenable in a clinical environment, as well as circumvented complicated targeting issues posed by competition from endogenous biotin, biotinylase cleavage of reagents or avidin immunogenicity problems.

Before Contrast   After Contrast

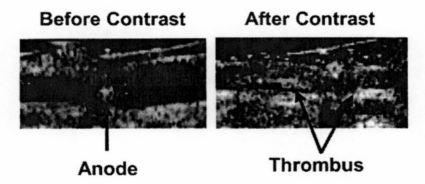

Anode   Thrombus

Figure 2. Femoral artery of dog with acute, platelet-rich thrombus created by electric injury before and after (45 minutes) fibrin-targeted nanoparticles. (Copyright 1996 Lippincott Williams and Wilkins.)

To understand the acoustic reflectivity of bound perfluorocarbon nano-particles, we developed and reported a first order mathematical model, i.e., an acoustic linear transmission model, which helped to elucidate the major principles governing the magnitude of acoustic reflectivity (Figure 3) *(12)*. In this model, the key elements influencing spectral ultrasonic reflectivity were the effective emulsion layer: density, speed of sound, and thickness. We extensively characterized the acoustic properties of a wide variety of potential perfluorocarbons and subsequently formulated a series of perfluorocarbon compounds of varying acoustic impedance *(13,14)*. In synopsis, these studies revealed that all liquid perfluorocarbon emulsions significantly increase target acoustic reflectivity when bound to a surface and that the magnitude of enhancement may be manipulated by formulating nanoparticles with perfluorocarbons of different acoustic impedance. Moreover, we determined that introduction of exogenous heating accentuated the acoustic contrast of nanoparticle targeted-tissues by lowering perfluorocarbon speed-of-sound, which augmented the acoustic impedance mismatch and increased the surface spectral reflections *(15)*. Essentially, the echogenicity of nanoparticles bound to a tissue surface was analogous to the reflectivity of light by a mirror, i.e., the more complete the coating of silver grains over glass surface, the clearer the reflected image.

$$R(k) = R_{12} + \frac{T_{12}R_{23}T_{21}e^{2ikd}}{1 - R_{21}R_{23}e^{2ikd}}$$

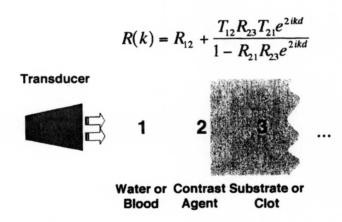

Transducer

1   2   3   ...

Water or Contrast Substrate or
Blood   Agent   Clot

Figure 3. Acoustic transmission line model provides a first-order mathematical approximation of ultrasound contrast achieved with targeted perfluorocarbon nanoparticles. Nanoparticles create a "layer"(2). "R" and "T" are reflection and transmission coefficients, respectively.

Complementary Acoustic Imaging with Information-Theoretic Receivers

Molecular imaging systems, regardless of the modality, must have complementary interaction between the chemistry of the contrast agent and the imaging hardware/software. For thrombus, where the concentration of fibrin epitopes is vast, the acoustic imaging of targeted nanoparticles is readily accomplished with traditional fundamental approaches. However, when biomarker targets have sparse density distributions, complementary acoustic receivers are required that are less dependent upon signal to noise (i.e., the size of the RF waveform) yet highly sensitive to the subtle changes in wave shape (i.e., the contours and ripple patterns). This is particularly true for angiogenic targets, where neo-vasculature often arises as microscopic islands along naturally reflective tissue interfaces (e.g., between tumor and muscle, Figure 4).

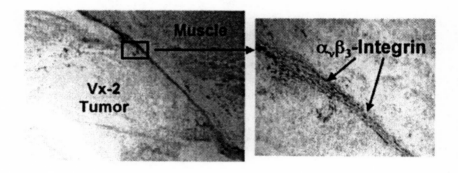

Figure 4. Angiogenic vessels induced by growing (3mm) Vx-2 rabbit adenocarcinoma. Neovasculature arising along the fascial interface between muscle and tumor. Dark brown staining of $\alpha_v\beta_3$-integrin with LM-609 antibody.

To accomplish sensitive discrimination of nanoparticles in small, highly localized deposits around tumors and between tissue planes, we implemented novel, nonlinear, thermodynamic/information-theoretic receivers, which allow in principle infinite stratification of the digitized acoustic waveform on a discrete time lattice. In simplistic terms, this family of detectors increases the received sensitivity to subtle differences in reflected acoustic power, improving contrast discrimination. Nonlinear entropy receivers were implemented first for single element acoustic microscopic imaging and today can be interfaced with commercial echocardiographic imagers. As discussed in more detail below in the context of magnetic resonance imaging, these advancements in acoustic imaging have been utilized to detect the expression of $\alpha_v\beta_3$-integrin on the induced neovasculature of nascent Vx-2 rabbit carcinomas. As is clearly noted in Figure 5, traditional RF processing produced random noise across the tumor image, however, implementation of a novel information-theoretic receiver, in this example based on Shannon's entropy, allowed the time-dependent, coherent accumulation of nanoparticulate contrast to be exquisitely appreciated. Histological examination of the tumor corroborated the distribution of angiogenesis based upon $\alpha_v\beta_3$–integrin expression within the tumor periphery, as depicted by nonlinear receivers. Thus, information-theoretic detectors provide the complementary contrast sensitivity required for perfluorocarbon nanoparticle detection, and they circumvent some of the potential, complicated imaging schemes currently employed to image targeted microbubbles.

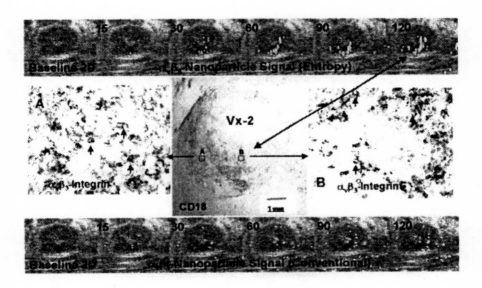

Figure 5. Vx-2 rabbit adenocarcinoma targeted with $\alpha_v\beta_3$-nanoparticles and imaged with conventional (bottom) and information-theoretic entropy (top) detectors at baseline, 15, 30, 60, 90, and 120 minutes post-injection. Immunohistochemistry of $\alpha_v\beta_3$-integrin in the Vx-2 tumor corroborating the presence and location of neovasculature.

MRI Molecular Imaging

Magnetic resonance (MR) is emerging as a particularly advantageous molecular imaging modality, given its high spatial resolution and its unique capability to elicit both anatomic and physiological information simultaneously without radiation exposure. In the earlier years of MR molecular imaging, an inability of targeted MR contrast agents to overcome partial volume dilution effects led to numerous failures (16), which were overcome with superparamagnetic (17-19) and "ultraparamagnetic" agents (3,20). Superparamagnetic agents take advantage of the wide-ranging effects of susceptibility artifacts produced by iron oxides, e.g., USPIO or MION. In general, these agents elicit strong signal effects, which appear as dark or "negative" contrast, particularly against a bright background. An alternative approach has been to develop "ultraparamagnetic" nanoparticles, which exert a tremendous influence on T1-weighted images due to high paramagnetic payloads, i.e., "bright-contrast" despite being bound to biomarkers at nanomolar concentrations.

As mentioned earlier, perfluorocarbon nanoparticles may incorporate large numbers of paramagnetic gadolinium ions onto their surface. Initial studies with this agent demonstrated that paramagnetic nanoparticles, targeted to the fibrin component of thrombi, *in vitro* and *in vivo*, could effectively overcome the partial volume dilution effect previously associated with the failure of most MR molecular imaging contrast agents (Figure 6). Moreover, the molecular relaxivity (i.e., r_1 per particle $(mM*s)^{-1}$) of these nanoparticles improved monotonically with increasing gadolinium surface concentrations. As with ultrasound imaging, the effectiveness of these "ultraparamagnetic" nanoparticles was first demonstrated for the molecular imaging of fibrin with its relatively high epitope density and prominent role as the *sine qua non* vascular marker of ruptured atherosclerotic plaque *(3,9)*. Subsequently, these paramagnetic particles were used to image the induction of neovasculature by exogenously implanted FGF pellets in a rabbit micropocket corneal model *(21)*, in nascent Vx-2 rabbit adenocarcinoma tumors (Figure 7) *(10)*, by human melanoma xenografts implanted into athymic mice (Figure 8) *(22)*, and a model of early atherosclerosis in hyperlipidemic rabbits (Figure 9) *(11)*.

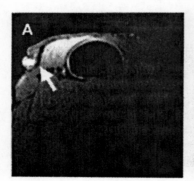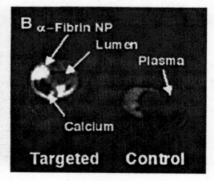

Figure 6. (A) In vivo T1w enhancement of canine thrombus in external jugular vein using fibrin targeted paramagnetic nanoparticles. (B) Ex vivo detection of ruptured plaque in carotid endarterectomy specimen with fibrin-targeted paramagnetic. nanoparticles. (Copyright 1996 Lippincott Williams and Wilkins.)

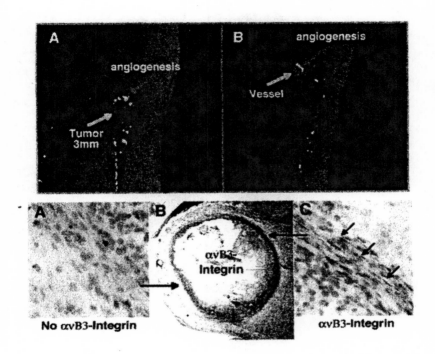

Figure 7. Top: Neovasculature of nascent Vx-2 tumor imaged with $\alpha_v\beta_3$-targeted paramagnetic nanoparticles (left), and the venous advential source of tumor neovasculature noted a few millimeters away. Bottom: Immunohistology of the Vx-2 tumor (imaged above) revealing asymmetric location of neovasculature adjacent to muscle plane (B) and high magnification insets showing the presence and sparse distribution of $\alpha_v\beta_3$-integrin expression within an imaging voxel. (See page 8 of color inserts.)

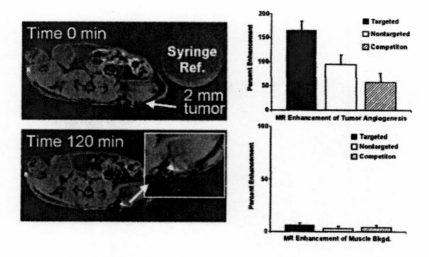

Figure 8. Left: Human melanoma tumor (~2mm) before and 120 min. after
$\alpha_v\beta_3$-*targeted paramagnetic nanoparticles (see 120 min. inset). Right:*
Histograms of percent signal enhancement 120 min. post treatment for mice
receiving targeted, nontargeted, or competitively-blockaded nanoparticle
therapy, Top: Signal derived from tumors; Bottom: Signal derived from
adjacent muscle. (Adapted from Schmieder et al. Magn Reson. Imaging 2004,
in review).

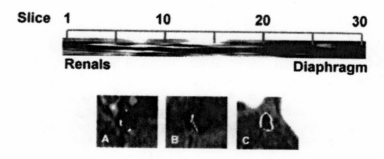

Figure 9. Top: In vivo spin echo image of aorta (long axis) from the renal
arteries to the diaphragm for one cholesterol-fed rabbit. Botton: Percent
enhancement maps from individual aortic segments at the renal artery (A), mid-
aorta (B) and diaphragm (C) 2 hours after treatment in a cholesterol-fed rabbit
given $\alpha_v\beta_3$-*targeted nanoparticles. (Copyright 1996 Lippincott Williams*
and Wilkins.)

Optimization of Nanoparticle Molecular Relaxivity

Targeted nanoparticle contrast agents may be improved by increasing the number and orientation of the paramagnetic chelates available to interact with the surrounding water milieu. As described above, nanoparticles have been routinely formulated with 10,000 to 100,000 gadolinium ions each with increasing molecular relaxivity and without induction of T2* effects from the high paramagnetic payloads *(9)*. Additionally, incorporation of the paramagnetic chelates onto the surface of slowly tumbling particles improves ionic relaxivity (i.e., the relaxivity per gadolinium ion $(mM*s)^{-1}$) over the free gadolinium chelate by two- to three-fold. Further improvements have been subsequently achieved by modifying the position of the paramagnetic chelate slightly beyond the stagnant, unstirred layer the particle.

The T_1 relaxivity (r_1) of nanoparticles formulated with a prototypic gadolinium diethylenetriaminepentaacetic acid bisoleate (Gd-DTPA-BOA) chelate was improved $(p<0.05)$ by reformulation with gadolinium diethylenetriaminepentaacetic acid phosphatidylethanolamine (Gd-DTPA-PE) and characterized at three fixed magnetic field strengths, 0.47 T, 1.5 T and 4.7 T (Table I) *(23)*. At each magnetic field strength, r_1 of the Gd-DTPA-PE formulation was approximately two-times greater than that measured for the Gd-DTPA-BOA agent. Variable temperature relaxometry measurements showed that r_1 of the Gd-DTPA-BOA emulsion was largely independent of temperature. In contradistinction, r_1 decreased at the lower temperature in the Gd-DTPA-PE emulsion. These temperature-dependence curves suggested that the water exchange rate with the paramagnetic ion was higher for the Gd-DTPA-PE chelate compared to Gd-DTPA-BOA. While at the higher temperature, the r_1 of Gd-DTPA-PE nanoparticles increased due to the faster water exchange and increased kinetic activity, the Gd-DTPA-BOA nanoparticles appeared to have somewhat restricted water access and did not benefit from the increased kinetic activity of water at the higher temperature. We have hypothesized that increased water accessibility results from the elevated position or improved complex rigidity of the chelate relative over the lipid surface for Gd-DTPA-PE nanoparticles. This postulate is further supported by the recent synthesis, formulation and characterization of two newer lipophilic chelates based upon methoxybenzyl-DOTA, with analogous results. Moreover, substitution of the macrocyclic DOTA chelate for the linear DTPA molecule markedly decreases the potential for gadolinium transmetallation from the particle, while retaining the improvements in relaxivity.

Table I. Relaxivity of Two Paramagnetic Emulsions at Three Field Strengths.

Magnetic Field (T)	Chelate	Relaxivity (mM*s)$^{-1}$ Ion-based (r_1)	Relaxivity (mM*s)$^{-1}$ Particle-based (r_1)
0.47	Gd-DTPA-BOA	21.3± 0.2	1,210,000 ± 10,000
0.47	Gd-DTPA-PE	36.9 ± 0.5	2,718,000 ± 40,000
1.5	Gd-DTPA-BOA	17.7 ± 0.2	1,010,000 ± 10,000
1.5	Gd-DTPA-BOA	33.7 ± 0.7	2,480,000 ± 50,000
4.7	Gd-DTPA-BOA	9.7 ± 0.2	549,000 ± 9,000

Gd-DTPA-BOA significantly different from Gd-DTPA-PE within field strength ($p < 0.05$).

Synergy Between Nanoparticulate Molecular Imaging and Targeted Drug Delivery

Targeted drug delivery has been a major goal of pharmaceutical scientists for decades but only recently, with the approvals of Gleevac™, Herceptin™, and Rituximab™, has this "Quest for the Holy Grail" led to viable therapeutic options in our clinic armamentarium. The primary benefit of homing therapeu-tics directly to a target site is the improved safety and effectiveness anticipated by lowering the dosage and concentrating the compounds at the pathological site. Unfortunately, targeted drug systems can not be perfect, and partial or poor therapeutic response to treatment can result from an ineffectiveness of the drug or an inadequacy of the delivered dose. Whereas traditional drug pharmaco-kinetics and pharmacodynamics can be monitored and correlated with therapeu-tic windows, circulating concentrations of targeted agents have little relationship with target tissue drug levels. The emergence of molecular imaging and its inte-gration into targeted drug delivery paradigm now provides a synergistic means to screen and stratify patients for therapy, to deliver treatment, to confirm and quantify target tissue levels, and to monitor effectiveness of the medical strategy. Thus, rather than waiting several weeks for a clinical outcome, doctors and patients will discern the day of treatment if the drug has accumulated at the target tissue within a therapeutic dosage window.

The inhibitory effects of classic antiproliferative agents incorporated into the targeted nanoparticle formulation on vascular smooth muscle cell prolifera-tion in culture have been reported. In contradistinction to other particulate agents, where active ingredients are sequestered within the aqueous or hydro-phobic core of the particle, we have embedded hydrophobic therapeutic agents into the outer surfactant monolayer of the nanoparticles, where they are constrained from quickly diffusing into the surrounding aqueous environment or inward into the perfluorocarbon core by insolubility. Therapeutic nanoparticles release drug primarily through interactions between the nanoparticle lipid surface and the opposing cell membrane of the targeted surface *(4)*. We have coined the term "contact facilitated lipid exchange" to describe the mechanism. This exchange of lipids between vesicles and cell membranes has been described and mathematically modeled by others, but this approach has not been employed heretofore as an effective method for drug delivery. Ordinarily, collision mediated lipid-exchange proceeds as a slow, concentration-dependent, second-order process. Ligand-directed binding of nanoparticles to cell surface receptors increases lipid-exchange by decreasing the activation energy barrier for the desorption of lipid molecules to the nearby cell membrane by van der Waals attraction. In other words, ligand binding of the nanoparticles minimizes the equilibrium separation of lipid surfaces and promotes the formation of collisional complexes. The increased frequency and duration of lipid surface interactions substantially enhance the net transfer of drug to the target cell membrane. As a result, therapeutic nanoparticles do not require cellular uptake or disruption of the particles in order to release their drug contents at the target site.

As an example, vascular smooth muscle cell cultures were exposed to tissue-factor targeted or nontargeted nanoparticles, incorporating paclitaxel at 0.0, 0.2, or 2.0 mole% into the surfactant layer (Figure 11). Control nano-particles, whether targeted or untargeted, exhibited no effect on smooth muscle cell proliferation. Tissue factor (TF)-targeted paclitaxel nanoparticles decreased cell proliferation ($p<0.05$) at both dosages, although the higher dosage was slightly more effective (74% versus. 81% decrease; $p>0.05$). Non-targeted pac-litaxel (Taxol™) nanoparticles exhibited no effect on cell proliferation. These data illustrate the contact dependence of drug-loaden nanoparticles with the target cell surface for effective delivery.

More recently, the effectiveness of this approach has been demonstrated *in vivo* for the delivery of the anti-angiogenic compound fumagillin in athero-sclerotic rabbits. In this study, a minute, one-time dosage of $\alpha_v\beta_3$–targeted paramagnetic nanoparticles incorporating fumagillin (30µg/kg) was found to be highly effective in pruning the expansion of the vasa vasorum in the hyper-lipidemic animals, whereas $\alpha_v\beta_3$–targeted paramagnetic nanoparticles without drug had no effect. In comparison with systemic therapy in the ApoE-deficient mouse model *(24)* with a water-soluble analogue of fumagillin, TNP-470, the effect was achieved faster (1 week versus 13 weeks) and with much less drug (30 µg/kg body weight versus 1.6 g/kg body weight). These early findings reflect the benefits of targeted therapy to have high pharmacological impact at lower total body dosages.

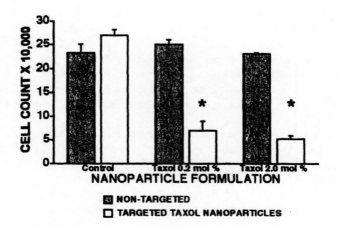

Figure 11. Percent cell counts of VSMC, 3 days following exposure to 0, 0.2, or 2.0 mole% of nontargeted or TF-targeted paclitaxel nanoparticles. Nontargeted nanoparticles have no therapeutic effect because interaction with target surface was transient versus TF-targeted nanoparticles with prolonged surface contact. (Copyright 1996 Lippincott Williams and Wilkins.)

Conclusion

Ligand targeted particles provide the opportunity to detect the expression of pathognomonic cell-surface molecular signatures present in nanomolar *(25)* quantities with ultrasound or magnetic resonance imaging. This unique platform technology has been conceptually demonstrated to improve the diagnosis of early atherosclerosis, vulnerable plaques, intracardiac thrombus and angiogenesis and could be studied in clinical trials for one or more of these applications within the next few years. The particulate nature of these agents creates an ideal platform for targeted drug delivery with the opportunity for rational therapeutic dosing, a feature unique to drug delivery coupled with molecular imaging. Molecular imaging agents, possibly in conjunction with rational targeted therapies, will likely alter the future of clinical medicine as these technologies continue to mature.

References

1. Weissleder R. Molecular imaging: Exploring the next frontier. *Radiology* **1999**, *212*, 609-614.
2. Lanza G.M., Wallace K., Scott M.J., Cacheris W., Abendschein D.R., Christy D., Sharkey A., Miller J.G., Gaffney P.J., Wickline S.A. A novel site-targeted ultrasonic contrast agent with broad biomedical application. *Circulation* **1996**, *94*, 3334-3340.
3. Lanza G.M., Lorenz C., Fischer S., Scott M.J., Cacheris W., Kaufman R.J., Gaffney P.J., Wickline S.A. Enhanced detection of thrombi with a novel fibrin-targeted magnetic resonance imaging agent. *Acad Radiol.* **1998**, *5* (suppl. 1), s173-s176.
4. Lanza G.M., Yu X., Winter P.M., Abendschein D.R., Karukstis K.K., Scott M.J., Chinen L.K., Fuhrhop R.W., Scherrer D.E., Wickline S.A. Targeted antiproliferative drug delivery to vascular smooth muscle cells with a magnetic resonance imaging nanoparticle contrast agent: Implications for rational therapy of restenosis. *Circulation* **2002**, *106*, 2842 - 2847.
5. Mattrey R.F. The potential role of perfluorochemicals (PFCs) in diagnostic imaging. *Artif. Cells. Blood Substit. Immobil. Biotechnol.* **1994**, *22*, 295-313.
6. Mattrey R.F., Scheible F., Gosink B., Leopold G., Long D., Higgins C. Perfluoroctylbromide: A liver/spleen-specific and tumor-imaging ultrasound contrast material. *Radiology* **1982**,*145*, 759-762.
7. Hughes M.S., Marsh J.N., Hall C.S., Fuhrhop R.W., Scott M.J., Lanza G.M., Wickline S.A. Optimization of site-targeted perfluorocarbon nanoparticle contrast in whole blood for molecular imaging applications. *J Acous. Soc Am.* **2004**, in press.

8. Lanza G.M., Abendschein D.R., Hall C.S., Marsh J.N., Scott M.J., Scherrer D, Wickline S.A. Molecular imaging of stretch-induced tissue factor expression in carotid arteries with intravascular ultrasound. *Invest Radiol.* **2000**, *35*, 227-234.

9. Flacke S., Fischer S., Scott M.J., Fuhrhop R.W., Allen J.S., McLean M., Winter P.M., Sicard G., Gaffney P.J., Wickline S.A., Lanza G.M. A novel MRI contrast agent for molecular imaging of fibrin: Implications for detecting vulnerable plaques. *Circulation* **2001**, *104*, 1280 -1285.

10. Winter P.M., Caruthers S.D., Kassner A., Harris T.D., Chinen L.K., Allen J.S., Lacy E.K., Zhang H., Robertson J.D., Wickline S.A., Lanza G.M. Molecular imaging of angiogenesis in nascent Vx-2 rabbit tumors using a novel alpha(v)beta3-targeted nanoparticle and 1.5 tesla magnetic resonance imaging. *Cancer Res.* **2003**, *63*, 5838-5843.

11. Winter P.M., Morawski A.M., Caruthers S.D., Fuhrhop R.W., Zhang H., Williams T.A., Allen J.S., Lacy E.K., Robertson J.D., Lanza G.M., Wickline S.A. Molecular imaging of angiogenesis in early-stage atherosclerosis with alpha(v)beta3-integrin-targeted nanoparticles. *Circulation* **2003**, *108*, 2270-2274.

12. Lanza G.M., Trousil R., Wallace K., Rose J.H., Hall C.S., Scott M.J., Miller J.G., Eisenberg P., Gaffney P.J., Wickline S.A. In vitro characterization of a novel, tissue-targeted ultrasonic contrast system with acoustic microscopy. *J Acoust Soc Am.* **1998**, *104*, 3665-3672.

13. Hall C.S., Lanza G.M., Rose J.H., Kaufman R.J., Fuhrhop R.W., Handley S.H., Waters K.R., Miller J.G., Wickline S.A. Experimental Determination of Phase Velocity of Perfluorocarbons: Applications to Targeted Contrast Agents. *IEEE Trans Ultrason Ferroelec Freq Contr.* **2000**, *47*, 75-84.

14. Marsh J.N., Hall C.S., Scott M.J., Fuhrhop R.W., Gaffney P.J., Wickline S.A., Lanza G.M. Improvements in the ultrasonic contrast of targeted perfluorocarbon nanoparticles using an acoustic transmission line model. *IEEE Trans Ultrason Ferroelectr Freq Control.* **2002**, *49*, 29-38.

15. Hall C.S., Marsh J.N., Scott M.J., Gaffney P.J., Wickline S.A., Lanza G.M. Temperature dependence of ultrasonic enhancement with a site-targeted contrast agent. *J Acous. Soc AM.* **2001**, *110*, 1677-1684.

16. Gupta H., Weissleder R. Targeted contrast agents in MR imaging. *Magn Reson Imag Clinics N Am.* **1996**, *4*, 171-84.

17. Weissleder R., Stark D.D., Compton C.C., Wittenberg J., Ferrucci J.T. Ferrite-enhanced MR imaging of hepatic lymphoma: An experimental study in rats. *AJR.* **1987**,*149*, 1161-1165.

18. Weissleder R., Elizondo G., Wittenberg J., Rabito C.A., Bengele H.H., Josephson L. Ultrasmall superparamagnetic iron oxide: characterization of a new class of contrast agents for MR imaging. *Radiology* **1990**, *75*, 489-493.

19. Moore A., Weissleder R., Bogdanov A., Jr. Uptake of dextran-coated monocrystalline iron oxides in tumor cells and macrophages. *J Magn Reson Imaging* **1997**, *7*, 1140-1145.

20. Sipkins D.A., Cheresh D.A., Kazemi M.R., Nevin L.M., Bednarski M.D., Li K.C. Detection of tumor angiogenesis in vivo by alphaVbeta3-targeted magnetic resonance imaging. *Nat Med.* **1998**, *4*, 623-626.
21. Anderson S.A., Rader R.K., Westlin W.F., Null C., Jackson D., Lanza G.M., Wickline S.A., Kotyk J.J. Magnetic resonance contrast enhancement of neovasculature with alpha(v)beta(3)-targeted nanoparticles. *Magn Reson Med.* **2000**, *44*, 433-439.
22. Schmieder A.H., Winter P.M., Caruthers S.D., Harris T.D., Chinen L., Williams T., Watkins M., Allen J.S., Wickline S.A., Lanza G.M. Molecular imaging of angiogenesis in human melanoma xenografts in nude mice by MRI (1.5T) with avB3-targeted nanoparticles. *Mol. Imaging* **2002**, *1*, 190.
23. Winter P.M., Caruthers S.D., Yu X., Song S.K., Chen J., Miller B., Bulte J.W., Robertson J.D., Gaffney P.J., Wickline S.A., Lanza G.M. Improved molecular imaging contrast agent for detection of human thrombus. *Magn Reson Med.* **2003**, *50*, 411-416.
24. Moulton K.S., Heller E., Konerding M.A., Flynn E., Palinski W., Folkman J. Angiogenesis inhibitors endostatin or TNP-470 reduce intimal neovascularization and plaque growth in apolipoprotein E-deficient mice. *Circulation* **1999**, *99*, 1653-1655.
25. Morawski A.M., Winter P.M., Crowder K.C., Caruthers S.D., Fuhrhop R.W., Scott M.J., Robertson J.D., Abendschein D.R., Lanza G.M., Wickline S.A. Targeted nanoparticles for quantitative imaging of sparse molecular epitopes with MRI. *Magn Reson Med.* **2004**, *51*, 480-486.

Author Index

Subject Index

316

319

Quadramet®, 284–286

role of bifunctional, in conjugate, 283f

samarium-153 Lexidronam injection, 284–286

skeletal targeted radiotherapy (STR), 287

therapeutic bone agents, 284–287

Chemoselective linking groups, dendrimers, 125–126

Chewing gum
menthol release, 275, 276f
preparation of, containing peppermint oil and yeast microencapsulated peppermint oil, 272–273
See also Microencapsulation

Chinese hamster ovary (CHO) cells
gene transfer efficiency of particles, 236, 238f
microparticle and lipid transfection efficiency in, 236, 238f
See also Double emulsion process

Cisplatin, tumor targeting, 34–35

Co-administration, macromolecular drug delivery, 71

Colloidal carriers, drug delivery and transport, 41

Colon adenocarcinoma cells
delivery of dipalmitoyl derivative of 5-fluorodeoxyuridine (FUdR-dP) to, 85
inhibition by FUdR-dP containing liposomes, 87
in vitro studies of FUdR-dP-containing liposomes, 86–87
in vivo studies of, 87
See also Human colon adenocarcinoma cells (Caco-2); Immunoliposomes

Competition reactions, relative reactivity of amines, 125–126

Complement protein binding

in vitro complement depletion assay, 100, 102, 108–112

protecting liposomes, 96–97
See also Polymer-protected liposomes

Conjugation
polymer as drug delivery vehicle, 138
polymer-drug, 254–256

Controlled drug delivery
potential advantages, 242–243
See also Solvent exchange method

Convergent synthesis approach, dendrimers, 124, 127

Cooperativity
adsorption, 97, 104–105
schematic of polymer blobs associating with surface at low and high, 113f
See also Polymer-protected liposomes

Copolymers. See Amphiphilic block copolymers (ABC); Amphiphilic graft copolymers; Thermosensitive polymeric micelles

Covalent modification, macromolecular drug delivery, 70

Cremophor® EL, surfactant for drug solubilization, 15

Critical aggregation concentrations, N-isopropylacrylamide (NIPAAm) copolymers, 44, 46t

Critical micelle concentration (CMC)
drug encapsulation by amphiphilic graft copolymers, 256–257
plasmid DNA condensation by cationic detergents, 208–209

Critical micelle temperature (cmt), N-isopropylacrylamide (NIPAAm) copolymers, 45, 46t

Crosslinking. See Amphiphilic graft copolymers

Cyclosporine A (CsA)
drug loading (DL), 59

324

336

Printed in the United Kingdom by
Lightning Source UK Ltd., Milton Keynes
141338UK00001B/91/A